Why Study Biology by the Sea?

CONVENING SCIENCE:
DISCOVERY AT THE MARINE BIOLOGICAL LABORATORY
A Series Edited by Jane Maienschein

For well over a century, the Marine Biological Laboratory has been a nexus of scientific discovery, a site where scientists and students from around the world have convened to innovate, guide, and shape our understanding of biology and its evolutionary and ecological dynamics. As work at the MBL continuously radiates over vast temporal and spatial scales, the very practice of science has also been shaped by the MBL community, which continues to have a transformative impact the world over. This series highlights the ongoing role MBL plays in the creation and dissemination of science, in its broader historical context as well as current practice and future potential. Books in the series will be broadly conceived and defined, but each will be anchored to MBL, originating in workshops and conferences, inspired by MBL collections and archives, or influenced by conversations and creativity that MBL fosters in every scientist or student who convenes at the Woods Hole campus.

Why Study Biology by the Sea?

EDITED BY KARL S. MATLIN,
JANE MAIENSCHEIN,
AND RACHEL A. ANKENY

The University of Chicago Press Chicago and London

The University of Chicago Press, Chicago 60637
The University of Chicago Press, Ltd., London

For more information, contact the University of Chicago Press, 1427 E. 60th St., Chicago, IL 60637.
Published 2020
Printed in the United States of America

29 28 27 26 25 24 23 22 21 20   1 2 3 4 5

ISBN-13: 978-0-226-67276-2 (cloth)
ISBN-13: 978-0-226-67293-9 (paper)
ISBN-13: 978-0-226-67309-7 (e-book)
DOI: https://doi.org/10.7208/chicago/9780226673097.001.0001

Library of Congress Cataloging-in-Publication Data

Names: Matlin, Karl S., editor. | Maienschein, Jane, editor. | Ankeny, Rachel A., editor.
Title: Why study biology by the sea? / edited by Karl S. Matlin, Jane Maienschein, and Rachel A. Ankeny.
Other titles: Convening science.
Description: Chicago : University of Chicago Press, 2020. | Series: Convening science: discovery at the Marine Biological Laboratory | Includes bibliographical references and index.
Identifiers: LCCN 2019042713 | ISBN 9780226672762 (cloth) | ISBN 9780226672939 (paperback) | ISBN 9780226673097 (ebook)
Subjects: LCSH: Marine laboratories—History. | Marine biology—Research—History. | Marine laboratories—United States—History. | Marine biology—Research—United States—History.
Classification: LCC QH91.6 .W49 2020 | DDC 578.77—dc23
LC record available at https://lccn.loc.gov/2019042713

♾ This paper meets the requirements of ANSI/NISO Z39.48-1992 (Permanence of Paper).

Contents

Foreword

NIPAM H. PATEL

Evolution has already solved most biological problems. Every organism is itself an experiment, and, given the enormous number of organisms on earth and the fact that life started at least 3.5 billion years ago, the number of experiments is astronomical. Through evolutionary processes, life has evolved to thrive in even the harshest conditions; has repeatedly mastered multiple solutions to locomotion on the land, in the air, and in the water; and can carry out reactions that amaze chemists. There are animals that have conquered regeneration and plants that have witnessed the birth and death of a dozen human generations.

Life on earth began its journey in the ocean, and, thus, we find that its oldest branches trace their history to a watery beginning and that diversity still reigns supreme within this realm. However, despite the fact that water covers more than 70 percent of our planet, the oceans remain a largely unexplored domain. We stare in amazement over their surface, but we know very little of what lies below. Over four thousand people have summited Everest, but the number who have reached the deepest parts of the oceans can be counted on one hand. Still, life teems in the oceans, and, thus, many of the solutions we strive to find lie there. Indeed, the history of biology is replete with discoveries that come from marine creatures, and many scientists got their start on the shore, picking their way through tide pools or watching plankton under a microscope.

Several essays in this book trace the history of a number of marine labs around the world that have played a critical role in all manner of biological discovery. Others detail specific scientific breakthroughs that were possible only because of the unique attributes of animals collected and raised at marine labs. Innumerable fundamental insights owe their origin to these marine creatures and the marine labs that pioneered the work being done with them. Furthermore, as we increasingly come to understand the interconnectedness of all life and the links between the physical environment and living organisms, we appreciate even more the importance of the oceans. The essays in this book document a rich history that we not only need to remember but also need to build on and use to guide us as we continue to investigate the mysteries of life.

Introduction

KARL S. MATLIN, JANE MAIENSCHEIN, AND RACHEL A. ANKENY

At the Marine Biological Laboratory (MBL) in 2016, a group of historians, philosophers, and life scientists gathered for the twenty-ninth annual MBL–Arizona State University History of Biology Seminar to ask, Why marine studies? Why have researchers since the mid-nineteenth century spent summers at the seaside? How did they move from working on the beach at the water's edge to the laboratory bench indoors? What questions motivated their work, and what opportunities appeared as a result of working by the sea?

As we explored marine studies from over a century, two themes emerged. First, the move from beach to bench became possible only with the development of institutions to house equipment, people, organisms, and the entire research enterprise. This happened at the end of the nineteenth century. Second, these institutions enabled a variety of different kinds of work to be carried out. Decisions from the past have informed the work done today.

As a result of the questions raised, we have developed this edited collection to explore the two themes of institutionalization and diversity of work at marine laboratories. We start with a focus on the MBL and its unique, longstanding commitment to research and education and their intersection. The MBL was inspired by the Stazione Zoologica in Naples, which preceded it but developed in different ways. And the MBL itself inspired the development of other institutions such as the one in Amoy, China, and has had important connections with others such as the Misaka Marine Biological Station in Japan. Essays in this collection look at each of these.

Marine institutions have embraced a wide range of studies, some involving marine organisms, and others exploring comparative studies with other, nonmarine organisms. Tools for exploration include microscopy, of course, and also other technologies and techniques. Sometimes researchers at the beach study intertidal zones or ecosystems or even coral reefs offshore. Then they bring life into the lab. Sometimes they observe living processes, and other times they use preserved materials to make images of what they have found. Molecules, genes, cells, organs, organisms, behaviors, complex systems, ecosystems, evolution, and so much more make up the realm of study at marine laboratories. Education through outstanding advanced courses occurs alongside research.

In this collection, we are purposefully focused on history. How did marine institutions develop, and what have they done over time? The institutions that still exist, including the MBL, have their own Web sites and discussions of what they are doing today, so we are letting them speak for themselves. Yet our historical analysis helps inform answers to the most relevant contemporary question, Why study biology by the sea? Experiences of the researchers, what they have learned in courses, and what they have discovered in the lab as they move from the beach to the bench and back to the beach again: these have all shaped what marine institutions are today. These institutions have tremendous capacity for facilitating different kinds of research, bringing together different groups of researchers, allowing sharing of ideas and ways of working, bringing new investigators into the mix alongside senior scientific leaders, and inspiring new thinking that helps keep science fresh. The foreword by MBL director Nipam Patel and the epilogue by Alejandro Sánchez Alvarado provide perspectives on why it matters for science today to focus on marine studies and to reflect historically.

PART ONE

Marine Places

ONE

Why Have Biologists Studied at the Seashore? The Woods Hole Marine Biological Laboratory

JANE MAIENSCHEIN

In the summer of 1893, Bashford Dean presented a public lecture at the Marine Biological Laboratory (MBL) in Woods Hole. He spoke about the marine biological stations of Europe, noting that European biologists had long embraced the value of studying biology at the seashore. He pictured the biologist first learning about marine organisms at an inland European laboratory, studying material shipped from a coastal station. Then the student would visit one of those stations first to learn and then to carry out research. Dean pointed out that, with rich research materials, equipment, and libraries, "the station becomes, in short, a literal emporium, cosmopolitan, bringing together side by side the best workers of many universities, tending, moreover, to make their observations upon the best materials sharper by criticism, most fruitful in results." He continued: "It has often been remarked how large a proportion of recently published researches [*sic*] was dependent, directly or indirectly, upon marine laboratories." He noted that "general interest in the advancement of pure science" had become important in Europe in the previous decade but that the same was not true of the United States. He sought to improve awareness among Americans, and he saw the MBL as a place to make the case for why biologists need marine stations (Dean 1894, 212).

This essay addresses the question of why biologists have studied at the seashore by looking briefly at previous answers such as Dean's. I first explore early studies at the seashore, then review Charles Kofoid's 1910 survey of marine stations. Historians extended Kofoid's story in a series of symposia and volumes around 1988, in connection with the MBL's centennial. This essay provides a summary of those studies and the resulting traditional picture of biology by the sea.

The essay then focuses on the MBL to show how Americans first embraced biology at the seashore. As the MBL's first director, Charles Otis Whitman, noted, the MBL offered a clear vision for an independent institution that brought together what he called *instruction and investigation*. From the beginning, the MBL leaders understood that students and researchers learn from each other and discover together and that they do so across the boundaries that normally separate biologists into different institutions and different disciplines with access to different tools and organisms. Having a building with running seawater, easy access to a wide diversity of research organisms, and a community of other curious biologists made the MBL and other marine institutions valuable. Coming together to carry out biology by the seashore has allowed new and otherwise unachievable biological advances for more than 130 years.

Despite this record of success, however, the viability of some marine institutions has recently come into question. In the United States, limited federal funding for biomedical research has made the traditional extended summer stays at marine laboratories a luxury that many scientists cannot afford. The positive aspects of year-round facilities by the sea are offset by the costs of maintaining them because many of the scientific and revenue-producing activities are in fact concentrated in the summer. In response to such challenges, institutions have hired full-time scientific staffs, hoping to support facilities with overhead charges on research grants, and have sought new conferences and courses to fill the off-season schedule. Fiercely independent institutions such as the MBL have found themselves responding to these sorts of pressures by affiliating with larger, deep-pocketed institutions (in MBL's case, the University of Chicago). While understandable, such changes have raised questions about how to preserve the unique, collaborative culture of marine stations that have relied on their independence.

These are not new problems; explorations of the history of selected institutions that have persisted over generations and the work done there help illuminate where we are today. Reflecting on these historical studies informs our understanding of why marine institutions have been and remain important.

Why Did Researchers Study at the Seashore?

As Dean showed clearly, by the end of the nineteenth century, discoveries at marine biological stations had contributed significantly to knowledge about the fundamental biology of organisms. Careful natural historical observations and experiments, typically during summer visits, continued to increase knowledge about such basic biological phenomena as fertilization, development, heredity, cells, regeneration, physiology, and the way in which external conditions influence biological processes. Technological developments in microscopy and imaging methods then enticed researchers to take their organisms from the beach indoors to the laboratory bench.

As with so many aspects of nature, the story of marine studies actually starts millennia earlier with Aristotle, who found marine life fascinating and informative for providing insights about living organisms. In his works focused on central phenomena of living systems, including *The Parts of Animals, The Generation of Animals, The History of Animals, The Movement of Animals,* and *The Progression of Animals,* he documented natural history, development, and distribution of marine as well as terrestrial organisms. Despite the obvious problems with these types of attribution, he is often referred to as the father of marine biology.[1]

Although others also studied marine organisms, it was not until the nineteenth century that significant numbers of investigators began to undertake "nature study" at the beachside, often as a way of enlightening and educating a general population about life (Armitage 2009; Kohlstedt 2015). Others looked into the ocean depths for pragmatic reasons, such as the need to know about the ocean bottom when laying transatlantic cable in the 1850s (Thomson 1873; Anderson and Rice 2006). Though oceanography remains separate from marine biology as a discipline, discoveries of marine creatures in the ocean depths raised questions about the processes of living systems. Other drivers for the study of marine life have come from commercial interests in fisheries or from imperialistic motives to establish what lands could be conquered, what resources could be commanded, and what information could be controlled (see, e.g., Osborne 1994). In short, diverse interests have taken researchers to the seashore.

Exploring expeditions in the nineteenth century involved observing but also collecting and documenting, with the result that major exploring expeditions typically carried along at least one naturalist and an artist. Naturalists such as Thomas Henry Huxley and Charles Darwin, for example, found their lives and their ideas about nature significantly affected by long expeditions across oceans as they developed evolutionary explanations of the rich organic diversity they observed and recorded.

Huxley's study of relatively simple marine life-forms led him to ideas about protoplasm as a material basis for life and to reflection on ways that study of individual development through ontogeny relates to evolutionary development of species or phylogeny (Huxley 1913, vol. 1 [discussing the *Rattlesnake* voyage, especially during the years 1846–50]). In Germany, Ernst Haeckel pursued similar questions about the connections of development and evolution through his biogenetic law, which held that ontogeny recapitulates phylogeny. His extensive study of marine organisms and his beautiful color drawings and paintings of marine life provided him the authority for his claims about both evolution and development (Richards 2008). Comparing development and morphology of different species suggested relationships among them. These studies took place at the sea but initially in the absence of marine stations or established institutions.

Aquariums provided another form of access to marine study, allowing observation of marine life without having to be actually at the seashore. Yet, until the late nineteenth century, there were only a few public aquariums, and only the wealthy could afford to keep the water sufficiently clean and aerated in their own aquariums at home (Barber 1984). Aquariums housed specimens for researchers as well as attracting public interest in marine studies, but they did not re-create nature, and they did not yet serve as laboratories (Nyhart 2009; de Bont 2015; Muka 2017). They were central to some—such as the Stazione Zoologica di Napoli (SZN)[2]—but not all marine stations.

A growing number of researchers began to move from the beach to comfortable rooms. They could study the natural history and environment outdoors, then move indoors to spread their collected specimens out on tables for further study while storing living organisms in seawater tanks. Researchers began to rent rooms where they could look closely with microscopes and other equipment that they could set up and use day after day and record their findings over longer periods of time than was convenient at the beach. They did not have to carry their specimens far and therefore could observe embryological development and physiological processes of the living organisms that would have been harder to study if the organisms had had to travel in jars for longer distances. Itinerant researchers—including notably Anton Dohrn and Ernst Haeckel, who rented rooms for research—soon began to long for even more permanent facilities where they could set up labs, leave their equipment, and return the next year to continue their research in reliable surroundings (Richards 2008; de Bont 2015).

By 1900, increasing numbers of researchers who aspired to become professional biologists had joined the migration to summer marine research sites and to a growing variety of institutions providing opportunities for seaside study. Small places for just one researcher and a handful of students existed alongside the magnificent building for the SZN founded by Dohrn in 1872 (Simon 1980; Groeben 1984; Benson 1988). Government facilities emerged as well, such as that at Plymouth, England (Southward and Roberts 1987). By 1893, so many places existed around the world with such different styles and purposes that Bashford Dean's MBL lecture may well have helped inspire the US government to commission a study to document and learn from what others were doing.

Kofoid's 1910 Survey of Marine Stations

The international development of marine stations caught the attention of the Bureau of Education in Washington, DC, which was then part of the Department of the Interior. The commissioner of education, Elmer Ellsworth Brown, requested a study of European stations with a published book to record the results. In his letter of transmittal for the resulting volume, he noted that both scientific research and traditional instruction are essential for educational programs. He explicitly intended the study to promote advancement of stations in America like those in Europe (Kofoid 1910, xi).

Charles Atwood Kofoid and his wife, Julia, took on the study, visiting European freshwater and saltwater marine stations through 1908 and 1909. Kofoid was born in 1865 in Illinois, graduated from Oberlin College in 1890, and received a PhD from Harvard University for research on cell lineage. With an appointment as assistant professor at the University of California, Berkeley, from 1903 until his retirement in 1936, he worked with William Emerson Ritter to develop the Marine Biological Association of San Diego, which later became the Scripps Institution of Oceanography. His biography in the National Academy of Sciences *Biographical Memoirs*, by the geneticist Richard Goldschmidt, explains his research commitments and his fascination with marine life, especially the plankton and protozoans (Goldschmidt 1951). It is also clear that the Kofoids enjoyed travel and discovering new places and new life-forms.

When he was asked to carry out the survey, Kofoid had already published the 1898 article "The Fresh-Water Biological Stations of America" in the *American Naturalist*. There he noted: "The fundamental purpose of all biological stations, both marine and fresh-water, is essentially the same. They serve to bring the student and the investigator into closer connection with nature, with living things in their native environment. They facilitate observation and multiply opportunities for inspiring contact with, and study of, the living world. They encourage in this day of microtome morphology the existence and development of the old natural history or, in modern terms, ecology, in the scheme of biological education" (Kofoid 1898, 391).

Kofoid went on to note that, while most marine stations were close to the sea, the United States had begun to offer some freshwater stations, especially in the Midwest. He pointed out that great potential for discovery lay in the mix of old-fashioned natural history description with experimental studies carried out by researchers and students working together.

For his larger study of stations in Europe and elsewhere, Kofoid visited each station for long enough to get a sense of its character. He documented the work being done, organisms studied, equipment used, and techniques developed. He also took photographs, included maps, and clearly paid close attention to the details of each station with the explicit intention that the European institutions could inspire development of more marine stations in the United States.

In the preface to the 1910 volume, Kofoid noted:

> Special attention has been given to the economic or applied scientific phases of their activities in the firm belief that biological stations and their methods of attack upon biological problems are destined not only to add greatly, and, in a unique way, to the advance of knowledge but are also of prime economic importance. They are laying the foundations for a scientific aquiculture [*sic*] that will make possible the conservation of the aesthetic and economic resources of lakes and streams from threatened pollution and destruction, and that will facilitate the reaping even-

tually of the annual harvest of the sea without destruction to its sources. (Kofoid 1910, xiii)

He then went on to describe nearly one hundred stations of various sorts, including the quite different, magnificent SZN, Plymouth Station in England, Station Biologique de Roscoff in France, and the then newly established station at Monaco.

Despite the differences in facilities and organisms studied, Kofoid identified key features of a successful station. It should be accessible, with available transportation, and near a city that can provide housing. A rich variety of flora and fauna, with pure water, are important for experimental study. Research stations benefit from the variety of organisms available in warmer waters, but that variety does not matter so much for teaching purposes. Aquariums to store organisms and a plumbing system to bring in local water both contribute to making research possible. When all these conditions are met, "the biological station is a unique agency in biological research, indispensable in the equipment of a nation for the upbuilding of leaders in biological teaching and in the development and expansion of the spirit of research" (Kofoid 1910, 6).

Kofoid's report made clear that the United States had opportunities to develop more marine stations, the message being that such development would benefit American biological research and education. Kofoid noted that over fifty Americans had visited facilities in Europe, which he felt demonstrated a demand for such research and facilities that was not being met in the United States. Americans could, and by implication should, establish more and diverse types of marine stations. In fact, a number of such stations soon appeared in the United States. While Kofoid's survey was certainly influential at the time, and many marine stations and programs did subsequently appear around the world (including the Misaki and Amoy Stations in Asia, on which see Luk, in this volume, and Ericson, in this volume), his arguments for the importance of marine stations and their research remain valid today.

Historical Studies of American Marine Stations

As the 1988 centennial of the MBL approached, historians began reflecting on the nature and role of marine stations in US biology. The MBL leadership collaborated with the SZN to hold a symposium in

1984 at the SZN's buildings on the nearby island of Ischia. The meeting brought together historians and experimental biologists to share reflections on a century of biology by the sea. Explorations of both research and teaching contributions provided insights into what was common between the two institutions and what differed and why. The discussions also raised questions about the future of marine institutions.

SZN director Antonio Miralto sounded an optimistic note about the important role marine stations could play as he asked, "What laboratories for what science?" He wrote: "I believe that science should fit within a broad, non-restrictive, non-deterministic cultural frame-work. Indeed, science should become the focal point of a process of cultural renewal; a point of reference for future generations; and a driving force in all fields, scientific and non." He argued that we must beware of creating isolated institutions and instead should encourage broad international exchange of ideas. Scientific laboratories need culture, he urged, including through visual arts and music: "As in the past, science can determine many aspects of the future of humanity. Men of science must look beyond the limits of their own research activities, and, by their culture and enlightenment, become intellectual leaders" (Miralto 1985, 203–4). Such a beautiful Italian sentiment had also appealed to the more practically minded Americans in establishing the MBL to promote cooperation, learning together, and sharing the culture of science.

The papers from this meeting of historians and scientists appeared in a special supplemental issue of the MBL's journal, the *Biological Bulletin* (see "The Naples Zoological Station and the Marine Biological Laboratory" 1985). In addition, soon thereafter the *Bulletin* reprinted a historical volume by Frank Lillie, the second MBL director (see Lillie 1944).

In 1986, the American Society for Zoology organized a session on the topic of "American marine biology and institutions" at its annual meeting. In his introduction to the session, Ralph Dexter noted the origins of US marine study and the growth to more than fifty marine institutions (Dexter 1980). My talk asked, "Why do research at the seashore?" and pointed to changes in questions asked and methods used over the decades of the MBL in particular. I noted the dual interests in working at the seashore both because that is where the marine organisms live and because it is pleasant for researchers to live there as well. Even as it became possible to ship more living organisms to, say, Kansas for research or teaching, researchers and students still chose to visit the seaside institutions (see Maienschein 1988). Discussion at the symposium made clear that a large percentage of the audience had themselves made the trip to study by the seashore at some point as students or researchers. Everyone who commented enthused about the experience, regretted not being able to make the trip more often, and expressed hope that marine stations would have a solid future continuing to attract future generations. The MBL stood as a favorite place that evoked nostalgic longings to return and reports of the impact time spent there had made on so many different people's biological research and teaching.

The MBL: Why Woods Hole?

The history of early US marine stations began with schools focused on natural history. Notably, Louis Agassiz and his son Alexander set up the Anderson School on the island of Penikese in a string of islands off the southern tip of Cape Cod. The school lasted only two seasons, and students and faculty noted a number of limitations of the remote location. Yet the group was enthusiastic about study at the seashore, and many of the Penikese students and assistants went on to play leadership roles in other institutions. Agassiz's student and assistant Burt G. Wilder provided an excellent sense of both what Agassiz hoped to accomplish and what happened there, who attended, and what Agassiz meant by the call he posted, "Study Nature, Not Books" (Wilder 1898). Among the students at Penikese were Charles Otis Whitman and Cornelia Clapp, who was the first person to arrive the first year at the MBL and who became the first librarian, the first female trustee, and an excellent scientist and teacher.

In 1880, Agassiz's student Alpheus Hyatt established a similarly short-lived summer laboratory at his home in Annisquam on Cape Ann, on the shore north of Boston. He attracted support from the Boston Society of Natural History and the Woman's Education Association of Boston, both of which later supported the MBL. As with the Anderson school, Hyatt's seaside lab offered instruction in natural history with a focus on the diversity of organisms. It attracted such students as Thomas Hunt Morgan, between his undergraduate studies and his graduate studies at Johns Hopkins. In addition, it was intended to offer research facilities and opportunities for advanced biology students and researchers (Maienschein 1985a). Although neither the Penikese or the

Annisquam school lasted long, both clearly indicated that the Boston area had a number of supporters with a lively interest in the sea for practical as well as research interests.

Another Agassiz student, William Keith Brooks, had served as an assistant both to Louis's son Alexander and to Hyatt at the Boston Society. In 1876, he moved to the new Johns Hopkins University, where he became professor of zoology and established research labs in Jamaica and then the Chesapeake Zoological Laboratory. Yet those remained small facilities for small groups of Johns Hopkins students and defined projects.

Woods Hole—at the southern tip of Cape Cod—became the home for permanent biological research in America. Woods Hole already housed the US Fish Commission (USFC), with a solid building and an active fisheries-oriented program. Despite his own smaller marine expeditions to other places, Brooks also sent his graduate students to the USFC station, established by Spencer Fullerton Baird in 1885 (Benson 1979). While the USFC focus remained on practical matters related to fisheries, Baird welcomed researchers working on a range of topics to use the lab. He pointed to the rich diversity of life at this location, where the warm waters of the Gulf Stream and the colder waters of the Newfoundland Current come together. In his centennial reflective essay, W. D. Russell-Hunter wrote about the considerable advantages of Woods Hole for reasons of biodiversity (as we call it today), water quality, and other ecological considerations (Conklin 1944; Russell-Hunter 1985).

At the USFC, Baird strongly supported the idea to develop an independent marine laboratory as a complementary, independent institution for both research and instruction just across the street from the USFC in Woods Hole. He looked forward to expanding the local biological community. In his history, Lillie recalled how the MBL had started when a group of Boston Society of Natural History members, along with the Woman's Education Association of Boston, decided in 1886 that the time had come to establish a marine biological laboratory dedicated to education and research. With Baird's support, they decided on Woods Hole. Their permanent and independent institution became the MBL (Lillie 1944). The MBL opened its doors in 1888 and began to grow in size and activity. The zoologist Winter Conway Curtis, who later testified in the Scopes trial, reported on the good times at the early MBL and USFC and the significant ways in which the two institutions reinforced each other (Curtis 1955; Galtsoff 1962).

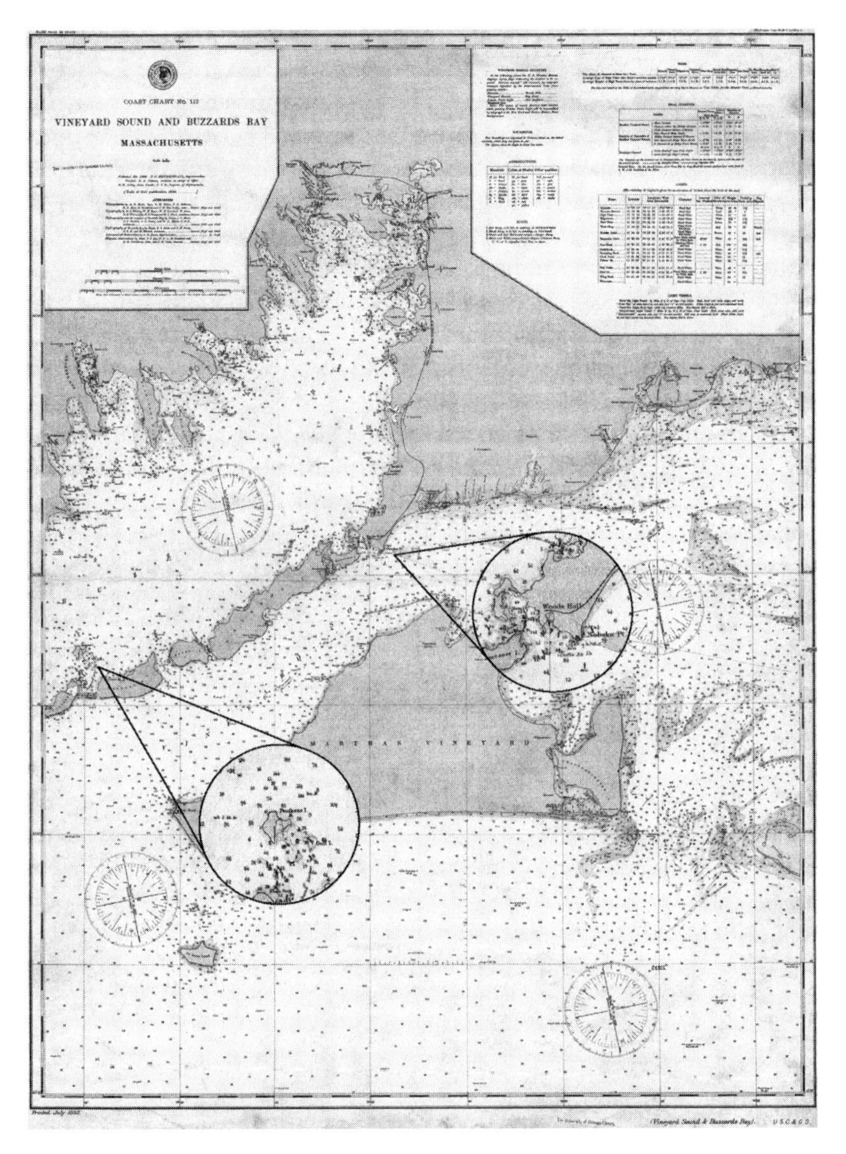

FIGURE 1.1. Map of the Woods Hole and Penikese Island area, Cape Cod, Massachusetts, 1892. Map courtesy of the University of Chicago Library (Map G3701.P5 svar.U4 no. 112 1892 MapCl).

From its opening in 1888 until 2013, the MBL remained ferociously independent, with Whitman writing strong letters insisting that it should never become affiliated with any one university.[3] When the University of Chicago opened in 1892, Whitman became the first chair of the Biology Department while also continuing in his unpaid position as MBL director. Despite several efforts starting in the 1890s to affiliate the two institutions, Whitman worried that the MBL needed national support rather than affiliation with just one group or one university. Ironically, and also perhaps quite appropriately, a century and a quarter later, in 2013, when the MBL Corporation addressed the question whether to become formally affiliated with the University of Chicago, the vote was overwhelmingly in favor (Maienschein 2014).

At the end of a railway line, Woods Hole provided a welcoming place for families to travel for the summer and enjoy the seaside research life together (Pauly 1988). Lillie met his future wife during the embryology course, and a number of other researchers pointed to the importance of their family connections with Woods Hole, with wives and children

FIGURE 1.2. Marine Biological Laboratory, Woods Hole, Mass., 1890s. Photograph by Howard Stidham Brode (1890–1958). https://history.archives.mbl.edu/archives/items/marine-biological-laboratorywoods-holl-mass.

FIGURE 1.3. Thomas Hunt Morgan, Edmund Beecher Wilson, Ross Granville Harrison, and Morgan's family apparently picnicking on the beach near Woods Hole, n.d. Photograph by Alfred Francis Huettner (b. 1884). https://history.archives.mbl.edu/search?search_fulltext=thomas+hunt+morgan%2C+E.B.+Wilson+andothers+having+a+picnic.

lobbying to return summer after summer. Thomas Hunt Morgan dedicated his 1901 *Regeneration* to his mother, thanking her for spending summers in Woods Hole and helping care for her grandchildren so that both he and his wife, Lilian Vaughan Morgan, could spend time doing research. A photograph shows Morgan (center rear), his mother, Ross Granville Harrison (left), Edmund Beecher Wilson (right, with teacup), and other family members enjoying one of many picnics at the beach. The life of the lab and life on the beach merged in many ways. Many MBL scientists report meeting collaborators on the beach or through their children who had become friends.

Marine Studies at the MBL

Having a permanent lab with buildings in which to store materials and equipment along with laboratory benches and meeting rooms made the MBL attractive for multiple purposes. In its early years, summer participants arrived with a mix of goals and needs. They got their feet wet watching the abundant life-forms in the intertidal zones, and they scooped and transported specimens to observe them in the lab. Often that meant putting on clumsy waders and carrying buckets, glass jars, and smaller tubes to collect specimens and take them away. Having shared teaching facilities near the labs for independent researchers helped inspire everyone to work and learn together. Observing the different organisms over time yielded a great deal of new information about marine invertebrates and small fish that then allowed researchers and students to compare what they found. The coexistence of investigation and instruction provided a strong foundation for the young MBL.

In his history, Lillie discussed the importance of the courses for bringing young people to learn about biological research alongside the more established teachers and independent researchers to create a mix of generations. Lillie noted: "There is a large body of students in this country enthusiastic about biology and eager to make a firsthand acquaintance with marine life; on the other hand, there are their instructors, many of whom are anxious to carry on their investigations in the summer under conditions furnished by the Laboratory but who are often without means to do so. The Laboratory brings them together in Woods Hole" (Lillie 1944, 86). This was true in 1888, and it remains true now.

The long-term courses included invertebrate zoology (which ran from 1888 to 1988 in various evolving forms), marine botany (which became marine biology from 1963 to 1967 and experimental marine botany between 1968 and 1979), physiology (from 1892 to the present), and embryology (from 1893 to the present) (for a more detailed discussion, see MacCord, in this volume). Lillie explained how the courses work, noting: "Naturally, the courses have been kept abreast of advancements in their fields and have been modified somewhat by successive instructors, but in their essential conceptions they have remained the same" (Lille 1944, 87). This is still true, with the addition of new, highly technical courses such as gene regulatory networks or computational approaches, special topics courses including the history of biology, and new secondary education courses.

Students and researchers could collect their own materials, but they also relied on the Supply Department for additional organisms. Under the guidance of Director George Gray, the growing department provided a boat and captain as well as a catalog of what was found and preserved specimens of many organisms. Since 1994, the George M. Gray Museum collection has been housed at Yale University and has

FIGURE 1.4. MBL botany class collecting specimens, 1895. Photograph by Baldwin Coolidge (1845–1928). https://history.archives.mbl.edu/archives/items/botany-course-photograph-1895-collecting-specimens.

added more specimens, photographs, and other materials to provide a rich record of New England marine biodiversity (Peabody 2018).[4]

Study of individual organisms focused especially on physiology, heredity, and development at first, then later also on neurobiology. These fields benefited from the ready availability of living material collected fresh from the sea. The wide diversity of organisms also allowed comparative studies. For several early summers, for example, Whitman assigned students and colleagues to study cell lineage of different organisms using similar methods so that they could map out and compare consistent and distinctive details of cell lineage in development. What were the exact patterns of cell division for each, and how did those patterns compare? Such research informed understanding of individual embryological development. It also allowed examination of evolutionary relationships, with the assumption that greater similarities and homologies of form or patterns reflected evolutionary closeness. And this approach brought the individual researchers into a community, comparing details, learning from each other, and raising new questions as they found reassuring similarities or surprising differences (Maienschein 1978).

The Woods Hole summer involved a lot of time observing outdoors and also embracing the new experimental ethos of biological discovery of the 1890s. For example, at the MBL in 1894, William Morton Wheeler presented a translation of Wilhelm Roux's introduction to his new *Archiv für Entwickelungsmechanik* on "developmental mechanics" with its call to carry out manipulative experimental research at the bench (Roux 1895). At the MBL, young scientists who became luminaries in biology took up this call, including Thomas Hunt Morgan, Jacques Loeb, Edmund Beecher Wilson, and many others who thrived under the laboratory's welcoming leadership of Charles Otis Whitman and then Frank Rattray Lillie.

As historians have argued vigorously, this experimental turn was not a matter of what the historian Garland E. Allen has called a "revolt from morphology." There was no rejection of descriptive methods or morphological studies but rather a combining of these approaches with study of physiology, embryology, evolution, and other aspects of what was being labeled *biology* (Allen [1975] 1978; Maienschein, Rainger, and Benson 1981; Caron 1988).

This is not the place to provide extensive details about all the research carried out at the MBL over its over 130 years and how it has changed over time. Some examples suffice to show the early focus, and Lillie (1944, chap. 6) provides much more detail up to that point, while annual reports give a good sense of what researchers were asking and what they were doing. Contributions that provided a foundation for courses and for further research included, for example, Whitman's research on cell lineage, along with extensive work on, respectively, *Clepsine* (leeches, by Whitman), *Crepidula* (slipper snail) and several ascidians (by Conklin, on whose work see Steinert [2016]), *Nereis* (by Wilson), and *Unio* (freshwater clam, by Lillie). Jacques Loeb studied parthenogenesis, regeneration, and fertilization and engaged in lively debates with Lillie and his student Ernest Everett Just about the mechanisms of fertilization that continue to inspire discussion about clashing assumptions informing biology. Morgan's study of regeneration gave way to his focus on heredity and chromosomes. Wilson's study of cells led to his classic *The Cell in Development and Inheritance* in 1896, with editions in 1900 and 1925, and it inspired the collective study of cells that led to Edmund Cowdry's edited volume *General Cytology* in 1924. Additional research in physiology, led by Loeb, and protozoology, led by Gary Calkins with a course from 1919 to 1940, added diversity to the research and educational offerings.

Since World War II, the MBL has added important study of ecosystems and the factors that shape a healthy environment. The Ecosystem Center makes up one of three research centers that also include the Eugene Bell Center for Regenerative Biology and Tissue Engineering and the Josephine Bay Paul Center for Comparative Molecular Biology and Evolution. Lively programs in cell biology, imaging, and biodiversity studies through the Marine Resources Department add to the current core areas of MBL research and education.

The MBL in Historical Context: 1888–2018

Director Whitman's initial vision to build a laboratory serving a diversity of researchers and students from different institutions remains intact. His passionate commitment that the MBL remain independent has been replaced by the economic reality that true independence is simply too expensive in the twenty-first century. Nearly every small private independent research lab has affiliated with a university or some other larger institution. The MBL has struggled financially from the beginning but has always managed to find a way to make things work.

Following the Naples model, in its first decades the MBL invited universities to subscribe to tables in exchange for the right to send some students for the summer. Independent investigators rented laboratory space and paid for any equipment or supplies that they needed. Students paid to attend courses. And the relatively modest wooden buildings did not cost much to maintain. Nonetheless, for the first half century, the MBL faced a budget deficit nearly every year because it kept expanding. The archival records make clear that Whitman felt that it was more important to grow than to worry about financial details. Fortunately, philanthropic donations made up the difference, and, in fact, Frank Lillie's brother-in-law Charles Crane made up a very high percentage of the shortfall each year. Crane also funded the first permanent brick building and supported establishment of a permanent library (Maienschein 1985b). Crane died in 1939, and, while Lillie had stepped down as MBL director in 1925, he remained president of the MBL board of directors until he retired in 1942. At that point, the MBL had to develop new funding models (Maienschein 1985b, 1989).

Fortunately for institutions like the MBL, by the 1940s the US federal government had begun to provide funding for scientific research. With external grants, researchers could take their funding with them to independent research stations such as the MBL. A broader range of researchers who had access to these funds could rent laboratory space and housing with grant dollars. Even though these researchers were not actually paying for the full cost of using space, the approach mostly worked because the courses were also bringing in funding and donations added to the mix. By juggling priorities, obtaining special funds to build new space that attracted more people, attracting donors and special earmarked public funding from the state and federal governments, and developing a year-round program, the MBL grew and managed to remain close to solvent but never with a significant financial buffer.

From very early, the MBL has had a national and international impact much larger than its resources would suggest possible. From the 1920s into the 1940s, for example, China sent a number of scientists to the MBL. They studied there and returned to establish marine stations in China (see Luk, in this volume). Lijing Jiang and Kate MacCord have documented this movement in their digital exhibit for the MBL History Project (Jiang and MacCord 2015).

Another connection—this one with Japan—is visible through a document left at the Misaki Marine Biological Station, which had been

FIGURE 1.5. Japan's emperor Hirohito and Sears Crowell at the MBL, 1975. Courtesy of MBL History Archives. https://hpsrepository.asu.edu/handle/10776/12863.

used as a Japanese naval facility during World War II. The document, written by the Japanese researcher Katsuma Dan and found by a US naval officer (see Ericson, in this volume), suggests that dedication to science should outweigh politics and war. Before taking up his position at the University of Tokyo and the Misaki Station, Dan and his American wife, Jean Clark Dan, had spent summers at the MBL studying cell biology and development. Their respect for the MBL inspired Dan's Japanese graduate student Shinya Inoué to pursue what became a distinguished career at the MBL, using exceptional light microscopic techniques to study cellular structure.

The Misaki Marine Biological Station was founded in 1896. The Japanese admiration for marine studies played out at the highest political levels with the emperor Hirohito's work. In 1975, when Hirohito was planning the agenda for a trip to the United States, he asked to visit the MBL. There, he declined afternoon tea and the proposed socializing because he preferred to spend his time looking through the microscope at the hydroids and related organisms that he had studied and summarized in *The Hydroids of Sagami Bay* (Hirohito 1988, 1995).

The MBL Today

In 2013, the MBL institutional situation changed. From its founding, the MBL Corporation had owned and technically managed the MBL. Corporation members were scientists who had spent time at the lab and paid annual dues to belong. Yet, by 2013, financial strains had become untenable. At a special meeting, the corporation overwhelmingly voted to make the University of Chicago the sole member and hence owner. With the long and rich connections between the university and the MBL, as documented in all of the histories of the MBL, the alliance appears to make sense (Maienschein 2014). Such a change, after more than 125 years, has also brought challenges. The initial mission to provide opportunities for instruction and investigation remains the guiding vision. But the institution has returned to a core focus on marine science, a commitment that had been lost at times when summer visitors brought diverse research projects to the lab.

Discussing the Chicago-MBL affiliation, the university president, Robert Zimmer, pointed to the lab's long success as a "convening power." The MBL draws people—because of the location, the rich history of discovery and education, and the opportunities it provides for networking and working with other scientific leaders. As MBL scientist Ron Vale put it, "how lucky can one get" to have the opportunity to spend time in this place with these people (Vale 2012).

In addition, the new mission emphasizes streams of strategic focus on new discoveries emerging from the study of novel marine organisms, encompassing research in regenerative biology, neuroscience, sensory physiology, and comparative evolution and genomics; the study of microbiomes and microbial diversity and ecology in a variety of ocean and terrestrial habitats; cutting-edge imaging and computation, illuminating cellular function and previously unknown biology; and organismal adaptation and resilience in the face of global change and rapidly changing ecosystems.[5]

The MBL's three centers and focus on marine organisms and related scientific programs, along with expansion of the year-round program, are attracting a dynamic group of scientists at all levels from distinguished researchers to energetic postdoctoral fellows to undergraduate assistants. These are all changes from the situation that prevailed before the 1970s. Then, only very few researchers stayed longer than the summer, individuals colonized small projects rather than far-reaching centers, and courses still included some undergraduates becoming familiar with biology (e.g., through the invertebrate biology course). Today, the year-round faculty is in the dozens, the centers provide new ways for individuals to connect ideas and explore cross-cutting projects, and the courses attract diverse international groups of enthusiastic and hardworking students.

The MBL has adapted. The future remains to be written, of course, and the affiliation with the University of Chicago has occurred so recently that its success remains to be seen. Yet the optimism for and energy devoted to innovation is palpable. Accepting the fact that the old models needed changing has brought new vitality and creativity to the MBL. New imperatives to study ecosystems or organisms perceived as being under siege by climate change, such as corals or saltwater marshes, suggest the need for changing priorities. An initiative is looking at ideas of regeneration across the scales of organisms, ecosystems, and microbial communities. The MBL hosts the National Xenopus Resource. And a new director has brought his own initiatives and research directions to inspire additional research and education. It will be fascinating to watch as the MBL and other marine stations continue to evolve and marine studies take new directions at the beachside and at the benchside. As the New York University physician, writer, and longtime MBL resident Gerald Weissmann puts it, the MBL is "the place where a lot of people fall in love with science" (Goldberg 1997).

Notes

1. See, e.g., https://marinebio.org/creatures/marine-biology/history-of-marine-biology.
2. See http://www.szn.it/SZNWeb/showpage/107?_languageId.
3. For Whitman's letters, see http://www.mblwhoilibrary.org/sites/default/files/Whitman.pdf (finding aid) and https://hpsrepository.asu.edu/handle/10776/11546 (the documents themselves).
4. See http://peabody.yale.edu/collections/invertebrate-zoology/george-m-gray-museum-collection.
5. See https://www.mbl.edu/strategic-themes.

References

Allen, Garland E. (1975) 1978. *Life Science in the Twentieth Century*. New York: Wiley. Reprint, Cambridge, Cambridge University Press.

Anderson, Thomas R., and T. Rice. 2006. "Deserts on the Sea Floor: Edward Forbes and His Azoic Hypothesis for a Lifeless Deep Ocean." *Endeavour* 30:131–37.

Armitage, Kevin. 2009. *The Nature Study Movement: The Forgotten Popularizer of America's Conservation Ethic*. Lawrence: University of Kansas Press.

Barber, Lynn. 1984. *The Heyday of Natural History, 1820–1870*. New York: Doubleday.

Benson, Keith R. 1979. "William Keith Brooks (1845–1908): A Case Study in Morphology and the Development of American Biology." PhD diss., Oregon State University.

———. 1988. "The Naples Stazione Zoologica and Its Impact on the Emergence of American Marine Biology." *Journal of the History of Biology* 21:331–41.

Caron, Joseph A. 1988. "'Biology' in the Life Sciences: A Historiographical Contribution." *History of Science* 26:223–68.

Conklin, Edwin Grant. 1944. "The Geography and Early History of Woods Hole." In *The Woods Hole Marine Biological Laboratory*, by Frank R. Lillie, 1–19. Chicago: University of Chicago Press.

Curtis, Winterton. 1955. "Good Old Summer Times at the M.B.L. and Rhymes of the Woods Hole Shores." *Falmouth Enterprise*, August 12, 19, 26, September 2, 9.

Dean, Bashford. 1894. "The Marine Biological Stations of Europe." *Biological Lectures of the Marine Biological Laboratory of Wood's Hole* 1893:211–34.

de Bont, Raf. 2015. *Stations in the Field: A History of Place-Based Animal Research, 1870–1930*. Chicago: University of Chicago Press.

Dexter, Ralph W. 1980. "The Annisquam Sea-Side Laboratory of Alpheus Hyatt, Predecessor of the Marine Biological Laboratory at Woods Hole, 1880–1886." In *Oceanography: The Past*, ed. Mary Sears and Daniel Merriman, 94–100. New York: Springer.

———. 1988. "Biology and Marine Biology Institutions Introduction: Origins of American Marine Biology." *American Zoologist* 28:3–6.

Galtsoff, Paul. 1962. *The Story of the Bureau of Commercial Fisheries Biological Laboratory, Woods Hole, Massachusetts*. Circular no. 145. Washington, DC: US Department of Interior.

Goldberg, Carey. 1997. "Summer in Woods Hole: Sea, Science and Synergy." *New York Times*, September 2.

Goldschmidt, Richard B. 1951. "Charles Atwood Kofoid, 1867–1947." *National Academy of Sciences Biographical Memoirs* 26:121–51.

Groeben, Christiane. 1984. "The Naples Zoological Station and Woods Hole." *Oceanus* 27:66–69.

———. 1985. "Anton Dohrn—the Statesman of Darwinism." *Biological Bulletin* 168 (suppl.): 4–25.

Hirohito, Emperor Showa. 1988. *The Hydroids of Sagami*. Tokyo: Biological Laboratory Imperial Household.

———. 1995. *The Hydroids of Sagami Bay, II: Thecata*. Annotated by Mayumi Yamada. Tokyo: Biological Laboratory Imperial Household.

Huxley, Leonard. 1913. *Life and Letters of Thomas Henry Huxley*. New York: D. Appleton.

Jiang, Lijing, and Kate MacCord. 2015. "China at the MBL, 1920–1945." MBL History Project digital exhibit. https://history.archives.mbl.edu/exploring/exhibits/china-mbl-1920-1945.

Kofoid, Charles A. 1898. "The Fresh-Water Biological Stations of America." *American Naturalist* 32, no. 378:391–406.

———. 1910. "The Biological Stations of Europe." *US Bureau of Education Bulletin*, no. 4, whole no. 440. Washington, DC: US Government Printing Office.

Kohlstedt, Sally Gregory. 2015. "Nature Study." In *Companion to the History of American Science*, ed. Georgina M. Montgomery and Mark A. Largent. Wiley Online Library. https://onlinelibrary.wiley.com/doi/abs/10.1002/9781119072218.ch36.

Lillie, Frank R. 1944. *The Woods Hole Marine Biological Laboratory*. Chicago: University of Chicago Press. Reprint, *Biological Bulletin*, vol. 174, no. 1S (January 1988). https://www.journals.uchicago.edu/toc/bbl/1988/174/1S.

Maienschein, Jane. 1978. "Cell Lineage, Ancestral Reminiscence, and the Biogenetic Law." *Journal of the History of Biology* 11:129–58.

———. 1985a. "Agassiz, Hyatt, Whitman, and the Birth of the Marine Biological Laboratory." *Biological Bulletin* 168 (suppl.): 26–34.

———. 1985b. "Early Struggles over the M.B.L.'s Mission and Money." *Biological Bulletin* 168 (suppl.): 192–96.

———. 1988. "History of American Marine Laboratories: Why Do Research at the Seashore?" *American Zoologist* 28:15–25.

———. 1989. *100 Years Exploring Life, 1888–1988*. Boston: Jones & Bartlett.

———. 2014. "A History of Rich Collaboration: The University of Chicago and Marine Biological Laboratory." Keynote lecture presented at the University of Chicago–MBL retreat.

Maienschein, Jane, Ronald Rainger, and Keith R. Benson. 1981. "Were American Morphologists in Revolt?" *Journal of the History of Biology* 14:83–87. Introduction to a special issue focused on Garland E. Allen's "revolt from morphology."

Miralto, Antonio. 1985. "What Laboratories for What Science?" In "The Naples Zoological Station and the Marine Biological Laboratory: One Hundred Years of Biology." *Biological Bulletin* 168, no. 3 (suppl.): 203–4.

Morgan, Thomas Hunt. 1901. *Regeneration*. New York: Macmillan.

Muka, Samantha. 2017. "Inland Oceans: Aquarium Technology and the Development of Marine Biological Knowledge, 1880–Present." PhD diss., University of Pennsylvania.

"The Naples Zoological Station and the Marine Biological Laboratory: One Hundred Years of Biology." 1985. *Biological Bulletin*, vol. 168 (suppl.).

Nyhart, Lynn K. 2009. *Modern Nature: The Rise of the Biological Perspective in Germany*. Chicago: University of Chicago Press.

Osborne, Michael A. 1994. *Nature, the Exotic, and the Science of French Colonialism*. Bloomington: Indiana University Press.

Pauly, Philip. 1988. "Summer Resort and Scientific Discipline: Woods Hole and the Structure of American Biology, 1882–1925." In *The American Development of Biology*, ed. Ronald Rainger, Keith R. Benson, and Jane Maienschein, 121–50. Philadelphia: University of Pennsylvania Press.

Richards, Robert J. 2008. *The Tragic Sense of Life: Ernst Haeckel and the Struggle over Evolutionary Thought*. Chicago: University of Chicago Press.

Roux, Wilhelm. 1895. "The Problems, Methods, and Scope of Developmental Mechanics." *Biological Lectures Delivered at the Marine Biological Laboratory in Woods Hole* 1894:149–90.

Russell-Hunter, W. D. 1985. "The Woods Hole Laboratory Site: History and Future Ecology." In "The Naples Zoological Station and the Marine Biological Laboratory: One Hundred Years of Biology." *Biological Bulletin* 168, no. 3 (suppl.): 197–99.

Simon, Hans-Reiner, ed. 1980. *Anton Dohrn und die Zoologische Station Neapel*. Frankfurt a.M.: Edition Erbrich.

Southward, A. J., and E. K. Roberts. 1987. "One Hundred Years of Marine Research at Plymouth." *Journal of the Marine Biological Association* 67: 463–64.

Steinert, Beatrice. 2016. "Drawing Embryos, Seeing Development." *The Node.* http://thenode.biologists.com/drawing-embryos-seeing-development/discussion.
Thomson, Charles Wyville. 1873. *The Depths of the Sea.* London: Macmillan.
Vale, Ronald. 2012. "How Lucky Can One Be? A Perspective from a Young Scientist at the Right Place at the Right Time." *Nature Medicine* 18:xvi–xviii.
Wilder, Burt G. 1898. "Agassiz at Penikese." *American Naturalist* 32:189–96.
Wilson, Edmund Beecher. 1896. *The Cell in Development and Inheritance.* New York: Macmillan.

TWO

Marine Biology Studies at Naples: The Stazione Zoologica Anton Dohrn

CHRISTIANE GROEBEN

"[Zoology] with all the results obtained up to now is nothing else but a product of individual initiative and serendipity. It is related to astronomy, mechanics and their way of scientific organization as Garibaldi's irregulars to a Prussian army Corps. . . . It lacks indispensable organization," Anton Dohrn wrote in his 1872 programmatic essay on the present state of the field of zoology and the foundation of zoological stations. While morphology—that is, systematics, anatomy, and embryology—was by then rigidly institutionalized, zoology, on the other hand, being the systematic and basic study of living organisms, required social recognition and participation, Dohrn argued, now that it had been given the "greatest modern research tool," namely, the "genetic principle" or Darwin's theory (Dohrn 1872b, 139–40, 137).

Very early in his scientific career Dohrn began to consider well-organized research facilities close to the seaside as essential to guaranteeing serious and continuous studies on marine organisms. Optimal research needed optimal structures. Through the foundation in 1872 of what has since become the Stazione Zoologica Anton Dohrn but is generally known as the Stazione Zoologica di Napoli (SZN),[1] Dohrn himself turned his theory into practice. This vision needed energy, imagination, courage, self-confidence, and persistence as well as scientific, political, and financial support at the right moment and in the right place. The biography of the Naples Station is an unlikely combination of all these factors.

Dohrn envisioned the SZN as an independent and international research institute for marine biology. Owing to the sustainable vision of its founder, the Naples Station has maintained its scientific autonomy throughout political and administrative changes. With World War I, the station ceased to be the private property of Dohrn's son and successor, Reinhard (1880–1962). In 1923, it was changed into an Italian nonprofit institution (*ente morale*); in 1982, it became a national research institute under the supervision of the Italian Ministry of Instruction, University and Research (Ministero dell'istruzione, dell'università e della ricerca). This implied a paradigm shift from guest research facility to research institute with in-house research programs but did not interfere with the station's original mission, that is, to study the fundamental and applied biology of marine organisms and ecosystems and their development in an integrated and interdisciplinary approach.

Anton Dohrn's Education

Felix Anton Dohrn (known as Anton; fig. 2.1),[2] the youngest of four children, was born in 1840 in Stettin (today Szczecin, Poland) to Carl August Dohrn (1806–92) and Adelheid Dohrn née Dietrich (1803–83). The Dohrn family belonged to the comfortably off Stettin bourgeoisie thanks to the commercial fortune of grandfather Heinrich Dohrn (1769–1852), a partner in an import business of colonial products and a cofounder (1817), a main shareholder, and the director of a sugar refinery, the Pommersche Provinzial Zuckersiederei, one of the first industrial enterprises in Prussia. Anton's father, Carl August, the only son of Heinrich, had studied law and commerce but did not show much interest in his father's business. On the advice of Alexander von Humboldt (1769–1859), Heinrich granted Carl August permission to travel and follow his many musical, literary, and scientific interests (Dohrn 1983, 30–81). As a collector of insects and the editor of the *Stettiner Entomologische Zeitung* (1840–87) and *Linnaea entomologica* (1846–66) and president (1843) of the Stettin entomological society, Carl August Dohrn became an internationally recognized entomologist.

The Dohrn children grew up in an open-minded atmosphere where it was more important to know one's Goethe or recognize a Beethoven

FIGURE 2.1. Anton Dohrn, Vevey, 1871. © Stazione Zoologica Anton Dohrn di Napoli, MAB SZN-AS, Lb.2.12. Reproduced with permission.

symphony than to be perfect in Greek grammar. The passion for music and literature, and Goethe in particular, was also part of Anton's heritage. He owed to his father his "intellectual protoplasm," as he gratefully admitted years later (Dohrn 1897, 34), although the father-son relationship had never been easy. In his early years, Anton was involved in his father's scientific activities as secretary and assistant editor. At the age of seventeen, he published his first paper (Dohrn 1858) and proudly told his father that a bedbug had been named after him (Anton) (*Notiobia* [*Diatypus*] *dohrnii* [Murray, 1858]) (Heuss 1991, 32). His future was already clear when he rather self-consciously told his teachers at the age of nineteen: "The study I selected holds me, body and soul. It is the sciences, particularly physiology and anatomy. Whatever can be achieved through love of the subject and natural talent, I hope to achieve. The external goal I try to reach is a professorship at a university" (Heuss 1991, 31).

In 1860, Dohrn started his studies at Königsberg University, moving thereafter to Bonn, Jena, and Berlin. His initial self-confidence in his studies and career soon faded away. Zoology, as he had been taught the subject in the early 1860s, failed to excite him. He was bored and was seriously thinking of becoming a book dealer. This situation changed radically when, in 1862, he moved to Jena and was introduced by Ernst Haeckel (1834–1919) to Darwin's work and theories.[3] As he told his Berlin friends and benefactors Fanny Lewald and Adolf Stahr:[4] "Through Darwin I felt hit by an excitement that reached my innermost being. All of a sudden, zoology proved itself for me to be an extraordinary source of human insight, and I gave myself over to it heart and soul. Even though the work is dull and tedious—Darwin and his theory give it a frame that could not be more magnificent. And so I started again to work myself through legs, antennas, bones and feathers."[5]

Dohrn promised himself to dedicate his further life to finding proofs for Darwin's theories in his own research and to creating this opportunity for others through the foundation of a zoological station. He felt ready "to change Darwinian theory into Darwinian practice."[6] His dissertation (completed in Breslau in 1865) dealt with the anatomy of three exotic bugs (Dohrn 1866a), "of course richly provided with Darwinian tendencies and point of views,"[7] but in the meantime his research interests had already shifted from insects to arthropods. In 1868, he was habilitated at Jena with a study on the embryology and genealogy of arthropods (Dohrn 1868b) and held his inaugural lecture on "Kant's relationship to the theory of descendance" (Müller and Wenig 1993).

Anton Dohrn's Meeting with Marine Life

In spring 1865, Haeckel asked Dohrn—who by that time was working on his PhD thesis—to join him on an expedition. Several places were discussed: Dalmatia, including southern Italy, Gibraltar or Alger, and Nice or Villefranche. "I would love to go south," Dohrn remarked,[8] but he also indicated that he would settle even for Iceland or the North Pole as long as he could learn something.[9] In the end, Haeckel took his assistants and students to Helgoland, the place where, in 1854, his teacher Johannes Müller (1801–58) had introduced him to the marine fauna and where he had decided to concentrate his research on marine biology.[10] Dohrn tried to go well prepared, taking along at least six reference books, two microscopes, one Brücke magnifying glass, and several nets.[11] They were a group of at least five,[12] but only Dohrn stayed for the entire period,[13] a fact appreciated by Haeckel, who also enjoyed Dohrn's pleasant company and permanent good spirits (Uschmann 1959, 65).

For Dohrn, this was the first contact with marine fauna—he worked on crustaceans (Dohrn 1866b)—and he also quickly became familiar with the technical difficulties associated with catching and preserving marine organisms for serious and extended observation. This situation spurred Haeckel and Dohrn to discuss how to improve these technical impediments to marine research. As Haeckel remembered in 1909 in a letter to Anton Dohrn's son Reinhard: "It has been forty-four years now that your father, one of my most excellent students, has collected and investigated together with me the marine fauna at Helgoland. When we both had to carry every day by our own hands a heavy chest with four big jars from the beach to the boat, the project of a zoological station was born."[14]

In summer 1867, Dohrn worked at the Hamburg Aquarium on the development of crayfish. Having seawater tanks right next to his workbench was quite a useful experience. He then moved to Millport (Scotland), where he also returned the following summer to study the development of Pycnogonids. Together with the amateur zoologist David Robertson (1806–96), he constructed a portable aquarium consisting of three tanks[15] and a very simple but innovative water-circulation system that allowed the observation of marine organisms over a certain stretch of time. From Millport, Dohrn and his aquarium moved for winter 1868–69 on to Messina, as most of Haeckel's students were encouraged to do because Haeckel himself had worked there quite successfully in 1859–60. In Messina, Dohrn met his friend and colleague from Jena the Russian zoologist Nikolai Micloucho-Maclay (1846–88), who was working there on the fish brain.[16] Dohrn installed his portable aquarium, and, through observation, he was able to answer, in an affirmative way, the widely discussed question of whether the transparent, paper-thin, flat, long-legged crayfish *Phyllosoma* was the larva of the spiny lobster *Palinurus* (Dohrn 1870). The "Prussian and the Russian sirs"—as they were soon called by the local fishermen—shared living and working quarters and discussed the hardships of lab work away from a lab, such as the lack of instruments, reference books, and discussion with peers and local experts about animal supply as well as language difficulties and lodging problems. Their results never seemed to match their efforts. A facility offering all the lacking features on site would really save effort and precious time, they argued. After seeing two Austrian frigates leaving the port of Messina for a circumnavigation, Maclay and Dohrn started to daydream about how convenient it would be to travel to faraway places and always find a zoological station on arrival (Dohrn 1872b, 144). Hence, they decided to cover the globe with a net of zoological stations connected one to the other just like the stations of a railway network, and they started with Messina.

It was at Messina that Anton Dohrn entered a new phase in his life: he started his career as a science manager. The previous four years at Jena, as assistant to Haeckel, had led him to accept the fact that a professorship was not what really suited his talents and potentials. Even more than Haeckel, his teacher Carl Gegenbaur (1826–1903) expressed doubts about his ability to pursue a serious, standard academic career, one including painstaking and consistent research (Uschmann 1959, 66). Dohrn's annelid theory of vertebrate origin as alternative to the amphioxus-ascidian theory supported by Haeckel and Gegenbaur contributed to further undermine his scientific credibility.[17] He admitted to friends: "Proper zoological work has elements that do not appeal to me and do not at all take into account an inner need, which is to occupy myself in a practical way, to make an impact on the outside world, to be of service to others." To his wife he wrote: "To create, to organize, to develop—this is my need, even passion" (Groeben 1985, 8, 4).

The initial warm friendship with Haeckel and the shared enthusiastic fight for public recognition of Darwin's theory had also cooled. After Dohrn's departure from Jena, the two never interacted again on a personal level.[18] This decision did not mean that Dohrn stopped doing research, but, beginning at Messina, his professional life split into two parallel tracks that complemented each other: doing and organizing research. His studies of the phylogeny of arthropods ended with his 1881 monograph on Pantopoda (Dohrn 1881b), whereas the research line he followed for the rest of his life started in 1875 with a study of the origin of vertebrates and the principle of succession of functions[19] (Dohrn 1875; Ghiselin 1994). Twenty-five detailed studies followed over the years. The quantity of new facts brought to light by Dohrn earned him an honored place in the field of animal morphology and in particular with regard to the most difficult of all questions within this field, the genesis of the vertebrate head (Boveri 1912, 463).[20]

Zoological Stations: Messina

With their efforts to do serious research in marine biology at Messina, Dohrn and Maclay continued a tradition that had started during the first half of the nineteenth century when German, Swiss, British, Russian, and Scandinavian naturalists started to travel to the Mediterranean to places such as Nice, Trieste, Naples, and Messina to explore the unknown marine world (Groeben 2008). Lazzaro Spallanzani (1729–99), Johannes Müller, and August David Krohn (1803–91), to name a few, had successfully worked at Messina, and the richness of the marine fauna in the strait of Messina was already well-known through numerous publications. In fact, by the 1860s, Messina was known as the "Mecca of the German Privat-Dozent" (Dohrn 1897, 7). These science tourists distinguished themselves from the culture tourists in that they combined scientific activity (such as working on the beach or exploring the unknown marine world) with comfortably organized visits to well-known Italian attractions in art and nature (Groeben 2008). Karl Ernst von Baer (1792–1876), Christian Gottfried Ehrenberg (1795–1876), Johannes Müller, Krohn, and Haeckel also belonged to this pioneer generation when useful firsthand information was practically nonexistent. When, for example, von Baer went to Trieste in 1846 to study sea urchin development, he arrived three months earlier than he needed to because he had no way to ascertain when the spawning period would occur. Timing was not the only problem. The fishermen were also not trained to look for specific, noncommercial organisms. And, when he finally had a few eggs to study in his makeshift lab-bedroom, the chambermaid threw them away thinking they were rubbish (Groeben 1993, 16, 43).

Dohrn belonged to the second generation of science tourists who could already rely on the experiences of their teachers; they could find information about places, methods, and availability of organisms, and they knew about local contacts, fishermen in particular. To travel well, cheaply, and successfully was not easy, Dohrn argued (Dohrn 1872b, 140); while "pleasure travelers" could rely on all kinds of advice and guidebooks, the "scientific traveler" could rely only on his or her own resources to make a trip successful, which always entailed a considerable loss of energy, time, and money. In addition, one had to face new and different linguistic and cultural environments, and, when exploring the fascinating marine world, it was sometimes difficult to concentrate on a single project. Dohrn spent almost five months in Messina, time enough to draw conclusions from his experiences and set his "creative imagination"[21] in motion to develop a scheme for constructing zoological stations. He decided to leave behind his equipment and research diary for others to come. Friends and the Swedish-Norwegian consul Julius Klostermann promised to take care of them. "The Zoological Station of Messina has now been established. I retain it an important progress for our science should we succeed in getting beyond the embryonic stage," he told his father (Groeben 1985, 8), and explained to Darwin in more detail:

> Having stayed now several times on the seashore for zoological studies, I have found how difficult it is to study Embryology without an Aquarium. This want has suggested me the idea of founding not only Aquariums, but Zoological Stations or Laboratories on different points of our European coast. Such a Station should consist of a little house of perhaps four rooms, an Aquarium connected with the sea and the house,—the Aquarium of perhaps 60 feet in Cubus,[22] where one might have streaming water,—a boat for dredging work, dredges, nets, ropes,—in short, all that is necessary for a marine Zoologist. Besides glasses larger tumblers, bottles; Acids and other chemical objects, and lastly a library. All this might be had at a not too high prize at such a place as Messina, where I thought of founding the first Station.[23]

Carl Vogt, Carl Theodor von Siebold (1804–85), Haeckel, and others were very much in favor of Dohrn's plan and offered their support to gain public acceptance and financing. Darwin replied: "As far as my judgment goes, I can feel no doubt that at present embryological investigations on the lower marine animals are of the utmost importance; & for this purpose your scheme offers obvious facilities."[24] Karl Ernst von Baer confirmed that he "recognize[d] in it [Dohrn's enterprise] a new era for investigations in natural history," in particular "on such a rich seabed as the Neapolitan" (Groeben 1993, 43, 105).

This first campaign in favor of zoological stations was concentrated on Messina, but, in spring 1870, Dohrn suddenly focused on Naples as the best place to build a zoological station. On his twenty-ninth birthday (December 29, 1869) he had expected to receive the same amount of money that had been given to his two elder brothers on the same occasion, but his father considered his project to be complete fantasy and not worth any investment. On a visit a few days later to the recently opened aquarium in Berlin,[25] he was impressed by the considerable number of visitors that it attracted. On the stagecoach to Jena, he suddenly realized that he could resolve his financial problem by building a big aquarium in the Mediterranean, the richest sea of Europe, where the revenues would easily also cover the running costs of a small laboratory. The obvious location for such an aquarium was Naples, one of the largest and most attractive cities in Europe with more than 500,000 inhabitants and about 30,000 tourists every year (Vogt 1871). A growing interest in marine life, the need for an unlimited supply of living marine organisms for research in morphology and embryology, Dohrn's own marine experiences, his championing of Darwinism, and his need to affirm himself with respect to his father and his teachers, all this converged in the creation of the Naples Station.

Zoological Stations: The Naples Station

From Dohrn's decision in 1870 to transfer his project to Naples until the opening of the Naples Station,[26] almost four years and an amazing variety of roadblocks, openings, misunderstandings, support, and coincidences passed, a story masterfully told by Anton Dohrn's biographer Theodor Heuss (Heuss 1991; see also Groeben 2010; and Florio 2015). Dohrn went to Naples in spring 1870 to start a site search and establish contacts with local city and academic authorities. It was not easy to convince the city council of the usefulness of a zoological station, but, in the end, he was given a plot of land, free of charge, right on the seashore in front of the exclusive city park Villa Reale (today Villa Comunale). He had to provide the building costs.

The greatest challenge that Dohrn had to face was, however, the construction of the aquarium and a system of seawater supply throughout the whole building. He availed himself of the expertise of the English technician William Alford Lloyd (1815–80), whom he had met while working in Hamburg. Lloyd, a trained bookbinder, had developed a passion for aquariums in the 1850s. He opened an "Aquarium Warehouse" in London in 1856 and invented and constantly improved the water-supply systems of public aquariums. Naples was the third modern aquarium built by Lloyd after Hamburg (1864) and the Crystal Palace (1871) (Groeben 2010, 152–54). After some difficulties with his first architect, Oscarre Capocci (1825–1904), a professor of architecture at the University of Naples, Dohrn asked his friend the sculptor Adolf von Hildebrand (1847–1921) to help with the facade of the building and hired the civil engineer Giacomo Profumo (?–1884) to supervise the construction. However, a recent study has shown that Dohrn himself should be considered the real architect of the Naples Station (Florio 2015). There was no model for such an intricate public and research space under one roof, but Dohrn knew exactly what he wanted and how he wanted it done. As he told his sister, it was going to be "a big establishment at the Mediterranean, as observatory and laboratory": "The house will be large-scale and spacious. The plans for the building, outside and inside, I have done myself."[27]

Before the foundation stone could be laid in March 1872, Dohrn's presence in Naples was interrupted by the Franco Prussian War (July 1870–May 1871). Dohrn enlisted and was assigned to an office in Kassel. At the same time, he continued to look for public support for his Neapolitan enterprise. In September 1870, during a fortnight's leave from the army, he was able to attend the annual meeting of the British Association for the Advancement of Science, held that year in Liverpool. There, he found a lively interest in his project among the participants. The general assembly unanimously voted in support of the following resolution, proposed by Dohrn: "The assembly declares that the foundation of zoological stations in different parts of the world will be of the most significant influence on the progress of biological science, and it considers the foundation of such a station in Naples as the first decisive step in this direction" (Heuss 1991, 107). The following year, a committee "for the purpose of promoting the Foundation of Zoological Stations in different parts of the World," of which Anton Dohrn initially was also a member, was nominated ([British Association for the Advancement of Science] 1872; see also Groeben 1982, 96 n. 43). This trip to England was also highly significant for Dohrn on a personal level: on September 26, he visited Charles Darwin, meeting him for the first time (Groeben 1982, 93–94).[28] Dohrn was surprised to find, not the sick-looking man he had expected, but "a tall, strong, grey bearded stature, full of life and cheerfulness and heart winning amiability." During their conversation, Dohrn noted, Darwin "took the most vivid interest in my Neapolitan plans, looking forward to seeing the Station well equipped," an interest he maintained until his very last letter to Dohrn. Two months before his death he wrote: "I am extremely glad to hear of the great success in all ways of your institution" (Groeben 1982, 94, 80).

Dohrn was also looking for consensus and support among peers in Germany. In September, he participated in the annual meeting of the Society for German Naturalists and Physicians[29]—more or less an equivalent to the British Association for the Advancement of Science—held that year in Rostock and also published an interesting eight-page leaflet with reports about and opinions of eminent German naturalists on the foundation of zoological stations (Dohrn [1871]). The physiologist Emil du Bois-Reymond (1818–96) envisioned experiments with the electric ray without the need to transport heavy equipment over the Alps. The anatomist Carl Gegenbaur underscored the fortunate choice of a site that would allow continuity of observations by multiple researchers, something that could be done only partially by individuals working in isolation. He considered well-equipped laboratories and a library to be important features of the station. Ernst Haeckel insisted on the need of laboratories on the beach in order to study the still-rather-neglected lower sea animals: "The most important and consequential discoveries on animal organization have been gained only through detailed investigation of the lower and lowest forms of animal life, which for the most part can be found only in the sea" (Dohrn [1871], 4). The physician, physiologist, and physicist Hermann von Helmholtz (1821–94) observed: "In particular the transformations of the various forms from each other that have become of highest interest for the whole development of the living creation have often been discovered strikingly late or rather incompletely" (Dohrn [1871], 5). The zoologist Rudolf Leuckart (1822–98) considered Dohrn to be a person who knew about the needs of science and capable of finding answers. He also appreciated the fact that the public aquarium would guarantee sustainability to the research facility, thus ensuring independence. The leaflet concludes with a supporting letter from Dohrn's friend Carl Vogt, who had tried several times—in Nice (1850–52), Naples (1863), and Trieste (1871)—to establish stations similar to what Dohrn intended to create in Naples. In fact, Dohrn always acknowledged Vogt as the "ancestor of the plan" (Groeben 1998).

Vogt had been correct about the need to find an answer to an urgent need in science—namely, to organize marine biological studies on a large scale—but he had chosen the wrong channels, focusing on political and administrative contacts. Dohrn took another line, advocating cooperation among scientists and the general public. Governments would not easily be induced to invest much money in the progress of science, he argued, but, if there was strong public pressure, the support of science would become a matter of science policy. He tried to achieve this public awareness through publications, at meetings, and by involving scientific, social, and political personalities. He also considered aquariums an essential and very substantial tool to keep the public interest alive. This recipe resulted in a research institution that administrated itself with public resources but was completely independent.

In the end, the first building, the central part of today's building, was ready in September 1873. Pumps and machines were kept in the basement, the public aquarium occupied the ground floor, while the upper two floors were reserved for research.[30] Dohrn felt that art (particularly music) and science were two parts of human culture, and he wanted them both to be present in his building. Initially, the second floor was therefore spilt up into two complementary parts. One big lab was equipped for research, while the room facing the south was dedicated to music. During summer 1873, the German painter Hans von Marées (1837–87) and Adolf von Hildebrand decorated the room with frescoes depicting scenes from Mediterranean life, but, instead of music, it soon (1876) came to house the library (Groeben 2000).

Already during the construction phase Dohrn had to make two decisions that shaped and guaranteed the future of the Naples Station. First, during his promotion campaign in favor of zoological stations as a requirement for modern biology, he realized that his original plan to offer hospitality to four guests at a time, including accommodation,[31] would be an inadequate answer to the increasing demand to use this research facility. He had to think big and dedicate all available space to research, which meant dispensing with an apartment for himself, in-house accommodation for his guests, and two of the four loggias. This change resulted in lab space for twenty scientists at a time.

The second major decision concerned finances. Dohrn succeeded in covering about 80 percent of the building costs (Dohrn 1876, 15–16) out of his own pocket (by then his father had changed his mind about his inheritance); the rest came from gifts and loans from friends and patrons, among them Frank Maitland Balfour (1851–82), Conrad Fiedler (1841–95), and Johann Nepomuk Czermak (1828–73). However, with an increase in research space, the income from the aquarium could no longer reasonably be expected to cover the running costs of the laboratory. As early as spring 1872, Dohrn started to think about renting out bench space or "working tables" to governments,[32] universities, and scientific institutions. Contracts defined the mutual obligations. For an annual fee, the contract partner had the right to send one scientist for one year to Naples, where he would find all that he needed for his research (Groeben 1993, 45). Thus, access to marine organisms was no longer an adventure or a courtesy extended to colleagues; instead, it became a rented service and was not linked to any program or specific research project. This system guaranteed Dohrn a regular income as well as public awareness of the Naples Station. The first tables were rented in 1873 by Strasburg University, the Netherlands, and Prussia, followed, in 1874, by Italy, Russia, Germany, and Cambridge University (Müller 1976, 155–87; see also Partsch 1980). Whoever wanted to work on marine organisms was welcome at Naples; there were only two conditions: (1) only scientists whose country had rented a table were admitted; one had to apply to the respective national contract partner to be granted the use of the "Naples table";[33] and (2) applicants had to be fully trained scientists, capable of working independently because no teaching or training was offered by the Station. Dohrn provided excellent working conditions, but the scientists were responsible for their own results. "Freedom in research" was the motto of Dohrn's life. The table system guaranteed scientists from different nations and research traditions open access to excellent working conditions and abundant research material. This table system was discontinued during the 1980s. By then, about ninety-four hundred scientists had spent research time in Naples.

The table system allowed Dohrn constantly to improve the facilities and adapt them to the growing number of applications as well as to the changing priorities in scientific research. By 1880, the station was already running out of space. A second building (1885–88) was added on the west side because the Departments of Preservation, Physiology, and Marine Botany had expanded and a lecture room for the training of navy officers and physicians in collecting and preserving marine organisms (Groeben 1990) was also needed. By the turn of the century, the expanding research in experimental biology required adequate space. The courtyard and the east wing were ready in 1906 to house the Departments of Physical and Chemical Physiology (fig. 2.2). A generous donation by Friedrich Alfred Krupp (1854–1902) and two subscription campaigns, one of which was headed by the German emperor, covered the substantial building costs (Müller 1976; Florio 2015).

Naples Station Organization

In spring 1876, after two years of regular activity, Dohrn summarized the scope and mission of his institute, stating:

> From the beginning of my venture I have attempted to create with the Zoological Station an institute that although taking into deepest consideration the present and

FIGURE 2.2. The Naples Zoological Station in 1950. Photograph by Franz Thorbecke. © Stazione Zoologica Anton Dohrn di Napoli, MAB SZN-AS, Lb.4.56. Reproduced with permission.

its needs should at the same time be far from serving exclusively contemporary currents in science. I have tried to include the advancement of the complete range of biology into the sphere of duties and impact of the institute, convinced that only this will guarantee the station a sustainable future. The zoological work of the sistematist and faunist as well as that of the anatomist, embryologist, and physiologist belong to this sphere. (Dohrn 1876, 13–14)

To create a flexible system of collaborators and services that would optimize the scientific output proved to be an ongoing challenge. Dohrn expected his scientific staff members or "assistants" first and foremost to assist the guest investigators. He also wanted them to fit well into the station's scientific community. The Baltic zoologist and Haeckel student Nikolaus Kleinenberg (1842–97), for example, whom Dohrn had convinced to join him in Naples, arrived in 1872 while the station was still under construction but left Naples in 1875 because he could not fit into daily routine schedules (Müller 1973).[34] The German zoologist Hugo Eisig (1847–1920), another of Haeckel's students, also joined Dohrn in 1872. Until 1909, he was deputy director and responsible for the administration of the laboratories. Paul Mayer (1848–1923) and Wilhelm Giesbrecht (1854–1913)[35] had first come as guest investigators and were then given permanent positions in 1878 and 1884, respectively. More assistants later joined for longer or shorter periods for reasons to be explained below. Each assistant had several areas of technical and administrative responsibilities such as the library, conservation, fisheries, laboratories, acquisitions, scientific collections, sales, or the aquarium.

In the beginning, Dohrn's technical staff consisted of ten people, all Neapolitans with one remarkable exception: Sophie Boutkewitch (?–1882), the widow of a Russian colonel. She was the first employee of the station and sold the tickets and guides[36] to the aquarium. The rest of the group was composed of fishermen, machinists, lab servants, a watchman, and a mason (Dohrn 1876, 17–18). This number also grew along with expanding activities, such as the opening of a department of marine botany in 1876 (Dohrn 1879a, 150). By 1884, Dohrn's full-time staff consisted of eleven scientists and twenty-four technicians (Vogt 1884).

A marine station is only as good as its supply of research material. Ships, tools, methods, technical expertise, and local knowledge were needed for a fast and efficient passage from beach to bench. A first step in this direction was the 14-meter/5-ton steamer *Johannes Müller* (May 1877), built in Great Britain, and donated by the Berlin Academy of Sciences and the Ministry of Agriculture. In 1883 it was followed by the 9-meter steam launch *Frank Balfour*, named after the English embryologist Francis Maitland Balfour (1851–82).[37] Smaller boats for diving and surface fishing completed the fleet. To improve the supply of requested organisms, fishermen and lab servants were trained with regard to the needs of the scientists. At the same time, records of when, where, and what (stages of) organisms could be found were constantly updated and published to help scientists plan their research stays effectively (Schmidtlein 1879a, 1879b).

Another challenge consisted in the problem of how to preserve organisms for future use when not immediately needed for the laboratories, the reference collections, or the public aquarium. In 1877, Dohrn implemented a preservation department and tasked the German chemist August Müller (1847–80) to study and improve the then-known methods of preservation. Müller's combination of experimentation and artistry produced splendid results. At Müller's untimely death, Dohrn nominated the twenty-year-old Neapolitan Salvatore Lo Bianco[38] (1860–1910) (fig. 2.3) his successor. Lo Bianco had worked as a lab servant since he was thirteen, time enough to learn amazingly much about marine organisms, their scientific names, and where they could be found. He developed the preservation methods to such

FIGURE 2.3. Salvatore Lo Bianco, 1889. © Stazione Zoologica Anton Dohrn di Napoli, MAB SZN-AS, La.119,37. Reproduced with permission.

perfection that the station could very successfully start to sell collections of preserved animals to museums, dealers, and university departments all over the world (Lo Bianco 1899; Raffaele 1910; Groeben 2002). His methods were kept secret for a long time, and scientists were not allowed to make their own collections while in Naples.

At any time of the year, scientists from different countries, disciplines, and backgrounds were present in Naples. Each contributed to the increase of knowledge, including favorite research methods, cutting and staining methods in particular. Research methods generally were discussed and constantly improved by Dohrn's collaborators. Testing methods and instruments soon led to close collaborations with companies producing instruments and chemicals, such as Zeiss (Jena) for microscopes, Jung (Heidelberg) for microtomes, and Merck (Darmstadt) for chemicals and dyes (Müller 1976, 255–71). As early as 1883, the American zoologist Edmund Beecher Wilson (1856–1939) stated: "For methods go to Naples" (Maienschein 1991, 102; see also Groeben 2005b).

Around 1876—by then the Station was an established "reality in brick and mortar" (Dohrn 1893a, 440)—Dohrn started to think about "the literary phase of [his] institute" (Dohrn 1893b, 440), which was realized in three publications. They all significantly increased the scientific output and visibility of the Naples Station. The first two, the journal *Mittheilungen aus der Zoologischen Station zu Neapel*[39] and the monograph series Fauna and Flora of the Gulf of Naples and Adjacent Waters, were complementary and corresponded to the mission and focus of the Naples Station, namely, to be the "centre for the biological exploration of the Mediterranean" (Dohrn 1879b, iii). According to the international character of the station, contributions in German, English, French, and Italian were admitted. The *Mittheilungen* offered staff members and guests the opportunity to publish short notes—often important in priority issues—on discoveries, research results, observations, or new methods and techniques. It also served to inform the public about recent activities at the station and included lists of guest investigators, of publications based on research done in Naples, and of the shipping of collections of preserved marine organisms and microscopic slides (Dohrn 1876, 1879a, 1881a, 1882, 1885, 1893a). The exchange of the *Mittheilungen* with specialized series from all over the world also considerably increased the library holdings.

The lack of reliable information was strongly felt when assistants and scientists had to identify what the fishermen brought in every morning; a revision of the systematics, anatomy, ontogeny, and ecology of the marine fauna and flora of the Gulf of Naples was much needed. The main object of the Fauna and Flora series was therefore "to create a firm basis for systematical knowledge" (Dohrn 1893b, 441). Each volume was supposed to describe one species as completely as possible in terms of its anatomical, embryological, ecological, and physiological aspects. The authors were carefully chosen by Dohrn from among collaborators and guest investigators. In several cases, he offered younger scientists a position as assistant for the time they needed to prepare their monographs. They then had to share other responsibilities with the other assistants but, in exchange, were given the opportunity to concentrate on site on their work without having to worry about supply of research material or career challenges. This explains the considerable mobility among Dohrn's scientific staff. True-to-nature illustrations were also still a rarity for marine organisms. For high-quality illustrations, Dohrn therefore hired two local artists, Comingio Merculiano (1845–1915) and Vincenzo Serino (1876–1945). The series was sold on a subscription basis and much appreciated for the beauty and accuracy of its plates (fig. 2.4). During Dohrn's lifetime, thirty-two volumes were published (Groeben and Müller 1975, 64–68), and eight more followed until 1980.

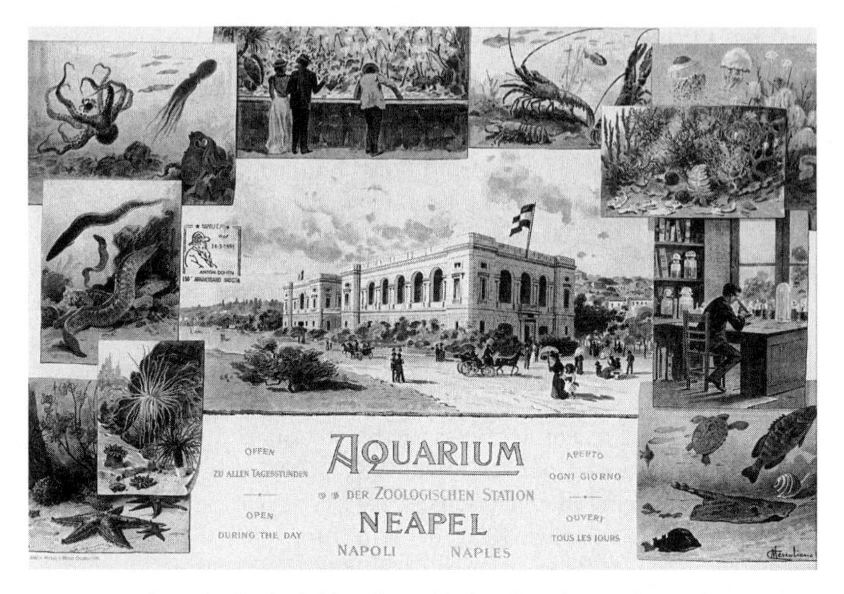

FIGURE 2.4. Poster by Comingio Merculiano with views from the aquarium tanks to attract tourists in four languages to visit the Naples Aquarium, 1902. *In the center*: the Naples Station. *On the right*: a scientist sitting at his "working table."

The third publication, the reference journal *Zoologischer Jahresbericht* (1879–1913), provided much needed, fast, and exhaustive information on recent publications in "all branches of animal knowledge, systematics, and anatomy as well as morphology and biology" (Carus 1880, iv). Once reviewed, the publications passed to the station's library. The *Zoologischer Jahresbericht* was soon appreciated in the zoological world for its completeness, reliability, and objective reporting. With the report for 1883 (1884–85), the editorial office passed from Julius Victor Carus (1823–1903) in Leipzig to Paul Mayer in Naples.[40] A constantly changing international group of experts was engaged to review the literature within their special subject areas and animal groups. Quite a number of reviewers belonged to the constantly increasing pool of guest investigators.[41]

From Bench to Book: Research Results from Naples

"Zoological stations are needed to increase the knowledge about geographic and bathymetric distribution, for physiology and embryology," Dohrn insisted.[42] They would also be useful for the "study of the life of animals in relation to all those conditions which regard the struggle for existence and the action on natural selection": "We have short notices on the habits of some fishes; but . . . of echinoderms, cuttle fish, jelly fish, polyps, &c., &c., our knowledge simply amounts to nothing" (Dohrn 1872a, 279). By 1915, the quantity and quality of knowledge about the Mediterranean marine fauna and flora had tremendously increased thanks to the scientific output of the Naples Station.

Dohrn once defined the Zoological Station as the "brainchild of his reflections on the problems of the theory of descendance."[43] Many anatomical, embryological, physiological, developmental, and morphological details needed to be studied before any serious phylogenetic conclusions could be drawn. This included the description of many new species and explained the remarkable number—about two thousand before 1915—of morphological-phylogenetic studies based on research done in Naples by Dohrn, his assistants, and many guest scientists.[44] Without going into details, it can be said that arthropods, hydrozoa, cephalopods, elasmobranchs, and polychaetes were the most studied groups and species.[45]

The natural input of ideas and methods from guest researchers and their lively interaction often led to developments not foreseen but always appreciated by Dohrn. His motto "freedom in research" was supposed to have just this effect. When Hans Driesch (1867–1941) stated, "It is fair to say that nine-tenths of all fundamental works in the field of modern zoology have been done at the Naples Zoological Station" (Driesch 1909, 514), he was referring not to the research line of functional descriptive morphology favored by Dohrn and many others but rather to the fact that the Naples Station had become "the birth-place of almost all of cell biology and experimental developmental physiology" (Driesch 1909, 515). This new experimental approach to the problems of development or *Entwicklungsmechanik* (mechanics of development) shifted embryology from being a subdiscipline of comparative anatomy to an experimental discipline along the lines of physics and chemistry.

This shift from phylogenetic to experimental embryology started in Naples with Theodor Boveri (1862–1915), who had been encouraged by his teacher Richard Hertwig (1850–1937) to study sea urchin development.[46] Boveri visited Naples for the first time in 1888. By then, the sea urchin had already entered the scene as one of the most suitable objects for the study of the processes of fertilization, from the morphological to the molecular level (Monroy and Groeben 1985; Monroy 1986; Müller 1996). The particular interest in sea urchins had started in 1875 when, during a research stay in Nice with his brother Richard, Oskar Hertwig (1849–1922) observed the fertilization of a sea urchin egg for the first time. Four years later, the two introduced sea urchins to the Naples community.[47] Sea urchins proved to be perfect research organisms for various reasons: they are abundant and easy to find and maintain in captivity; they rapidly produce many eggs that are resistant, transparent, and big enough to be observed under the microscope. Through his ingenious "hybrid merogone" experiments, Boveri proved, in 1889, that only the nucleus controls development (Boveri 1889; Laubichler and Davidson 2008), and, in 1901–2, he demonstrated, through his classic polyspermy experiments, that chromosomes are qualitatively different (Boveri 1902).

Between 1891 and 1909, Hans Driesch, certainly one of the most interesting and influential guest researchers, and Curt Herbst (1866–1946) spent almost every winter in Naples. Driesch worked on sea urchin development, following Wilhelm Roux's experimental research methods. Roux himself never visited the Naples Station—he too was one of Haeckel's students—but his theories and methods were directly imported to Naples by Driesch.[48] Repeating Roux's frog experiments with sea urchins, Driesch observed that, when the first two cells of the embryo are separated from one another, each can give rise to an entire sea urchin larva. He came to the conclusion that there must be a vital principle beyond the physicochemical forces that is responsible for the integrity of the embryo and the regeneration of lost parts.

Two methods discovered at Naples contributed considerably to the obtaining of reliable results in this innovative research line. In 1895, working in Naples next to Driesch on the development of ctenophore eggs, Thomas Hunt Morgan (1866–1945) discovered that a tripolar nuclear spindle would sometimes form when eggs were gently shaken after polyspermy fertilization (Morgan 1896). In 1898, while studying the influence of inorganic substances on the development of organisms, Curt Herbst discovered that calcium free seawater caused blastomeres to separate spontaneously (Herbst 1900).

Boveri and Driesch had started their scientific careers with studies in descriptive morphology and moved to experimental zoology through their work in Naples. This was also the case for quite a number of other guest investigators (Müller 1996). In 1892, Edmund Beecher Wilson (1856–1939) returned to the Naples Station.[49] Driesch was there at the same time. Owing to the influence of Driesch, Boveri, and Jacques Loeb (1859–1924), Wilson lost his initial interest in morphological questions and turned to experimental embryology. His experiments with amphioxus eggs, which repeated Driesch's sea urchin experiments, resulted in his classic paper on the mosaic theory of development (Wilson 1893). According to Wilson, Driesch and Roux had started a new era in the history of embryology. Jacques Loeb spent two research periods at the Naples Station between 1889 and 1891,[50] shortly before Driesch's first visit. Loeb suggested a purely chemical explanation of life and enthusiastically welcomed mechanics as a new principle guiding research. In Naples, he studied, together with the British zoologist Theo. T. Groom, heliotropism in nauplia of *Balanus perforatus*. He also became interested in the phenomena of regeneration, which seemed to confirm his strictly mechanistic-materialistic viewpoint. In numerous experiments, he investigated the conditions for organ regeneration and the possibilities of controlling and regulating morphogenesis by external influences (Müller 1996, 108; Loeb 1891, 1892). Loeb's spectacular results highly impressed the younger generation of zoologists and the ability to get animal morphogenesis under control attracted quite a number of scientists to Naples to study regulation and regeneration, where they availed themselves of the abundant supply of ascidians, planarians, and hydrozoans. Loeb's and Driesch's research on regeneration was taken up by Thomas Hunt Morgan during his second stay in Naples in summer 1900 (Morgan 1901) and also by Charles Manning Child (1869–1954) during winter 1902–3 (Child 1903).

Undoubtedly, the Naples Station has had an impact on various research disciplines and national research traditions by promoting an experimental approach to understand the processes of life. Two factors contributed to this. First, the Gulf of Naples offered a year-round unlimited supply and a rich variety of "marine Guinea-pigs,"[51] and, second, the Naples Station was able constantly to adapt its research facilities to the needs of its patrons and changing priorities in science.

By the end of his life, Dohrn could look back at a well-running independent organism,[52] one based on postindustrial principles where the parts interacted smoothly to create an entity. He can be considered an early manager of science or, as he defined himself, the "statesman of Darwinism" (Groeben 1985, 15; De Masi and Gentile 1991). At that point, more than twenty-two hundred scientists from Europe and the United States had already worked in Naples, published their results, written and reported on the station, and/or sent their students there, thus contributing to enlarging the network of knowledge and international support for the Naples Station as the place to work in marine biology.

The Naples Experience

Spending research time in Naples was always a challenge. On the one hand, one found familiar setups in the lab, met colleagues, and continued scientific discussions; on the other hand, one had to face a different cultural and linguistic environment and also the temptation to get lost in the overwhelming tourist attractions of the Naples area. In such a context, the role of host was important, unobtrusively advising and guiding so that guests felt at ease both at work and at leisure. In this sense, Anton Dohrn (fig. 2.5) and, after him, his son Reinhard (fig. 2.6) were the perfect hosts. They offered as much hospitality as was needed, welcoming guests at their home, the Casa Dohrn in Naples, and at Villa Dohrn on the island of Ischia.[53] On Thursday afternoons there was *jour fixe* at Casa Dohrn where one could play or listen to music or discuss the origin of the vertebrate body with Anton Dohrn or Russian literature with the lady of the house, Marie Dohrn née von Baranowska. The house had a very special atmosphere, and its destruction by bombing on August 4, 1943, was an irreplaceable loss to many. Margret Boveri (1900–1975),[54] the daughter of Theodor Boveri and Marcella O'Grady Boveri (1863–1950) and close friend of the Dohrn family, remembered:

> The Casa, the family, and the station formed an indissoluble unity. Much of the station's atmosphere had its origin in this remarkable house, where the traces of past and present, of serene southern lifestyle and northern initiative, of eastern intensity of sentiments and German reflectiveness, had almost tangibly condensed. In the extensive hospitality of the Casa scientists met with travelers who came from all corners of the world and docked in Naples, diplomats on their way to new positions, writers, actors, painters who looked for refreshment and rest from their work, state officials, industrialists, bankers, royalties. . . . Maxim Gorki, with his intensively blue eyes, coming from his exile in Sorrento, drank tea with lemon. (Boveri 1964, 35)

Dohrn lived for and from music; it was his elixir. Whenever possible, he invited musicians to give concerts at Casa Dohrn after their official concerts. The violinist Joseph Joachim (1831–1907), for example, performed several times there. Guests from the station were always welcome, and several of them were not only lovers of music but also

FIGURE 2.5. Anton Dohrn in 1899. © Stazione Zoologica Anton Dohrn di Napoli, MAB SZN-AS, Lb.2.4. Reproduced with permission.

performers, such as Edmund Beecher Wilson or the German behavioral physiologist Erich von Holst (1908–62). Hans Driesch remembers:

It was a great pleasure to be allowed to witness at the hospitable Casa Dohrn how he [Dohrn] together with a musically equal colleague such as Rudolph Bergh[55] would become enthusiastic while remembering musical beauties: the two men would then rather sing, hum, and warble than talk. Unforgettably stuck in my memory are two musical evenings at the Dohrn home in spring 1897: he had invited the

FIGURE 2.6. Reinhard (Rinaldo) Dohrn, son and successor (1909–54) of Anton Dohrn, as director of the Naples Station. Foto Schafganz, Bonn, 1955.

Halir[56] Quartet to Naples. They had to play for him from morning to night. (Driesch 1909, 515)

Social life at the Station included female colleagues and featured regular teatime each afternoon on the loggia and evenings spent in the old city. Excursions on the station's steamer *Johannes Müller* were frequently mentioned in letters, diaries, and memories as particularly pleasant and fascinating highlights. For many of the guests, it was the first time that they had seen the fascinating variety of marine organisms that came up with each haul:

Such an excursion on the *Johannes Müller* . . . with Dohrn and the resourceful conservator Lo Bianco belongs to the most precious and beautiful memories. How exciting when the map is tested and the announcement is made: now we are going to bring up with the steam dredge purple sea urchins, then red *Cynthia papillosa*, and finally big white bumpy ascidians (*Phallusia mammillata*). The immediate observation of the barely caught still perfectly alive animals, the sorting by the expert eye of the experienced conservator, the eventually following debate on the classification offers every time a wealth of suggestions and knowledge. To this by no means inferior is the aesthetic pleasure of such excursions on the sea. (Tschermak 1914, 26)

Very often a stay at Naples turned out to be a perfect combination of research results and pleasant memories or, as the American anatomist Caroline McGill (1879–1959) perfectly summarized in 1910, "a mix of work and play" (Groeben 2005a, 115). In her 1975 survey, the Czechoslovakian sociologist Olga Skalovà asked guest investigators to rank the factors that had contributed to making their research stays at Naples satisfactory or successful. The highest values were given to the following working conditions: (1) supply of experimental marine material; (2) supply and operation of the library; and (3) skilled work by the fishing personnel. On the personal level, respondents rated highest the "overall atmosphere for creative activity" and the freedom to devote all personal efforts to research, free from teaching and administrative duties (Skalovà 1975, 62).

Many guest researchers considered their research stay(s) in Naples to be the highlight of their scientific careers. In Naples, the neurophysiologist John Zachary Young (1907–97) appreciated the freedom to pursue whatever curious subject aroused his interest. Neuroscience might have gone differently had he not made sections of a yellow spot at the hind end of the stellate ganglion of *Eledone*, which years later led to the discovery of the epistellar body (Young 1985, 154). In 1980, on the one hundredth anniversary of the birth of Reinhard Dohrn, the British biologist Maurice Wilkins (1916–2004) mentioned highlights of discoveries and encounters made at the Naples Station during its long history. He had come to Naples in 1951 to obtain sperm of *Sepia* in order to investigate the possibility that the genes were arranged regularly in crystalline array in the sperm heads using X-ray diffraction. An X-ray diffraction pattern of crystalline DNA that he showed at a conference stimulated James D. Watson (b. 1928) to begin his collaboration with Francis Crick (1916–2004) that led to the description of the double helical structure of DNA.[57] Wilkins fully agreed with the statement of an unidentified "distinguished zoologist": "A zoologist ought, absolutely, to have had his Naples 'experience'" (Wilkins 1983, 7–8).

From Guest Research Facility to Research Institute

From 1872 to 1915, the Naples Station had been the private property of Anton and then Reinhard Dohrn. After the First World War, the legal status of the station was turned into that of an Italian nonprofit corporation (*ente morale,* 1923–82) under the administration of a board of administration with Reinhard Dohrn as director, followed in 1954 by his son Peter (Pietro) Dohrn (1917–2006).[58] It has been Reinhard's merit to guide the station through two world wars and to keep the international research tradition alive. Aggregation points of research continued to be physiology of the embryo, developmental mechanics, and regeneration. The particular richness of the fauna and flora of the Gulf of Naples stimulated a new ecological approach, focusing on food chains and the organism's relation to its biological and physical/chemical environment (Fantini 2002, 20–25).

After World War II, Reinhard Dohrn was well aware of the fact that a choice needed to be made whether to stay small and eclectic or join big science. "It is urgent now to work on the economic and political connections and to reinsert the station into the always stronger organized business of science without, however, sacrificing its inner essence. This is Peter's task."[59] Peter Dohrn had joined the Naples Station in 1945 as assistant. His family's dedication to the station, his training in the medical and natural sciences, and his fluency in five languages was considered a guarantee for respecting the past and changing the future. In 1954, he was appointed director and legal representative of the SZN, while Reinhard as director emeritus remained in charge of international relations. Scientific events, new laboratories, the transformation of Villa Dohrn in Ischia from a guesthouse to a laboratory for marine ecology (1963–65), and the creation of the Anton and Reinhard Dohrn Foundation (1955; today simply the Dohrn Foundation)[60] changed the image of the SZN from a passively hosting to an actively proposing institution.

However, by the end of the 1970s, growing financial and institutional constraints became a major difficulty for the scientific life of the Naples Station. A new institutional and financial leap was needed. In 1982, the SZN became a national research institute (*ente pubblica di ricerca non strumentale*) under the supervision and control of the Ministry of Public Instruction and was renamed after its founder the Stazione Zoologica Anton Dohrn. The SZN maintained, however, autonomy concerning all decisions related to the running of the institute and its scientific policy. New statutes were approved in 1985. The most important innovation concerned the president, to whom a more active responsibility was attributed.

For one hundred years, the SZN had focused on guest research. It was the first president under the new statute, Gaetano (Nino) Salvatore (1987–1997), who implemented a paradigm shift on the institutional level, reshaping the institute from mainly a service-centered research facility into a research institute. Because of his scientific status, his international links, and his management abilities, Salvatore was able to secure scientific and financial support from national and international scientific organizations.

An important addition was institutionalized in 1999 with the introduction of an international PhD program in collaboration with Italian universities and the Open University, London. In 2018, the SZN community consisted of 150 staff members and about the same number of PhD students and collaborators. The mission of the SZN continues to be to conduct basic research in biology with a focus on marine organisms and their biodiversity, using an integrated and interdisciplinary approach. Research at the SZN today is organized in four departments, two focusing on scientific research (biology and evolution of marine organisms; integrative marine ecology) and two on technological research (research infrastructure for marine biological resources; marine biotechnology).

In 2011, new statutes were approved with some significant innovations with the intent of streamlining the structure even more along the lines of Italian research policies. In particular, the administrative board members were reduced to three, namely the president, one member nominated by the Ministry of Instruction, University and Research, and one member elected by the national scientific community of reference.[61] Collaboration and interaction with the private sector were also no longer excluded. While in 1985 the SZN was defined as a "special scientific institution," the definition given in the most recent statute (2017) now fixed its lifelong mission and tradition as "national institute of marine biology, ecology, and biotechnology."[62]

Anton Dohrn's idea of establishing a network of marine biological stations (Dohrn 1872b, 158) and offering research hospitality has recently been revived in the European Marine Biological Resource Centre (EMBRC), an EU-funded consortium of thirteen marine biological stations and institutes across the European Research Area, providing services to support research on marine biological resources. In June 2018, the EMBRC was adopted as a European Research Infrastructure

Consortium (EMBRC-ERIC) and started its operational phase. The SZN hosts and coordinates the Italian national research unit of EMBRC (EMBRC^{IT}).

The Naples Station has a long research tradition focused exclusively on marine biology that cannot be evaluated through simple arithmetic. That this was strongly felt in the international community became evident when in 2010 the Italian government published a list of institutes to be suppressed or integrated into a different administrative system as part of a measure meant to achieve financial stabilization and economic competitiveness—among them the SZN. An international petition "Save the Stazione Zoologica Anton Dohrn," asking to maintain the SZN as an autonomous structure, was signed in a few days by more than thirty-six hundred supporters from sixty-five nations worldwide, showing that the station is not only valued for its tradition and achievements but also considered indispensable for maintaining today's level of excellence in marine biology research.

Conclusions

By the mid-1850s, pioneers in seaside research such as Johannes Müller, Carl Vogt, and Ernst Haeckel had drawn attention to the richness of marine life in the Mediterranean and also to the difficulties of obtaining results that went beyond the observation and description of hitherto unknown organisms. Haeckel and Dohrn managed at Helgoland with a wooden box and glass jars, Vogt kept his "scientific suitcase"[63] always ready for the next trip, and Alexander Kowalewsky (1840–1901) had to invent a way to obtain ctenophore eggs because the much-praised Müller nets for plankton fishing proved to be inadequate (Groeben 2008, 141). It was a period of makeshift tools and of time, energy, and money consuming individual research trips. On the basis of his personal experiences, Anton Dohrn created an independent international research facility in Naples where unlimited supply of research material was guaranteed, technical problems could be solved, peer discussion was possible, new methods could be tested, and the latest research results could be consulted in an unrivaled research library.

Dohrn was a complex and remarkable personality. He had a strong ego, revered creativity, and despised authority (Ghiselin 1994, 4), and he seems to have been the right man at the right moment in history to achieve his goal, namely, creating a research facility where large-scale and continuous studies of marine fauna and flora and of the processes of life could be done. It was in fact in Naples that international scientific collaboration in the modern sense was invented, collaboration based on the quick and free communication of ideas, methods, techniques, and instruments and on exchange and personal contacts between scientists of different cultural traditions. As Theodor Boveri aptly summarized: "Dohrn's Station first made the study of marine life practical" (Boveri 1912, 454). Several stations in other countries followed the example and standard set by Naples, adapting, however, the general principle to local needs, among them the Misaki Marine Biological Station in Japan (1887), the Marine Biological Laboratory in Woods Hole in the United States (1888), and the Plymouth Marine Laboratory in Great Britain (1888). The Naples Station has been described as Mecca for zoologists (Whitman 1883, 94) and as a continuous zoological congress (Boveri 1912, 467), but what Dohrn wanted to and did achieve, and what he certainly considered to be the highest recognition of his success, was Darwin's statement: "You did a great service to science" (Groeben 1982, 42).

The organism created by Dohrn in 1872 underwent—as has been shown—several administrative changes; it could react to challenges in science management and was able to survive the vicissitudes of two world wars. Hosts, statutes, research topics, and approaches have changed, but the focus on the marine environment and its inhabitants never has. Understanding and protecting the ocean as a complex ecosystem requires a continuous interaction of beach and bench research. The biodiversity in the Gulf of Naples and surrounding areas is still rife with surprises and challenges, be it the study of sea turtles to promote the protection of this important flagship species, the discovery of an oasis of biodiversity at great depth in the Dohrn Canyon (Gulf of Naples) holding, for example, giant clams thought no longer to exist as well as many species of deep white corals,[64] the observation of the phenomena of natural marine acidification on the Island of Ischia, or the discovery of kleptopredation, a yet unknown method used by sea slugs preying on cnidarians.[65]

When at the end of 1872 Anton Dohrn tried to convince members of the Berlin Academy of Sciences to express themselves in favor of the Naples Zoological Station—the Prussian government had asked for an expert opinion before granting financial support—the German naturalist, zoologist, comparative anatomist, geologist, and microscopist Christian Gottfried Ehrenberg (1795–1876) commented that, if Dohrn really intended to investigate the Neapolitan fauna with such an armory of manpower and equipment, in five to ten years there would be nothing left there to investigate (Boveri 1912, 458). Today, we know for sure that he was wrong.

Notes

The following abbreviations have been used throughout the notes: BSB = Bayerische Staatsbibliothek München, Handschriftenabteilung; EHA = Ernst-Haeckel-Archiv Jena; MAB SZN-AS = MAB Stazione Zoologica di Napoli Anton Dohrn—Archivio Storico.

1. Before 1982, the institution founded by Anton Dohrn never had a proper name or acronym, and hence its name has always been translated differently in various languages. In 1982, the institute became a public research institute under the supervision of the Italian Ministry of Scientific and Technological Research and was given the name of its founder: Stazione Zoologica Anton Dohrn.
2. Dohrn, named Felix after his godfather Felix Mendelssohn-Bartholdy (1809–47), was called only Anton throughout his life. On Dohrn, see Oppenheimer (1978), Groeben (1991, 2013a), and Heuss (1991).
3. Dohrn was one of the twenty-five students attending Haeckel's first course on Darwin's theory of evolution.
4. Fanny Lewald (1811–99) and Adolf Stahr (1805–76). She was a writer, a feminist, and the hostess of a popular literary saloon. He was a feared literary critic.
5. Anton Dohrn to Adolf Stahr and Fanny Lewald, April 19, 1866, Jena (BSB, Ana 525, Ba.1178). Quotations in English from publications and correspondence originally in German are my translations.
6. Dohrn to his father in 1867, quoted in Heuss (1991, 63).
7. Dohrn to Haeckel, February 17, 1865, Berlin (EHA Jena, A 3291).
8. Dohrn to Haeckel, August 3, 1865, Stettin (EHA Jena, A 3304).
9. Dohrn to Haeckel, August 5, 1865, Stettin (EHA Jena, A 3305).
10. Krausse (1987, 26–27). On the research tradition at Helgoland, see Florey (1995).
11. Dohrn to Haeckel, August 5, 1865, Stettin (EHA Jena, A 3305).
12. Haeckel, Dohrn, Richard Greef (1829–92), Mathijs Salverda (1840–86) from Delft, and Pietro Marchi (1833–1923) from Florence (Zissler 1995).
13. August 18–October 2, 1865.
14. Haeckel to Reinhard Dohrn, October 2, 1909, Jena (MAB SZN-AS, A.1909.A.Dohrn.H.).
15. The system was easy to mount and dismantle, even though a bit heavy to carry along (about 100 kilograms). The floor and the side walls were of slate, the main walls of thick glass panes. The central tank measured 90 × 60 × 35 centimeters. For a detailed description, see Müller (1976, 78–79).

16. Sources give him also as Nicholas Miclucho, Miklucho or Miklouho. At Messina, Maclay introduced Dohrn to the Russian-Polish family von Baranowski. The oldest daughter, Marie (1856–1918), would become Dohrn's wife in 1874. For Micloucho-Maclay, see Uschmann (1959, 66–67) and (Groeben and Müller (1975, 57).
17. For details on the scientific disagreement of Haeckel and Gegenbaur with Dohrn's hypothesis that vertebrates derive from annelids, see Kühn (1950, 60–61), Uschmann (1959, 83–87), and Maienschein (1994).
18. This had far-reaching consequences. Haeckel was one of the most prominent and influential teachers of zoology in Germany, whereas Dohrn offered top research facilities at Naples in marine biology, but students from Jena were encouraged to go to Messina, not to Naples. There never was any institutional collaboration.
19. According to Dohrn: "The transformation of organs occurs through the succession of functions, the bearer of which remains one and the same organ. Each function is a resultant of many components, of which one forms the main, or primary function, while the others represent subsidiary, or secondary functions. The sinking of the main function and the rising of a subsidiary function changes the total function. The subsidiary function gradually becomes the main function, the total function becomes another one, and the result of the entire process is the transformation of the organ" (Ghiselin 1994, 67).
20. On Dohrn as a scientist, see Kühn (1950), Oppenheimer (1978), and Ghiselin (1994).
21. Anton Dohrn to his wife, Marie Dohrn, August 1, 1886, Naples (BSB, Ana 525, Bd. 372).
22. About 18 m^3.
23. Dohrn to Darwin, December 30, 1869, Stettin (Groeben 1982, 25–26). At that point, Dohrn was thinking of other places such as Venice, Gibraltar, Portugal, Ireland, Ceylon, Australia, and the Cape of Good Hope (Groeben 1982, 91 n. 16).
24. Darwin to Dohrn, January 4, 1870, Down, Beckenham, Kent, Darwin Correspondence Entry 7071, accompanying letter to Darwin to Dohrn, January 4, 1870 (Groeben 1982, 29).
25. Opened on May 11, 1869, under the direction of Alfred Brehm (1829–84), whom Dohrn had already met at Hamburg in 1865 on his way to Helgoland.
26. The official founding date is March 1872, and the first scientist arrived in 1873, but the continuous flow started only in 1874, which is, therefore, sometimes given as the opening date.
27. Dohrn to Anna Wendt, March 30, 1870, Naples (BSB, Ana 525, Bd.06).
28. Dohrn and Darwin's correspondence started in 1867 with Darwin acknowledging receipt of Dohrn's publication on the morphology of Arthropoda (Dohrn 1868a) and finished with a letter from Darwin of

February 13, 1882, in which he repeats his invitation to pay him a visit whenever Dohrn would return to England.

29. Gesellschaft deutscher Naturforscher und Ärzte, founded in 1822.
30. For a detailed description, see Dohrn (1876, 1–9) and Kofoid (1910, 7–32).
31. See the sketch in a letter to Adolf Stahr of April 13, 1870, reproduced in Groeben (2000, 37).
32. Depending on the nation, represented by different state departments or ministries such as the Ministry for Public Affairs (Italy, Russia, Romania), Interior (The Netherlands), Cultural Affairs (Austria), or Cultural, Educational and Medical Affairs (Prussia).
33. Exceptions were made for special guests who would—Dohrn hoped—encourage their governments to rent a table at Naples as well. This system, however, excluded several generations of biologists from countries without table agreements such as France and Sweden.
34. Kleinenberg later held a professorship first (1879) at Messina, then (1895) at Palermo.
35. On Wilhelm Giesbrecht, see Steiner (in this volume).
36. Beginning in 1880, detailed and well-illustrated guides to the aquarium of the Naples Station were published in English, German, Italian, and French; they all went through several editions until the 1940s.
37. Balfour had worked at the Naples Station in 1874, 1875, and 1877.
38. Found also as Lobianco.
39. The journal was published as *Mittheilungen aus der Zoologischen Station zu Neapel, zugleich ein Repertorium für Mittelmeerkunde* from 1879 to 1921 (vols. 1–22). It continued as *Pubblicazioni della Stazione Zoologica di Napoli* from 1916 to 1977 (vols. 1–40) and then *History and Philosophy of the Life Sciences* (1979–) and *Marine Ecology* (1980–).
40. Some volumes were published in collaboration with Wilhelm Giesbrecht (the reports for 1882–85) and Julius Gross (1869–1933; the reports for 1912–13).
41. For a detailed list of reviewers for the *Zoologische Jahresbericht*, see Müller (1976, 196d–196i).
42. Dohrn to Thomas Henry Huxley, April 4, 1870, Naples (BSB, Ana 525, Ba.473).
43. "Die Zoologische Station ist das Kind meines Nachdenkens über die Probleme der Deszendenztheorie." Dohrn to Carl Ludwig, April 13, 1883 (MAB SZN-AS, Ca II, 129–32).
44. The following section is mainly based on Müller (1975, 1976, 1996), Ghiselin (1994), Groeben and Ghiselin (2001), and Fantini (2002).
45. For details, see Müller (1976, 349–413).
46. For Boveri, see Baltzer (1962). Boveri worked at Naples eight times (1888, 1889, 1896, 1901–2, 1905, 1910, 1911–12, 1914).
47. Oskar and Richard Hertwig worked at Naples from April 11 to May 11, 1879, as guest investigators no. 117 and 118.

48. Driesch had been particularly fascinated by the programmatic *Die Entwicklungsmechanik der Organismen, eine anatomische Wissenschaft der Zukunft* (Roux 1890).
49. January 11–April 16, 1892. Two more research visits followed in 1899 and 1903.
50. October 10, 1889–May 1, 1890, and October 31, 1890–April 25, 1891.
51. A term coined by Georg Grimpe (1889–1936), referring to cephalopods (Grimpe 1928, 382).
52. As Dohrn told his wife in August 1886: "I found chaos before me and have created out of that both a practical organism: the Station, and a theoretical one, the 'Urgeschichte der Wirbelthiere' [Origin of vertebrates]" (Groeben 1985, 16–17).
53. Built in 1906 at the entrance to the port of the island of Ischia in the Gulf of Naples, the house served from the 1920s to the 1940s as a guesthouse for the Dohrn family, their friends, and guest investigators; it was remodeled as a lab for Benthic ecology in the 1960s (Richter 1969).
54. German journalist. In 1911–12 and 1914, she had accompanied her parents to Naples, where she also went to school; from 1926 to 1929 she worked as secretary to Reinhard Dohrn.
55. Rudolph Sophus Bergh (1859–1924), a Danish zoologist and composer, had worked at Naples in 1898 and 1903.
56. Karel Halir (also Karol Halíř; 1859–1909), the Czech violinist and conductor. See also Fantini (2015).
57. On Nobel laureates at the Naples Station, see Groeben and de Sio (2006).
58. For details, see Groeben (2013b).
59. Reinhard Dohrn to Theodor Heuss, 1955, in Partsch (1980, 90).
60. For more than fifteen years the Fondazione Antonio e Rinaldo Dohrn (FARD) was extremely active, providing grants and fellowships to young scientists, sponsoring publications, and implementing the archives and historical studies (FARD 1961, 1967).
61. Consisting, in 2011, of 584 scientists from 53 Italian universities and institutions.
62. Istituto Nazionale di Biologia, Ecologia e Biotecnologie Marine. Statuto della Stazione Zoologica Anton Dohrn, come into force on October 28, 2017, par. 1. http://www.szn.it.
63. A suitcase with compartments for a microscope, nets, and jars that had already accompanied him to Nice, the North Cape, Ireland, and Italy (Vogt 1876).
64. See http://www.szn.it/index.php/en/news/hot-topics/1982-marine-litter-and-deep-sea-biodiversity-oases.
65. For further details, see http://www.szn.it/index.php/it/ricerca.

References

Baltzer, Fritz. 1962. *Theodor Boveri: Leben und Werk eines grossen Biologen (1862–1915).* Stuttgart: Wissenschaftliche Buchgesellschaft.

Boveri, Margret. 1964. "Reinhard Dohrn: Ein Leben für die Zoologische Station." In *Dem Andenken von Reinhard Dohrn: Reden, Briefe und Nachrufe*, ed. Heinz Götze, 21–42. Berlin: Springer.

Boveri, Theodor. 1889. "Ueber partielle Befruchtung." *Sitzungsberichte der Gesellschaft für Morphologie und Physiologie in München* 4:64–72.

———. 1902. "Ueber mehrpolige Mitosen als Mittel zur Analyse des Zellkerns." *Verhandlungen der Physikalisch-medizinischen Gesellschaft Wuerzburg*, n.s., 35:67–90.

———. 1912. "Anton Dohrn [Graz, 1910]." *Science* 36:453–68.

[British Association for the Advancement of Science]. 1872. "Report of the Committee, Consisting of Dr. Anton Dohrn, Professor Rolleston, and Mr. P.L. Sclater, Appointed for the Purpose of Promoting the Foundation of Zoological Stations in Different Parts of the World." *Report of the British Association for the Advancement of Science, Edinburgh 1871*, 192. London.

Carus, J. Victor. 1880. "Vorwort des Redacteurs." *Zoologischer Jahresbericht* 1:iv.

Child, Charles Manning. 1903. "Form Regulation in Cerianthus." *Biological Bulletin* 5, no. 5:239–60.

De Masi, Domenico, and Paolo Gentile. 1991. "Un congresso permanente: Anton Dohrn e la Stazione Zoologica di Napoli." In *L'emozione e la regola: I gruppi creativi in Europa dal 1850 al 1950*, ed. Domenico De Masi, 29–57. Rome: Editori Laterza.

Dohrn, Anton. 1858. "Hemipterologisches." *Stettiner entomologische Zeitung* 19:163–64.

———. 1866a. "Zur Anatomie der Hemipteren." *Stettiner entomologische Zeitung* 27:321–52.

———. 1866b. "Zur Naturgeschichte der Caprellen." *Zeitschrift für wissenschaftliche Zoologie* 16:245–52.

———. 1868a. "On the Morphology of the Arthropoda." *Journal of Anatomy and Physiology* 2:80–86.

———. 1868b. *Studien zur Embryologie der Arthropoden: Habilitationsschrift.* Leipzig: Wilhelm Engelmann.

———. 1870. "Untersuchungen über Bau und Entwicklung der Arthropoden, VI: Zur Entwicklungsgeschichte der Panzerkrebse (Decapoda loricata)." *Zeitschrift für wissenschaftliche Zoologie* 20:248–71.

———, ed. [1871]. *Kurzer Abriss der Geschichte, sowie Gutachten und Meinungsäusserungen hervorragender Naturforscher über die Gründung der Zoologischen Stationen.* Jena: Fr. Frommann. Reprinted in Hans-Reiner Simon, ed., *Anton Dohrn und die Zoologische Station Neapel* (Frankfurt a.M.: Edition Erbrich), 13–20.

———. 1872a. "The Foundation of Zoological Stations." *Nature* 5, no. 119: 277–80.

———. 1872b. "Der gegenwärtige Stand der Zoologie und die Gründung zoologischer Stationen." *Preussische Jahrbücher* 30:137–61. Reprinted in Hans-Reiner Simon, ed., *Anton Dohrn und die Zoologische Station Neapel* (Frankfurt a.M.: Edition Erbrich), 23–46.

———. 1875. *Der Ursprung der Wirbelthiere und das Princip des Functionswechsels: Genealogische Skizzen*. Leipzig: Wilhelm Engelmann.

———. 1876. *Erster Jahresbericht der Zoologischen Station in Neapel*. Leipzig: Wilhelm Engelmann.

———. 1879a. "Bericht über die Zoologische Station während der Jahre 1876–1878." *Mittheilungen aus der Zoologischen Station zu Neapel* 1, no. 1:137–64.

———. 1879b. "Vorwort." *Mittheilungen aus der Zoologischen Station zu Neapel, zugleich ein Repertorium für Mittelmeerkunde* 1, no. 1:iii–viii.

———. 1881a. "Bericht über die Zoologische Station während der Jahre 1879 und 1880." *Mittheilungen aus der Zoologischen Station zu Neapel* 2, no. 4:495–514.

———. 1881b. *Die Pantopoden des Golfes von Neapel und der angrenzenden Meeres-Abschnitte*. Fauna und Flora des Golfes von Neapel III. Leipzig: Wilhelm Engelmann.

———. 1882. "Bericht über die Zoologische Station während des Jahres 1881." *Mittheilungen aus der Zoologischen Station zu Neapel* 3, no. 1:1–14.

———. 1885. "Bericht über die Zoologische Station während der Jahre 1882–1884." *Mittheilungen aus der Zoologischen Station zu Neapel* 6, no. 1:93–148.

———. 1893a. "Bericht über die Zoologische Station während der Jahre 1885–1892." *Mittheilungen aus der Zoologischen Station zu Neapel* 10, no. 4:633–74.

———. 1893b. "Publications of the Zoological Station at Naples." *Nature* 48, no. 1245:440–43.

———, ed. 1897. *Das 25 jährige Jubiläum der Zoologischen Station zu Neapel am 14. April 1897*. Leipzig: Breitkopf & Härtel. Reprinted in Hans-Reiner Simon, ed., *Anton Dohrn und die Zoologische Station Neapel* (Frankfurt a.M.: Edition Erbrich), 61–104.

Dohrn, Klaus. 1983. *Von Bürgern und Weltbürgern: Eine Familiengeschichte*. Pfullingen: Günther Neske.

Driesch, Hans. 1909. "Zur Erinnerung an Anton Dohrn." *Suddeutsche Monatshefte* 6, no. 2: 513–18.

Fantini, Bernardino. 2002. *The History of the Stazione Zoologica Anton Dohrn: An Outline*. Naples: Stazione Zoologica Anton Dohrn.

———. 2015. "Music and Biology at the Naples Zoological Station." *History and Philosophy of the Life Sciences* 36, no. 3:346–56.

FARD. 1961. *Fondazione "Antonio e Rinaldo Dohrn": Statuto—Elenco donatori—Resoconto delle attività svolte durante il quinquennio 1956–1960*. Naples: Stazione Zoologica di Napoli.

———. 1967. *Fondazione Antonio e Rinaldo Dohrn: Rapporto di attività, 1961–1965*. Naples: Giannini.

Florey, Ernst. 1995. "Highlights and Sidelights of Early Biology on Helgoland." *Helgoländer Meeresuntersuchungen* 49:77–101.

Florio, Riccardo. 2015. *L'architettura delle idee: La Stazione Zoologica Anton Dohrn di Napoli*. Naples: Artstudiopaparo.

Ghiselin, Michael T. 1994. "The Origin of Vertebrates and the Principle of Succession of Functions: Genealogical Sketches by Anton Dohrn: 1875: An English translation from the German, Introduction and Bibliography." *History and Philosophy of the Life Sciences* 16:3–96.

Grimpe, Georg. 1928. "Pflege, Behandlung und Zucht von Cephalopoden für zoologische und physiologische Zwecke." In *Handbuch der biologischen Arbeitsmethoden*, ed. Emil Abderhalden, sec. 5, pt. 9:331–402. Berlin: Urban & Schwarzenberg.

Groeben, Christiane, ed. 1982. *Charles Darwin, 1809–1882, Anton Dohrn, 1840–1909: Correspondence*. Naples: Macchiaroli.

———. 1985. "Anton Dohrn—the Statesman of Darwinism." *Biological Bulletin* 168 (suppl.): 4–25.

———. 1990. "The *Vettor Pisani* Circumnavigation (1882–1885)." *Deutsche Hydrographische Zeitschrift, Erg.-Heft B*, 22 (suppl. B): 220–34.

———. 1991. "Dohrn, Felix Anton." In *Dizionario Biografico degli Italiani*, 11:380–82. 92 vols. to date. Rome: Istituto dell'Enciclopedia Italiana.

———, ed. 1993. "Correspondence, Karl Ernst von Baer (1792–1876), Anton Dohrn (1840–1909)." *Transactions of the American Philosophical Society* 83, pt. 3:1–156.

———. 1998. "'Le précurseur du plan': La contribution de Carl Vogt à la fondation des stations marines." In *Carl Vogt: Science, philosophie et politique (1817–1895)*, ed. Jean-Claude Pont, Daniele Bui, Françoise Dubosson, and Jan Lacki, 287–312. Chêne-Bourg: Georg Éditeur.

———. 2000. *The Fresco Room of the Stazione Zoologica Anton Dohrn: The Biography of a Work of Art*. Naples: Gaetano Macchiaroli Editore.

———. 2002. "The Stazione Zoologica: A Clearing House for Marine Organisms." In *Oceanographic History: The Pacific and Beyond*, ed. Keith R. Benson and Philip F. Rehbock, 537–47. Seattle: University of Washington Press.

———. 2005a. "Briefe aus Neapel: der Blickwinkel der Gastforscher." In *Der Brief als wissenschaftshistorische Quelle*, ed. Erika Krausse, 103–24. Berlin: Verlag für Wissenschaft & Bildung.

———. 2005b. "Catalysing Science: The *Stazione Zoologica di Napoli* as a Place for the Circulation of Scientific Ideas." In *Stätten biologischer Forschung/ Places of Biological Research*, ed. Christiane Groeben, Joachim Kaasch, and Michael Kaasch, 53–64. Berlin: Verlag für Wissenschaft & Bildung.

———. 2008. "Tourists in Science: 19th Century Research Trips to the Mediterranean." *Proceedings of the California Academy of Sciences* 59, suppl. 1, no. 9:139–54.

———. 2010. "'Sotto sarà una pescaria, sopra una piccola università': La Stazione zoologica Anton Dohrn." In *L'acqua e la sua vita*, ed. Pietro Redondi, 151–202. Milan: Guerrini.

———. 2013a. "Felix Anton Dohrn." In *Enciclopedia italiana di scienze, lettere ed arti: Il contributo italiano alla storia del pensiero: Ottava appendice*, ed. Istituto della Enciclopedia italiana, 597–601. Rome: Istituto della Enciclopedia Italiana.

———. 2013b. "Stazione Zoologica Anton Dohrn." *eLS 2013*. doi 10.1002/9780470015902.a0024932.

Groeben, Christiane, and Fabio de Sio. 2006. "Nobel Laureates at the Stazione Zoologica Anton Dohrn: Phenomenology and Paths to Discovery in Neuroscience." *Journal of the History of the Neurosciences* 15:376–95.

Groeben, Christiane, and Michael T. Ghiselin. 2001. "The Zoological Station at Naples and Its Impact on Italian Zoology." In *Giovanni Canestrini, Zoologist and Darwinist*, ed. Alessandro Minelli and Sandra Casellato, 321–47. Venice: Istituto Veneto di Scienze, Lettere ed Arti.

Groeben, Christiane, and Irmgard Müller, eds. 1975. *The Naples Zoological Station at the Time of Anton Dohrn*. Naples: [Stazione Zoologica di Napoli].

Herbst, Curt. 1900. "Über das Auseinandergehen von Furchungs- und Gewebezellen in kalkfreiem Medium." *Wilhelm Roux's Archiv für Entwicklungsmechanik* 9:424–63.

Heuss, Theodor. 1991. *Anton Dohrn: A Life for Science*. With an introduction by Karl Josef Partsch and a contribution by Margret Boveri. Berlin: Springer.

Kofoid, Charles Atwood. 1910. "The Biological Stations of Europe." *US Bureau of Education Bulletin*, no. 4, whole no. 440. Washington, DC: US Government Printing Office.

Krausse, Erika. 1987. *Ernst Haeckel*. Leipzig: B. G. Teubner.

Kühn, Alfred. 1950. *Anton Dohrn und die Zoologie seiner Zeit*. Pubblicazioni della Stazione Zoologica di Napoli, vol. 22. Naples: Stazione Zoologica de Napoli.

Laubichler, Manfred D., and Eric H. Davidson. 2008. "Boveri's Long Experiment: Sea Urchin Merogones and the Establishment of the Role of Nuclear Chromosomes in Development." *Developmental Biology* 314:1–11.

Lo Bianco, Salvatore. 1899. "The Methods Employed at the Naples Zoological Station for the Preservation of Marine Animals." Translated by Edmund Otis Hovey. *Bulletin of the United States National Museum* 39, pt. M:1–42.

Loeb, Jacques. 1891. *Untersuchungen zur physiologischen Morphologie der Thiere, I: Ueber Heteromorphose*. Würzburg: Georg Hertz.

———. 1892. *Untersuchungen zur physiologischen Morphologie der Thiere, II: Organbildung und Wachstum*. Würzburg: Georg Hertz.

Maienschein, Jane. 1991. *Transforming Traditions in American Biology, 1880–1915*. Baltimore: Johns Hopkins University Press.

———. 1994. "'It's a long way from *Amphioxus*': Anton Dohrn and Late Nineteenth Century Debates about Vertebrate Origins." *History and Philosophy of the Life Sciences* 16, no. 3:465–78.

Monroy, Alberto. 1986. "A Centennial Debt of Developmental Biology to the Sea Urchin." *Biological Bulletin* 171:509–19.

Monroy, Alberto, and Christiane Groeben. 1985. "The 'New' Embryology at the Zoological Station and at the Marine Biological Laboratory." *Biological Bulletin* 168 (suppl.): 35–43.

Morgan, Thomas Hunt. 1896. "A Study of a Variation in Cleavage." *Archiv für Entwicklungsmechanik der Organismen* 2, no. 1:72–80.

———. 1901. "The Factors That Determine Regeneration in Antennnularia." *Biological Bulletin* 2:301–5.

Müller, Gerhard H., and Klaus Wenig. 1993. "'Kants Verhältnis zur Descendenztheorie': Anton Dohrns Probevorlesung 1868 in Jena." *Nuncius* 8, no.2:521–53.

Müller, Irmgard. 1973. "Der 'Hydriot' Nikolai Kleinenberg, oder: Spekulation und Beobachtung." *Medizinhistorisches Journal* 8:131–53.

———. 1975. "Die Wandlung embryologischer Forschung von der deskriptiven zur experimentellen Phase unter dem Einfluß der Zoologischen Station in Neapel." *Medizinhistorisches Journal* 10, no. 3:191–218.

———. 1976. *Die Geschichte der Zoologischen Station in Neapel von der Gründung durch Anton Dohrn (1872) bis zum ersten Weltkrieg und ihre Bedeutung für die Entwicklung der modernen biologischen Wissenschaften.* Habilitationsschrift, Universität Düsseldorf, Mathematisch-naturwissenschaftliche Fakultät.

———. 1996. "The Impact of the Zoological Station in Naples on Developmental Physiology." *International Journal of Developmental Biology* 40:103–11.

Oppenheimer, Jane M. 1978. "Dohrn, Felix Anton." In *Dictionary of Scientific Biography* (vol. 15, suppl. 1), ed. Charles Coulston Gillispie, 122–25. New York: Scribner.

Partsch, Karl Josef. 1980. *Die Zoologische Station in Neapel: Modell internationaler Wissenschaftszusammenarbeit.* Göttingen: Vandenhoeck & Ruprecht.

Raffaele, Federico. 1910. "Salvatore Lo Bianco, n. 10 Giugno 1860, m. 9 Aprile 1910: Commemorazione." *Bollettino della Società di naturalisti di Napoli* 23:109–12.

Richter, Gotthart. 1969. "The Ecology Department at Ischia (Punta S. Pietro) of the Zoological Station of Naples." *Pubblicazioni della Stazione Zoologica di Napoli* 37 (suppl.): 16–24.

Roux, Wilhelm. 1890. *Die Entwicklungsmechanik der Organismen, eine anatomische Wissenschaft der Zukunft.* Vienna: Urban & Schwarzenberg.

Schmidtlein, Richard. 1879a. "Beobachtungen über Trächtigkeits- und Eiablage-Perioden verschiedener Seethiere." *Mittheilungen aus der Zoologischen Station zu Neapel* 1, no. 1:124–36.

———. 1879b. "Vergleichende Uebersicht über das Erscheinen grösserer pelagischer Thiere während der Jahre 1875–1877." *Mittheilungen aus der Zoologischen Station zu Neapel* 1, no. 1:119–23.

Skalovà, Olga. 1975. "An Analysis of Geographical Mobility of Scientists and Their Communications as a Component of Their Working Conditions with Regard to the Naples Zoological Station: Contributi per la storia della Stazione Zoologica I." *Pubblicazioni della Stazione Zoologica di Napoli* 39 (suppl. 2): 1–126.

Tschermak, Armin von. 1914. *Die Zoologische Station in Neapel.* Berlin: Ernst Siegfried Mittler & Sohn.

Uschmann, Georg. 1959. *Geschichte der Zoologie und der Zoologischen Anstalten in Jena, 1779–1919.* Jena: G. Fischer.

Vogt, Carl. 1871. "Eine zoologische Beobachtungs-Station in Triest: An Prof. Oskar Schmidt in Graz." *Neue Freie Presse*, November 22.

———. 1876. "Die Zoologische Station in Neapel." *Frankfurter Zeitung*, no. 156.

———. 1884. "Die Zoologische Station in Neapel." *Vom Fels zum Meer* 2:365–78, 523–29.

Whitman, Charles Otis. 1883. "The Advantages of Study at the Naples Zoological Station." *Science* 2 (June 27): 93–97.

Wilkins, Maurice H. F. 1983. "[Reinhard Dohrn]." In *Reinhard Dohrn, 1880–1962: Reden, Briefe und Veröffentlichungen zum 100. Geburtstag*, ed. Christiane Groeben, 5–10. Berlin: Springer.

Wilson, Edmund Beecher. 1893. "Amphioxus and the Mosaic Theory of Development." *Journal of Morphology* 8, no. 3:579–638.

Young, John Zachary. 1985. "Cephalopods and Neuroscience." *Biological Bulletin* 168 (suppl.): 153–58.

Zissler, Dieter. 1995. "Five Scientists on Excursion—a Picture of Marine Biology in Helgoland Before 1892." *Helgoländer Meeresuntersuchungen* 49:103–12.

THREE

The First Marine Biological Station in Modern China: Amoy University and Amphioxus

CHRISTINE YI LAI LUK

In October 1925, Edwin G. Conklin, a prominent American biologist, visited the Peking Union Medical College (PUMC) in China. On hearing of Conklin's upcoming visit, Gist Gee, in his capacity as the secretary of the Peking Natural History Society, conveyed the following message to him: "The society would be glad if it could receive a lecture in the PUMC auditorium about the aims of biological training that would include an outline of the Marine Biological Laboratory at Woods Hole, Massachusetts; this latter might 'give encouragement to the effort to establish a marine biological laboratory in China'" (Haas 1996, 181). Conklin is known among historians of biology for helping lay the foundations of experimental biology in twentieth-century America. Not only was he an active participant in creating the "Hopkins" model;[1] he was also an influential figure at the Marine Biological Laboratory (MBL) and Princeton University (Maienschein 1991). His reputation was acknowledged by his host, who had nominated him to be a corresponding member of the Peking Natural History Society during a special dinner held in his honor. But it is compelling to note that, among all the topics that Conklin could have covered during his visit, Gee specifically asked him to talk about the institutional approach of the MBL, looking ahead to creating a marine biological laboratory in China. This choice of topic had far-reaching implications for the unfolding of marine biology in modern China.

This essay examines the founding of China's first marine biological laboratory at Amoy University between 1930 and 1935. Through describing the institutional approach to building an embryonic Chinese Woods Hole, my purpose is to unravel the tension between locality and transnationality arising from marine biological research in the historical context of Republican China. Amoy was chosen primarily because of its abundant supply of marine species, particularly amphioxus, which was in great demand for teaching the evolutionary origin of vertebrates and for researching the Chinese characteristics of "man's oldest vertebrate ancestor." This growing need to supply teachers and researchers of biology with zoologically important animals is a common thread that connects the Chinese biological community with its American counterparts.

In what follows, I first give a historical overview of biology in modern China, describing the general context in which marine biological study took off in Republican China; then I explore the motivations behind the ambition to build China's first marine biological station by examining historical actors' rationales for seeking to promote marine biology in modern China. Next, I discuss how Amoy University emerged as the birthplace of marine biological laboratory research and the types of summer research that took place there and correlate the locality with the types of marine-based research undertaken between 1930 and 1935. My overall argument is that, while the locality of marine stations is crucial (Muka 2016), the transnationality of marine research is equally important. China's first marine biological station attracted interest from both the Chinese and the American scientific communities owing mostly to the abundance of Amoy-supplied amphioxus—a marine organism that combines elements of both indigenous characteristics and transnational commonality. Unlike biological research surrounding terrestrial organisms, which is often a local attempt blending classicism with nationalism (Jiang 2016), biological investigation of marine species swims between poles of national sovereignty and transnational synergism.

Marine Biology in Modern China

Although curiosity in China about ocean life dates back to ancient times, a systematic study of marine organisms began only in modern China. Tchang Si (張璽, also spelled Zhang Xi),[2] one of the most distinguished marine zoologists in modern China, asserted that it was only in the early years of the Republic of China that marine biological stations were constructed for the classification and investigation of sea creatures. In an article published in the *American Zoologist*, he wrote: "In China, there were no far sea explorations nor marine biological stations until the early years of the Republic of China" (Tchang 1946, 599). Although an absence of marine stations does not necessarily mean an absence of marine biological investigation, Tchang's remark was congruent with the expanding scope of biological and geological surveys during the Republican period. It is worth revisiting the context and significance of scientific research in Republican China as marine biology emerged from the same historical context that gave rise to many other scientific disciplines.

In the late nineteenth century, as China's imperial order began to crumble and alongside the demise of the Qing dynasty, many thinkers and reformers set out to search for new ways to understand the world and manage the country (Schwartz [1964] 1983). As the traditional values of Confucian learning came under direct attack, many looked elsewhere for inspiration. Modern science and technology provided an important framework for intellectuals attempting to conceptualize the new world order and come up with strategies for social change (Wang 1995; Wang 2002). China's adoption of Western technology began with the construction of the first arsenal, the first steamship, and the first machine-tool manufacturer in the 1870s (Elman 2006). After realizing that building modern warships and gunboats alone would not save the country from military defeats, many started to look outside the realm of military technology and into basic science for ideas to strengthen the nation. Sanitation and public health (Rogaski 2004), geological surveys (Shen 2014), and mammalian fossils (Schmalzer 2008) represent some of the Chinese efforts to generate scientific understanding of the natural realm.

Within this broader context in which Western science was generally revered and feverishly pursued, some leading scientists began to advocate the advancement of marine biology in the cause of nation building. Bing Zhi (秉志, also spelled Ping Chih), a cofounder of the

Biological Laboratory of the Science Society of China, was of the view of "cultivating morality through biology" (Jiang 2016). Bing, a Cornell PhD holder in entomology, wrote popular essays on the relationship between biology and university education, biology and women's education, biology and people's livelihood, and so on. A central theme underlying these essays is that biology holds the key to national survival since understandings of the objective world and subjective beings converge in biological knowledge. Lijing Jiang nicely summarizes Bing's view of biology as crystallized in the motto, "no biology, no nation" (Jiang 2016, 190).

In view of the significance of biology for nation building, the next question that arose was how to promote biology, and one way was through developing marine biology. In an article titled "In Support of Building a Marine Biological Laboratory" (倡設海濱生物實驗所說), Bing offered reasons why it was necessary to build a marine biological laboratory in China at that time. Since that article is one of the main pieces linking marine biology to biology and nation building, it is worth quoting at some length:

> Biology matters to applied industry and education. No one who studied biology doesn't know about it. There are two ways to develop biological study, the first of which is to emphasize the teaching of biology at schools through hiring teachers and experts to conduct specialized research, to arouse scholarly interest, and to cultivate specialists; the second way is to investigate sea creatures with coastal surveys to sort out the relationship between biology, applied industry, and teaching. Anything about the life history and distribution of marine living organisms is worthwhile for specialists to study in order to enrich scholarship and contribute to human knowledge. These two means are indispensable for promoting biology. (Bing 1923, 307)[3]

In other words, the rationale is that biology deserves financial support from the state because it is important for moral education and has applied values. Bing suggested that marine biology is one of the two indispensable ways to develop biology, and therein lies the significance of marine biological research. In the rest of the article, he gave examples about how collecting marine biological specimens relates to classroom learning. In addition to using specimens for illustration and observation, he emphasized the role of actual fieldwork: "School learning benefits from coastal experiments; otherwise, the training is lopsided at the expense of seeing the full picture of biology" (Bing 1923, 307). He then mentioned the trend of setting up marine biological laboratories at coastal locations in Europe, America, and Japan that enabled marine biologists to take up summer surveying trips. He stressed the need to emulate this worldwide trend: "If we want to promote marine biology (in our country) for the sake of benefiting applied industry (i.e., fishery) and education, we need to imitate the actions of Europe, America, and Japan" (Bing 1923, 308). Among all the possible localities in Japan and Europe/America, he highlighted the Naples Zoological Station in Italy and the MBL at Woods Hole (烏子吼耳) in America as shining examples from which the Chinese could learn.

Bing went to some lengths to discuss Woods Hole, beginning with details about his two excursions to the MBL during his academic sojourn in the northeast part of the United States. He was very impressed with Woods Hole's success, which he hoped his Chinese compatriots could emulate. It is unsurprising why a foreign biologist like Bing Zhi would find Woods Hole attractive as the MBL had always been a "world biological laboratory," a place for prominent scientists from all over the world to gather each summer to pursue the best science (Maienschein 1988, 1989). But what is new is the expressed interest in bringing the Woods Hole model to China, hoping that a budding "Chinese Woods Hole" would take root in the specific political soil of early twentieth-century China.

Although Bing was not the first nor the only Chinese biologist visiting the MBL in the early twentieth century,[4] he was likely the first one who translated "Woods Hole" into Chinese and introduced it to the average Chinese audience. A year before he popularized the Woods Hole concept among the Chinese public, he wrote another *Kexue* (Science) article in which he discussed his summer collecting trips in Shanghai. In it, he stressed the significance of conducting a nationwide survey of coastal life to overcome what he perceived to be the increasing Japanese dominance in the supply of biological materials. He noted that most marine samples used in Chinese schools at the time were purchased from Japan and were thus Japanese specimens. He continued: "[Even though] there were similarities between Japanese species and ours, there were also substantial differences, and we know specimens coming from other places could not exhaust our range of indigenous species. Moreover, the Japanese decorated generic, valueless specimens and sold them to our people at an expensive price. Our countrymen should do the collecting themselves." He further observed: "Our nation is situated at the temperate zone, blessed with an abundant supply of seafood in the southeastern part of the nation. Yet we have to depend on Japan for seafood provision" (Bing 1922, 84). From the nationalist rhetoric reproduced above alone, one would get the impression that Bing was just a nationalistic instigator. Yet what underscored his proposition was the notion that China must make up its own mind to take marine studies seriously. The competition from Japan was a contextual factor, but the actual motivation came from Bing's larger vision to "expand our scope of understanding of new marine species and contribute to scientific knowledge." To this end, the competing efforts from Japan or other countries could not replace "the imperative of undertaking marine research in China" (Bing 1922, 98).

Comparing China first to Japan and then to Woods Hole makes it evident that Bing saw China as capable of rising to the highest level, comparable to that of the leading countries in terms of biological excellence. At least in the realm of marine biology, he hoped that China could one day take up the role of leadership, rather than just catching up with its Western counterparts. His vision for developing marine biology in China was picked up three years later by Gist Gee, who invited E. G. Conklin to deliver a lecture about the MBL, as mentioned above.

An Embryonic Chinese Woods Hole

It was not accidental that it was Gee who asked Conklin to discuss the MBL approach at PUMC in 1925. Before Conklin's talk, Gee had been an advocate of the teaching of biology in China. An esteemed American missionary-teacher living in China between 1901 and 1935, Gee was said to be "universally respected among the teachers of science in China" (Haas 1996, 12). More importantly, he was celebrated as the first person introducing a modern biological education to China in addition to contributing to the institutional development of the teaching of biology and of biological research in modern China (Li and Luo 2009). As an avid biologist-naturalist with a passion for biological specimens from Southeast China, Gee had called for the investigation of China's indigenous fauna and flora. He urged the Nationalist government to sponsor a biological survey similar to the geological survey in order to understand the biological resources of China: "[The Chinese] government must organize a biological survey and must provide museum facilities for the accumulation of a national collection. . . . It is a cause for shame . . . that when we want accurate information about things already known . . . we have to go to the museums of the West to examine their collections and their records" (Haas 1996, 13).

Gee felt that it was essential for the Chinese to embark on a domestic investigation of the fauna and flora of China even though some Western naturalists, particularly the British, had already accumulated some understanding of China's indigenous animals and plants (Fan 2004). Being a non–Chinese national, he was nevertheless aware of not just the inconvenience but also the profound sense of shame that it inflicted on Chinese researchers to have to consult the biological collections of museums in the West. It was uncommon for a foreigner to utter a nationalist-leaning sentiment like this. As Gee's biographer suggested: "What was unusual was that Gee, an American missionary, let go an opportunity to promote natural history as the occupation of Westerners, in favor of showing it as something a Chinese person had command of" (Haas 1996, 102). It was Gee's deep sympathies for Chinese things that led him to side with the Chinese approach to natural history.

Among all the creatures that subsisted in and around Chinese territory, what captured Gee's fascination was what lay beneath the water. A leading expert on freshwater sponges, Gee produced forty-four papers on this topic, with most of his research based on sponges collected from Lake Taihu in Southeast China. His research interest in sponges spread to other nonterrestrial organisms and fueled his desire to sponsor the institutionalization of marine biological investigation. Gee was of the view that "marine and freshwater biological stations could be established through existing educational institutions to study 'problems of great economic importance' with only minimal equipment" (Haas 1996, 13). He felt that tapping into existing resources at local universities was conducive to marine biological investigation. His idea of creating a university-based marine biological station foreshadowed the construction of a modern marine biological laboratory at the University of Amoy in 1930.

Yet it is not very clear why Gee picked Amoy, among other possible coastal locations, as a favorable place to institutionalize marine biology. The next section explores factors that underlined his arguments for choosing Amoy as the birthplace of a marine biological laboratory in modern China.

Amoy and Amphioxus

In an article surveying China's research progress in marine zoology, Tchang Si noted: "Amoy is the most important due to its having the University of Amoy, and being the original place of the Marine Biological Association of China" (Tchang 1946, 596). But when and why did Amoy emerge as a top location for marine biological research?

After Conklin's 1925 visit, Gee took more concrete actions to realize his ambition of building the first marine biological laboratory in China. In a 1929 letter addressed to Max Mason, a senior officer at the Rockefeller Foundation in New York City, he conveyed his abiding wish to create "a simple marine biological laboratory in some suitable place on the China coast where the teachers and advanced students of biology might gather during their vacations and continue their work, both of teaching and research, upon living salt water materials." He went on to suggest that "Amoy seems to be a place where something of this kind might be tried out."[5]

Gee had his eyes on Amoy for a number of reasons. Amoy was chosen primarily because of its rich marine biological resources and easy access to the water, which provided plenty of opportunities for collecting marine fauna and flora. In addition, it was known as a storehouse of amphioxus. The abundance of this primitive vertebrate captured the attention of S. F. Light, an American biologist teaching at the University of Amoy at the time. Light contributed an article to *Science* that began as follows: "This note is to announce the discovery of an apparently inexhaustible supply of amphioxus near the University of Amoy" (Light 1923, 57). What piqued his interest was the fact that "on this little strip of Chinese coast somewhere around a billion amphioxi are caught and consumed each year." Light's finding was so influential that George Sarton, the founder of the History of Science Society, incorporated his article in his fifteenth annotated bibliography of the history and philosophy of science, appended a personal note: "Most of us have been brought up with the idea that the amphioxus is a relatively rare animal. Light describes 'fisheries' of amphioxus near his own university, where about a billion amphioxi are caught each year! The Chinese name of the lancelet is wen shen yü, meaning fish of the god of the literature (literary composition fish). It is also called silver spear fish and carrying pole fish. Light tells the story connected with the first of these three names" (Sarton 1924, 193).

Intending his bibliography for a Western audience, Sarton was apparently quite surprised when he learned that amphioxus was so abundant at Amoy that the Chinese could make a fishing industry out of what elsewhere was viewed as a precious marine fauna. Amoy's large reserve of amphioxus was a compelling feature that appealed to both historians and scientists. As Gee wrote in 1932: "The one fact that this center is near the very large amphioxus fishing grounds make [*sic*] it unique and if it is as rich in other interesting forms as our present knowledge seems to indicate, it will certainly make a very splendid station."[6] Apparently, the prolific presence of amphioxus was an outstanding factor when it came to selecting Amoy as the site of the marine biological station.

In addition to its rich natural resources, Amoy's institutional support was also a helpful factor. Lim Boon Keng (林文慶), founding president of Amoy University, noticed the growing interest in marine biological work in the early 1920s. In a letter addressed to H. C. Zen (任鴻雋, also spelled Ren Hong-jun), a founding member of the Science Society of China and staff director of the China Foundation for the Promotion of Education and Culture (the China Foundation), he described the institutional resources that his university could offer in support of summer marine research activities: "With our Biology Building located on the shore of Amoy Harbor, the laboratory and library facilities in workable condition and the Amoy region rich in marine materials for study, the University of Amoy stands in a very favorable position to supply the same need. I feel that time has come for us to open up our laboratories for summer work."[7]

Not only was Amoy University adjacent to the shore with easy access to salt water; it was also equipped with a functional biological laboratory and a library to facilitate marine biological research. But what underpinned the institutional support above everything else was Lim's eclectic philosophy, as discussed by Lee Guan Kin in her research examining his eclectic approach blending Western science with Confucianism (Lee 1991). As Lee shows, Lim's unique background as an overseas Chinese with medical training from Edinburgh structured his reformist belief in engineering Chinese society through Western science and medicine. During his sixteen-year tenure as president of Amoy University, his investment in science and engineering, perceived as undermining the humanities, met with some forms of resistance from the May Fourth intellectuals teaching at Amoy University such as Lu Xun, Lin Yutang, and Gu Jiegang (Lee 2009).

Lim's affinity to biological and medical sciences among all the scientific disciplines offered at Amoy University was quite remarkable. In the keynote speech delivered at the tenth anniversary of Amoy University (Lim 1931), Lim likened the university to a living organism: the buildings were the body, the teaching staff and administrative council the nervous organs, and the students the vegetative function. But it is important to point out that his interest in biological science was by no means limited to using biology-related metaphors; rather, his agenda in promoting biology at Amoy was manifested in concrete terms. In other words, he was not interested in just talking about biology; he also took actions to advance biology at Amoy University.

Lim's expressed interest in biology manifested in the physical layout of Amoy University. As shown in figure 3.1, the biology building is one of the two campus buildings dedicated to the learning and teaching of science and is located next to the chemistry building. Both buildings were situated close to the beach and Amoy Harbor, with easy access to public transportation (provided by a bus road). Given the importance of university buildings in Lim's keynote speech, this proximity is hardly a coincidence. In fact, the chemistry and biology buildings were among the first erected on campus following the school's inception in 1922. Inside the biology building, the first floor was allocated to the Department of Zoology, with half the space reserved for zoological laboratories and the other half dedicated to the museum (see fig. 3.1).

The provision of space for biological research and collections underlined the institutional commitment to marine biological study at Amoy University. These spatial facilities were an important source of evidence to lend support to Lim's articulation of Amoy's distinctiveness. As Lim noted in the *First Annual Report of the Marine Biological Association of China* (in a section titled "Amoy; or, The Island That Remembers the Mings"): "Owing to the easy access from Amoy to all parts of the interior of the province, and also owing to the facilities of an excellent library and of the science laboratories of the University it can be easily appreciated that the locality offers many attractions to those interested in making zoological collections. The University zoological museum shows at a glance the wealth of material available on the spot. The flora is also fairly extensive if the geological structure is taken into consideration" (Lim 1932, 52).

In "Amoy; or, The Island That Remembers the Mings," Lim delineated the history, topography, climate, population, customs and religions, political and diplomatic relations, education, fauna and flora, and fishing industry of Amoy. This ethnographic sketch gave a portrait of Amoy's natural and man-made resources. Not only did Lim introduce readers to Amoy's rich marine stock, but he also emphasized amphioxus as Amoy's most attractive marine species: "The most interesting marine product is the primitive *Branchiostoma belcheri var. amoyensis*—better known as amphioxus or lancelet. This species of the cephalochorda [*sic*] occurs in great abundance at Liu-Wu-Tien, where a

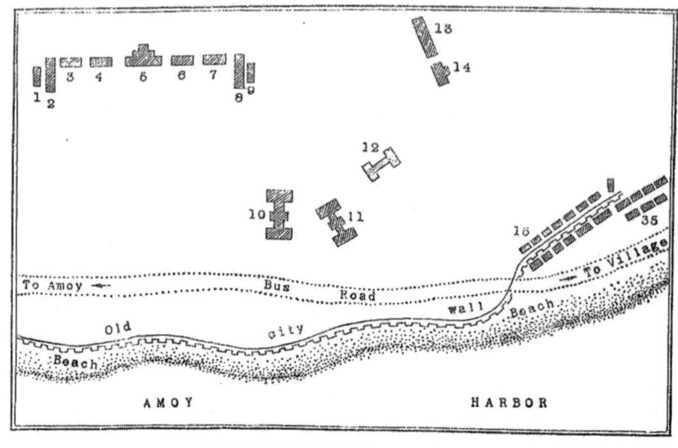

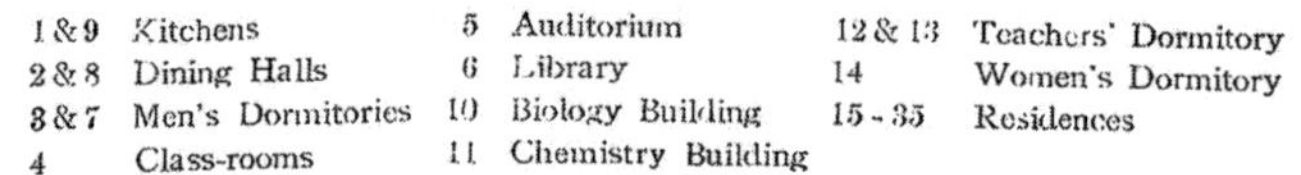

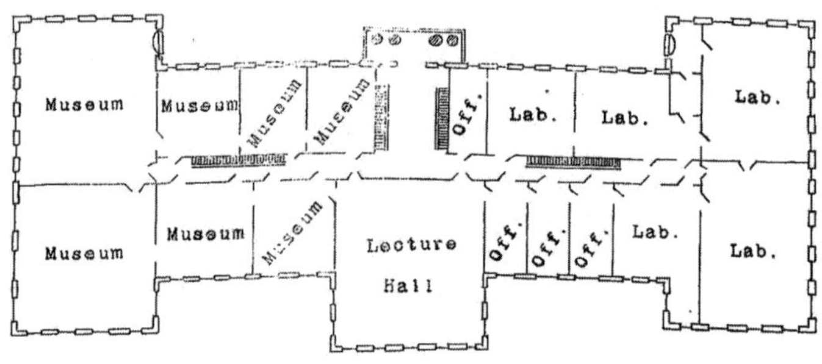

FIGURE 3.1. *Upper panel*: Layout of Amoy University. The biology building and the chemistry building were the only science buildings on campus and were constructed near the beach and Amoy Harbor, with easy access to public transportation. *Lower panel*: Plan of the first floor of the biology building at Amoy University. Adapted from the *First Annual Report of the Marine Biological Association of China* (1932). Courtesy of Rockefeller Archive Center.

special fishing industry has long developed for its capture. Both in the fresh and the dried form, it is sold in large quantities in the market" (Lim 1932, 52).

Lim's characterization of Amoy's amphioxus as "*B. belcheri var. amoyensis*" reflects his Amoy-centric view. By incorporating Amoy into the scientific terminology of amphioxus, Lim sought to highlight the indigeneity of Amoy for newfound nonterrestrial species. The Amoy-centric approach is evident in the new forms of marine fauna introduced by the Zoological Department of Amoy University in 1932, such as *Acromitus tankahkeei* (a jellyfish named after Tan Ka Kee, the patron-founder of Amoy University), *Asterina limboonkengi* (a starfish named after Lim Boon Keng), *Lemdana limboonkengi* (a nematode named after Lim Boon Keng), and *Platyrhina limboonkengi* (a type of ray named after Lim Boon Keng). Not only did these new monikers bear the names of the patron and the president of Amoy University; they were also featured in a special issue of *Amoy Weekly* celebrating the university's fifteenth anniversary (*Amoy Weekly* 1933). These new monikers bear witness to the fact that Amoy was beginning to play an important role in biological studies and was fulfilling Lim's ambition to make Amoy University a center of marine biological research and teaching.

Following Lim's administrative leadership and the worldwide trend of studying amphioxus at that time (Maienschein 1994), Chinese and Western biologists began to take Amoy's amphioxus seriously as a research subject. One of the first notable Western biologists to do so was Alice Boring, a PhD holder from Bryn Mawr who went to PUMC to teach biology (Ogilvie and Choquette 1999). A specialist in amphibians and reptiles, Boring and her colleague Li Hui-lin coauthored a study of the morphological characteristics of Chinese amphioxus. Through comparing American and European amphioxi specimens ordered from MBL and those collected from the first summer marine biological survey conducted at Amoy University in 1930, Boring and Li concluded that the structural dissimilarities in muscular characteristics and behavioral responses as well as external anatomical features suggested that the Chinese amphioxus (*B. belcheri*) represented a different species from the American amphioxus (*B. caribeum*) and the European amphioxus (*B. lanceolatum*) (Boring and Li 1932). Boring's provocative idea sparked more follow-up research on Chinese amphioxus. Shortly thereafter, Tchang and Koo compared the amphioxi samples collected from Tsingtao's Kiaochow Bay and the Amoy coast and determined that the differences in fin chambers and myotomes set Tsingtao's amphioxus apart from Amoy's (Tchang and Koo 1936).

If one takes a closer look, it is not difficult to discern a general pattern linking all the above-mentioned Amoy-based marine fauna. From Boring and Li's pathbreaking 1932 study of Chinese amphioxus vis-à-vis Euro-American amphioxus to Lim's 1932 description of *B. belcheri var. amoyensis* and the inventory of new marine species reported by the *Amoy Weekly* in 1933, it seems that all the new research surrounding

Amoy's marine fauna took place in or around 1932. It is to these gravitational events that took place around 1932 that we now turn.

The Summer Surveys at Amoy

The previous section discussed Lim's ambition to make Amoy a model place for China's marine biological survey. With Amoy's easy access to the coast, an abundant supply of amphioxus, and institutional commitment from the president of Amoy University, Amoy was uniquely positioned to take over the country's marine biological survey. Lim was hoping to use the survey as a platform to make Amoy into a research hub for other types of nationwide investigations. As he wrote to H. C. Zen of the China Foundation on February 21, 1930: "The survey of the Amoy fauna and flora by a chosen group of specialists, which this year is to be one of the main undertakings, will serve as a model for similar undertakings on a smaller scale in other parts of China. I consider such a study a very important one at the present stage of our biological work in China."[8]

But Lim's grandiose plan to make Amoy a center of excellence for China's biological studies required funds to back it up. At the time, the Rockefeller Foundation was famous for its medical-missionary work in China, especially its founding of PUMC. As discussed before, Gee leveraged his connection with PUMC to obtain more foundation funding to support biological investigation in China. Indeed, he was the main proponent and the contact person at the foundation for the entire Amoy marine biological station project. However, other than Gee, most senior foundation officers were not terribly supportive of this initiative. It was felt that the foundation's priority was to bring Western medicine to poverty-stricken China; biology and natural history may have been relevant to biomedical teaching, but they were simply not an area of focus: "Natural history in China was not on foundation president George Vincent's list for support" (Haas 1996, 188). To be clear, it was not the case that the Rockefeller people were uninterested in marine biology. To the contrary: "The Rockefeller Foundation at one point or another supported almost every marine biology station in the world but never any in China" (Haas 1996, 188). It was the combination of China and marine biology, and modern science in general, that drew opposition from the foundation's senior officers. Warren Weaver, the director of the division of natural sciences between 1932 and 1955, was unsympathetic to China's scientific competence: "Weaver questioned

Chinese capabilities for modern science and whether modern science would benefit China" (Haas 1996, 250).

Despite the strong skepticism of its decision makers, the Rockefeller Foundation did allocate a sum of up to $5,000 to sponsor the summer marine biological survey to be conducted at Amoy University, provided that matching funds from other sources were available. The foundation's financial support was secured only after active lobbying by Gee, who had petitioned for it from the very beginning. In board minutes dated April 28, 1931, it was noted: "During the fall of 1929 Mr. N. Gist Gee, the Foundation's representative in China, became interested in the formation of a marine biological institute at some biological laboratory. . . . The officers of the Rockefeller Foundation indicated interest in the undertaking, but did not suggest participation, preferring to have the project inaugurated by outside sources and thus to ascertain the demand for such an institute."[9] It is thus clear that Gee was the main champion of building China's marine biological institute. Were it not for his advocacy, it would be hard to imagine anyone from the foundation taking any interest in the matter. Second, the minutes conveyed how financially conservative the foundation was when it came to supporting China's marine biology. In spite of preliminary interest, it was recommended that direct involvement be avoided and that other groups be allowed to take over the project. In other words, in 1929, the American philanthropic organization was lukewarm about this proposal and unwilling to pay for it.

Advancing China's biological enterprise might not be an institutional priority for the Rockefeller Foundation, but it surely was for the China Foundation, which had offered generous funding for the Biological Laboratory of the Science Society of China since its inception in 1924 (Wang 2002). For example, the foundation allotted $5,000 to support the undertaking of the first summer marine biological investigation, beginning in 1930. As a result, the Marine Biological Association of China (中華海產生物學會, MBAC) was founded at Amoy University between July 14 and August 24, 1930, when eleven leading biologists and their representatives from across the country gathered to conduct a marine survey at China's first modern seaside laboratory.

The research agenda of the first summer survey revolved around the forms and functions of amphioxus, like nearly all the investigations conducted at the site concerned some aspects of amphioxus. Robert K. S. Lim, President Lim's son and a prominent neurobiologist, set out to examine the muscular and nervous systems of amphioxus, while others studied its glands and secretion, respiration, and nutritional value and the ecology around the organism. Amphioxus was, he found, the converging factor for the first summer survey commissioned by the MBAC—an institution dedicated to "the encouragement of investigations in marine biology" and "the training of workers in methods of investigation and in recognition of Chinese forms" (*First Annual Report of the Marine Biological Association of China 1932*). Although amphioxus was not unique to China, the intensive summer study session created the conditions that led to Boring and Li's 1932 study of the Chineseness of Amoy's amphioxus, which in turn inspired subsequent studies of amphioxus fisheries around Amoy by T. Y. Chen, chairman of the Zoology Department at Amoy University between 1930 and 1943. Although Chen received his doctorate from Columbia University under the tutelage of the famous geneticist T. H. Morgan, in China he was considered, then and now, a leader in marine biology rather than genetics. In two reports (Chen 1935; Chen and Chin 1930), he suggested that amphioxus is an important food source for local people and a valuable experimental organism for evolutionary biologists; thus, the amphioxus fishery is a special type of fishery with local characteristics. These fisheries-cum-marine-biological surveys of the 1930s highlighted the economic as well as the scientific value of amphioxus for the local fisherfolk and the transnational biological community.

Yet this transnational scientific network forged over a shared curiosity about nature did not survive wartime contingencies. The outbreak of the Second Sino-Japanese War in 1937 brought the development of the embryonic Chinese Woods Hole to an abrupt halt. Between 1937 and 1946, Amoy University was relocated away from the coast to the interior. The once-elaborate biology building on Amoy campus was bombed by the Japanese army, which in 1939 also confiscated all the university's equipment and laboratory supplies. The biology building was rebuilt in 1954 as Amoy University became nationalized after the establishment of the People's Republic of China in 1949. Even though Amoy's marine research facilities did not survive the Pacific War and the subsequent civil war, the cultural imagery and material practice surrounding amphioxus did. Amphioxus is now popularized as an endangered (but delicious) species in China and a symbol for protecting the local way of life in Xiamen—the current romanization of Amoy. Adults are now teaching Xiamen's children to sing a folksy song about amphioxus in their rapidly disappearing local dialect, and traditional methods and tools for catching amphioxus are broadcasted on television programs. More recently, Chinese researchers joined the phylogenetic bandwagon and offered genomic data obtained from amphioxus to enrich the debate on evolution and development (Gee 2018).

Conclusion

The emergence of marine biology in China can be traced to Republican China as leading biologists at the time began to connect marine biological study with the objective of nation building. Marine biology was a loosely defined discipline structured around the locality of marine biological stations. The location of these stations is quite important as they represent the meeting points at which biologists from different places convened every summer. The construction of the first marine biological station at Amoy University exemplified a host of factors embedded in the localness of marine biology. But Amoy's attractiveness as the fishing ground for amphioxus also reflects the transnationality of marine biological research. As a key experimental organism for the teaching of and research into vertebrate evolution, amphioxus embodies the duality of locality and transnationality. The scarcity of amphioxus in the United States and its abundance in China fueled the interest surrounding Amoy as a top choice for building an embryonic Chinese Woods Hole. While China's development of marine biology cannot be reduced to this one marine animal, it is through amphioxus that we can better understand the local and transnational characters of marine biological research and teaching in early twentieth-century China.

Notes

This paper was first presented at the annual meeting of the History of Science Society held in Atlanta between November 3 and 5, 2016. I thank Lijing Jiang and Subo Wijeyeratne for their contribution to the panel "Transnationalism and Transformation of Science in Modern East Asia." I am also grateful to Zuoyue Wang for serving as the panel's commentator and Sigrid Schmalzer for serving as the panel's chair. In writing this essay, I benefited from discussions with Kjell Ericson, Lijing Jiang, Guo Jinhai, Zuoyue Wang, Subo Wijeyeratne, and Gray Williams. I appreciate the constructive comments on and suggestions regarding an earlier draft offered by Rachel Ankeny, Jane Maienschein, and Karl Matlin. Any errors that remain are mine.

1. The term *Hopkins model* refers to the integrative approach in American biology at the turn of the twentieth century in which morphological

tradition merged with physiological traditions at the Johns Hopkins University. For details, see Maienschein (1991).

2. For Chinese individuals, family name (surname) precedes given name (first name). I use traditional Chinese characters whenever it is possible and am sensitive to the time and place of historical actors. For instance, I use 張璽 rather than 张玺 because simplified Chinese characters were not available when the paper cited as Tchang (1946) was delivered. For American individuals, I adopted the Western convention of giving the family name last.
3. Quotations from Chinese sources were translated into English by the author.
4. One of the MBL's virtual exhibitions collects short snippets of some Chinese biologists studying at the MBL between 1920 and 1945: http://history.archives.mbl.edu/exploring/exhibits/china-mbl-1920-1945.
5. China Medical Board Collection, Research Group 1, Series 601, box 38 (hereafter China Medical Board Collection), folder 312, Rockefeller Archive Center, New York.
6. China Medical Board Collection, folder 314.
7. China Medical Board Collection, folder 312.
8. China Medical Board Collection, folder 312.
9. China Medical Board Collection, folder 312.

References

Amoy Weekly. 1933. "Xiada zhoukan: xiamen daxue shi wu zhounian jinian zhuanhao" (廈大周刊: 廈門大學十五週年紀念專號; *Amoy Weekly*: A special issue celebrating the fifteenth anniversary of Amoy University). *Amoy Weekly* (廈大周刊) 22, no. 15 (April): 1–67.

Bing Zhi (秉志; Ping Chih). 1922. "Xinyou xiaji caiji dongwu biaoben jishi" (辛酉夏季採集動物標本記事; A report of collecting animal specimens in the summer of 1921). *Kexue* (科學) 7, no. 1: 84–98.

———. 1923. "Chang she haibin shengwu shiyansuo shuo" (倡設海濱生物實驗所說; In support of building a marine biological laboratory). *Kexue* (科學) 8, no. 3:307–10.

Boring, Alice M., and Li Hui-lin. 1932. "Is the Chinese Amphioxus a Separate Species?" *Peking Natural History Bulletin* 1931–32, no. 6, pt. 3:9–18.

Chen Tse-Yin (陳子英; Ziying Chen). 1935. "Fujian sheng yuye diaocha baogao" (福建省漁業調查報告; Fisheries survey report of Fujian Province). *Zhongguo jianshe* (中國建設0) 11, no. 1:29–76.

Chen T. Y. and Chin T. G. (金德祥). 1930. "Xiamen wenchangyu yuye diaocha" (廈門文昌魚漁業調查; Survey of amphioxus fisheries in Amoy). *Zhongguo jianshe* (中國建設7), no. 3:115–23.

Elman, Benjamin A. 2006. *A Cultural History of Modern Science in China*. Cambridge, MA: Harvard University Press.

Fan Fa-ti. 2004. *British Naturalists in Qing China: Science, Empire, and Cultural Encounter*. Cambridge, MA: Harvard University Press.

First Annual Report of the Marine Biological Association of China. 1932. Amoy: Marine Biological Association of China.

Gee, Henry. 2018. *Across the Bridge: Understanding the Origin of the Vertebrates*. Chicago: University of Chicago Press.

Haas, William. 1996. *China Voyager: Gist Gee's Life in Science*. New York: M. E. Sharpe.

Jiang Lijing. 2016. "Retouching the Past with Living Things: Indigenous Species, Tradition, and Biological Research in Republican China, 1918–1937." *Historical Studies in the Natural Sciences* 46, no. 2:154–206.

Lee Guan Kin. 1991. *The Thought of Lim Boon Keng: Convergency and Contradiction between Chinese and Western Culture* (in Chinese). Singapore: Singapore Society of Asian Studies.

———. 2009. *Lim Boon Keng and Xiamen University* (in Chinese). Singapore: Centre for Chinese Language and Culture, Nanyang Technological University, and Global Publishing.

Li Ang (李昂) and Luo Gui-huan (羅桂環). 2009. "Qi tianci dui zhongguo shengwuxue shiye de gongxian" (祁天賜對中國生物學事業的貢獻; N. Gist Gee and his contribution to the development of biology in China). *Chinese Association for the History of Science* (中華科技史學會學刊) 13 (December): 15–25.

Light, S. F. 1923. "Amphioxus Fisheries Near the University of Amoy, China." *Science* 58, no. 1491 (July): 57–60.

Lim Boon-keng (林文慶). 1931. "Xiada shi zhounian jinian de yiyi" (廈大十週年紀念的意義; The meanings of celebrating the tenth anniversary of Amoy University). *Amoy Weekly Tenth Anniversary Special Issue*, April, 5–11.

———. 1932. "Amoy; or, The Island That Remembers the Mings." In *First Annual Report of the Marine Biological Association of China*. Amoy: Marine Biological Association of China.

Maienschein, Jane. 1988. "History of American Marine Laboratories: Why Do Research at the Seashore?" *American Zoologist* 28, no. 1:15–25.

———. 1989. *100 Years Exploring Life, 1888–1988: The Marine Biological Laboratory at Woods Hole*. Boston: Jones & Bartlett.

———. 1991. *Transforming Traditions in American Biology, 1880–1915*. Baltimore: Johns Hopkins University Press.

———. 1994. "It's a Long Way from 'Amphioxus' Anton Dohrn and Late Nineteenth Century Debates about Vertebrate Origins." *History and Philosophy of the Life Sciences* 16, no. 3:465–78.

Muka, Samantha. 2016. "Marine Biology." In *A Companion to the History of American Science*, ed. Georgina M. Montgomery and Mark A. Largent, 134–46. Malden, MA: Wiley.

Ogilvie, Marilyn Bailey, and Clifford J. Choquette. 1999. *A Dame Full of Vim and Vigor: A Biography of Alice Middleton Boring, Biologist in China*. Amsterdam: Harwood.

Rogaski, Ruth. 2004. *Hygienic Modernity: Meanings of Health and Disease in Treaty-Port China*. Berkeley and Los Angeles: University of California Press.

Sarton, George. 1924. "Fifteenth Critical Bibliography of the History and Philosophy of Science and of the History of Civilization." *Isis* 6, no. 2:135–251.

Schmalzer, Sigrid. 2008. *The People's Peking Man: Popular Science and Human Identity in Twentieth-Century China*. Chicago: University of Chicago Press.

Schwartz, Benjamin. (1964) 1983. *In Search of Wealth and Power: Yen Fu and the West*. Cambridge, MA: Harvard University Press.

Shen, Grace Yen. 2014. *Unearthing the Nation: Modern Geology and Nationalism in Republican China*. Chicago: University of Chicago Press.

Tchang Si. 1946. "Progress of Investigations of the Marine Animals in China." *American Naturalist* 80, no. 795:593–609.

Tchang Si and Koo Kwang-chung. 1936. "Description of a New Variety of *Branchiostoma belcheri* Gray from Kiaochow Bay, Shantung, China." *Contributions from the Institute of Zoology of the National Academy of Peiping* 3, no. 4:76–114.

Wang, Hui. 1995. "The Fate of 'Mr. Science' in China: The Concept of Science and Its Application in Modern Chinese Thought." Translated by Howard Y. F. Choy. *Positions* 3, no. 1:1–68.

Wang, Zuoyue. 2002. "Saving China through Science: The Science Society of China, Scientific Nationalism, and Civil Society in Republican China." *Osiris*, 2nd ser., 17 (Science and Civil Society): 291–322.

FOUR

The Misaki Marine Biological Station's Dual Roles for Zoology and Fisheries, 1880s–1930s

KJELL DAVID ERICSON

Since 1886, the seaside town of Misaki, located on the Miura Peninsula due south of Tokyo, has hosted facilities that founding director Mitsukuri Kakichi introduced in English as a "marine biological station" (Mitsukuri 1887; see fig. 4.1).[1] The Misaki Marine Biological Station (or Laboratory, as it also came to be known) opened under the auspices of Japan's first zoology program and first imperial university in Tokyo.[2] Ever since, historians have put Misaki at the center of zoology as a Japanese academic discipline. Isono Naohide's history of Misaki, easily the most comprehensive study published to date, narrates the station's activities in terms of "the birth of zoology in Japan" (Isono 1988). Yet marine biological stations have not existed solely inside the academic disciplines of zoology or biology. At Misaki, marine biological work was part and parcel of fisheries work.

Fisheries problems shaped what Raf de Bont has called the Euro-American "station movement" of the late nineteenth century (de Bont 2014, 11, 25). In the United States, the early history of seaside research features the US commissioner of fish and fisheries Spencer Fullerton Baird's

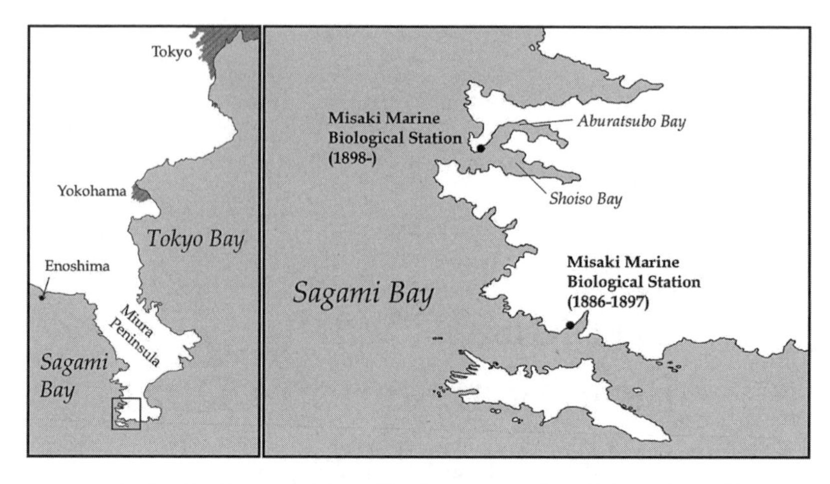

FIGURE 4.1. The Misaki Marine Biological Station's two locations (*right*) and the wider region (*left*). The maps, along with the shaded Yokohama and southwestern Tokyo metropolitan areas, have been redrawn from Yatsu (1926). Special thanks to Kristine Ericson for map design.

efforts to bring scientists to Woods Hole, Massachusetts, from 1871 on (Allard 1967; Maienschein 1989, 19–21). In different parts of the world, biological stations opened with plans, and sometimes with state and industry funding, to study marine organisms amid widespread interest in applied or economic biology. Such was the case in Japan, where Mitsukuri's notion of a fisheries-focused applied zoology was central to the Misaki Marine Biological Station's first decades. As the Columbia University zoologist Bashford Dean later eulogized Mitsukuri Kakichi, "in many regards he reminded one of Spencer F. Baird" (Dean 1909, 353).

The meanings of fisheries work were transformed at the Misaki Marine Biological Station in the early twentieth century. It was during this period when fisheries science (*suisangaku*)—encompassing elements of applied zoology and botany, oceanography, and industrial management—began to share space at Misaki with the zoology program (Hibiya 1981). Fisheries science developed within Japan's state fisheries bureaucracy, guided to a remarkable extent by officials who had studied zoology under Mitsukuri. The new field served as a pedagogical model overseas, including at America's first fisheries program at the University of Washington.[3] Just as importantly, fisheries science entered Japan's imperial research universities and the coasts where their students and faculty studied.

Misaki's connections to fisheries science have not received sustained attention in any language.[4] It is becoming apparent, however, that the divergent historiographic trajectories of the natural sciences from agricultural, forestry, and fisheries sciences are artifacts of tendencies to consider pure biology apart from other kinds of applied research (Phillips and Kingsland 2015). Reconsidering what it meant for Misaki to be a marine biological space outside the disciplinary frameworks of zoology or biology illuminates how multiple styles of research and teaching interacted at the water's edge.

In four sections, this essay traces the convergence and divergence of zoological and fisheries work during Misaki Marine Biological Station's first half century, a period spanning from the 1880s to the 1930s. The first section introduces Mitsukuri Kakichi's efforts to institutionalize zoology as a state-supported scientific discipline connected to applied fisheries concerns. The idea of a seaside laboratory modeled on European and American stations appeared amid ongoing debate about how to organize scientific knowledge in Japan's post-1868 Meiji state. The second section follows several of Mitsukuri's zoology protégés who became involved in the formulation of fisheries science as a new discipline. Fisheries science emerged as a pedagogical system for training students in scientific fisheries practices of catching, cultivating, and processing fish. It was, moreover, inextricable from the work of surveying waters across the expanding Japanese empire. The third section considers how Misaki changed when the discipline of fisheries science joined the zoology program from the early 1910s on. For a quarter of a century (and in fact beyond), the College of Science's zoology program and the College of Agriculture's fisheries science program operated in parallel and overlapping ways. The fourth section retells the 1920s and 1930s rise of an indoor, physiologically focused experimental zoology at Tokyo Imperial University and its consequences for both zoology and fisheries science. Zoologists who found themselves outside the experimental turn made new claims about the applications of systematic and morphological research to problems on land and in the sea. Some found a home in fields like fisheries science. But, with the opening of a dedicated fisheries science research station outside Misaki in 1936, zoology and fisheries science moved further apart.

By the end of World War II, the Misaki Marine Biological Station's history had come to look more like the early twentieth-century rise of a globalized (but in some ways distinctly American) kind of experimental zoological laboratory—a seaside place where scientific discovery took place indoors. Such an interpretation, however, overshadows Misaki's deep connections to fisheries problems as a constituent part of applied zoology and to the multistranded field of fisheries science itself. Focusing on the place of fisheries issues in Misaki's first fifty years sheds light on the unstable boundaries of zoology—and marine biology—in early twentieth-century Japan.

Applied Zoology and the Founding of Misaki Station

In Japan, an institutional basis for zoology—*dōbutsugaku* or "the study of animals"—took shape after the 1868 establishment of a new Japanese political regime under the Meiji emperor that replaced more than two and a half centuries of rule by the Tokugawa shogunate over 250–odd domains. Zoology found a home in the Meiji capital of Tokyo, the new name for the Tokugawa capital of Edo.

Why should state funding be devoted to the study of animals? This question was not unique to Meiji Japan, but it remained a pressing concern for Mitsukuri Kakichi from the 1880s until his death in 1909. After nearly eight years of study at institutions including Yale (as an undergraduate) and Johns Hopkins (for his PhD), Mitsukuri returned to Japan in 1881 to take a position as the third professor of zoology at Tokyo University. He followed two Americans, Edward Sylvester Morse and Charles Otis Whitman, who had each come and gone after serving two-year terms (Isono 1987).

Mitsukuri was well aware that knowledge about nature (including the terminology employed when discussing natural phenomena) had been transformed in nineteenth-century Japan. One lexical concoction was *dōbutsugaku*, a gloss of *zoology* coined in the early 1870s by a naturalist and Meiji official named Tanaka Yoshio. Tanaka followed on earlier work on Dutch botanical treatises by the scholar Udagawa Yōan (1798–1846), who in the 1822 publication *The Botany Sutra* (*Botanika kyō*) had resurrected a term from Chinese antiquity to describe a category that referred to animated things or animals: *dōbutsu* (Tanaka 1874; Miller 2013, 25–30; Marcon 2015, 257–59; Carr 1979, 47–49). Mitsukuri praised Yōan—a scholarly contemporary of his grandfather Mitsukuri Genpo[5]—as the leader of a "new school of natural history" who had, in rejecting the study of materia medica or *honzōgaku*, broken ground in the conceptualization of the natural world (Mitsukuri 1907, 987–88). Looking back from the early twentieth century, he acknowledged Tokugawa-era practices of natural history as a basis on which new scientific disciplines had developed in Tokyo during the 1870s and 1880s. He noted that botany (*shokubutsugaku*) had taken over institutions associated with the *honzōgaku* tradition, most notably the venerable Koishikawa Botanical Garden (Mitsukuri 1907, 1000–1001).[6] Even so, he saw Japanese natural history—both *honzōgaku* and the comparatively recent mid-nineteenth-century formulation of *hakubutsugaku* or "the study of myriad things" (Miller 2013, 31)—as fields based on approaches that were ill suited to modern Japan. In place of pre-1868 varieties of natural history came disciplines led by people like the Cornell-trained botanist Yatabe Ryōkichi (1851–99) and the Hopkins-trained Mitsukuri himself. The focal point of the new sciences was Tokyo's Imperial University.[7]

Justifying the existence of Japan's first and, for the next several decades, only zoology program led Mitsukuri to stress the economic importance of the embryological and morphological research in which he had specialized overseas.[8] He promoted zoology as a discipline of immediate use to the self-consciously developing post-1868 Meiji state, presenting it in terms of the pure (*junsei*) and applied (*ōyō*) categories that had gained prominence in late nineteenth-century Japan. Yet he downplayed the conceptual separation of pure and applied science, arguing that "today's theory should be tomorrow's practice" (Mitsukuri 1895, 376–77).[9] He identified the space for realizing such a synthesis along the shore.

Appeals to fisheries management undergirded Mitsukuri's arguments for the construction of a zoological laboratory in Japan. During the early 1880s, Mitsukuri apprised Japanese fisheries boosters of American plans for a permanent laboratory at Woods Hole then being explored by Spencer Baird's US Fish Commission: "A coastal experimental station is the single most vital thing for fisheries science and zoology. . . . From what I have heard, the Fish Commission of the United States central government has decided to build a large research station that will join forces with several universities. If we too were to have such a commendable thing as this, it would be of no small benefit for fisheries and academe" (Mitsukuri 1883, 30). He enlisted the young lecturer Ishikawa Chiyomatsu, one of the first group of Tokyo University zoology students who had studied under Edward S. Morse, to join him on visits to possible locations for a marine station. The pair made several trips to the Miura Peninsula, an area farther south and more time consuming to reach than the Enoshima coastline that Morse had previously used as a temporary summer laboratory (Isono 1988, 32). Alluding to Morse's earlier activities, Mitsukuri and Ishikawa laid out the case for a station in 1884: "If there were no permanent laboratory in our country, we would have to look anew for a building to house the laboratory and do other preparations as soon as we arrived at the beach. We would have no choice but to fritter away our limited time on such things" (Mitsukuri and Ishikawa 1884, 253).

Mitsukuri was one of many late-nineteenth-century biologists who sought to model seaside laboratories on the world's oldest and best-known zoological station, Anton Dohrn's Stazione Zoologica in Naples (de Bont 2014; Groeben, in this volume). Letters sent to Dohrn between 1884 and 1886 indicate that Mitsukuri regularly sought his advice about the layout and workings of the imagined Misaki Station (Isono 1988, 38; Isono 1999). Mitsukuri based his vision of Misaki not only on Naples, which he had visited while studying at Cambridge, but also on the institutions of American seaside research with which he was even more familiar. If anything, his Johns Hopkins adviser William Keith Brooks's Chesapeake Zoological Laboratory provided a more consistent reference point for his public appeals in Japan. His rhetoric owed much to Brooks, who had devoted years of research in the Chesapeake to the study of edible oyster reproduction and cultivation (Keiner 2009). For both, marine creatures existed as organisms of embryological study and as living foundations of regional and national economies. A Japanese marine station would be a place where zoological and fisheries work would not just coexist but also possibly be indistinguishable. "For example, the abalone fishery not only serves as food for the Japanese people, but they are also exported to foreign countries in very large amounts," Mitsukuri and Ishikawa wrote. "Therefore, their cultivation is of the utmost importance. In order to do this, we must clarify their developmental process and behavior. When we have understood their development, we will be able to intervene and propagate them. However, these things will be difficult to do if there is no research by the coast" (Mitsukuri and Ishikawa 1884, 254). The argument for research into abalone, then one of Japan's largest marine product exports to Qing China, reflected the influence of aquacultural practices on Mitsukuri's research and teaching.[10] Craft knowledge of aquatic creatures facilitated Mitsukuri's embryological research. Embryology, in turn, offered the possibility of controlling the reproduction and development of aquatic creatures for private and national gain.

As Isono Naohide has shown, support for the new station depended on Mitsukuri's access to personal and bureaucratic channels unavailable to other would-be institution builders at Tokyo's Imperial University. Most notably, at the time, Mitsukuri's elder brother Kikuchi Dairoku headed the university's College of Science, which included the zoology and botany programs.[11] In early 1885, the Imperial University took control of a small patch of Kanagawa Prefecture–owned land in Misaki.

The site was the former grounds of a seventeenth-century Tokugawa ship inspection office (*funa-bansho*) that had checked boats coming into and out of the early modern capital in Edo. Though the university was struggling financially at the time, it agreed to fund the new station with a special allocation in the 1886 budget of 1,500 yen, an amount that would cover the costs of a modest two-story wooden laboratory building.[12] Construction lasted from June to December 1886 and was overseen by two architects from the Ministry of Education (Isono 1988, 32–46). The Misaki Marine Biological Station (Misaki rinkai jikkensho) opened the following year as part of the Imperial University's College of Science. But the station did not stay in the same place for long. Citing concerns including the cramped facilities, Director Mitsukuri and Iijima Isao, a Morse student who had returned from study in Germany in 1886 to join Mitsukuri on the zoology faculty, oversaw the station's relocation a few miles north along the Miura Peninsula to the former site of the Miura clan's castle. From 1898 on, the station's home was on a hillside facing the narrow bay of Aburatsubo (Isono 1988, 74).

The fishing port of Misaki played a significant role in the operations of the marine biological station. Misaki grew as part of the sprawling hinterwaters that supplied marine products to markets in the early modern metropolis of Edo (Wilson, 2015–16). By the late nineteenth century, Misaki fishers (in particular boat fishermen) had also begun to find occasional employment from Western merchants such as the Yokohama-based Englishman Alan Owston, who plied waters near the Miura Peninsula in order to acquire specimens for sale to collectors and to institutions like the British Museum (Isono 1988, 76). With the establishment of the marine biological station, a few fishermen parlayed their expertise into positions as professional collectors. The best known among this latter group was Aoki Kumakichi, hired as Misaki's first full-time collector in 1898 (Isono 1988, 78). Aoki, or "Kuma-san" as many called him, gained a reputation as a collecting and systematizing savant, a once-unlettered fisherman who came to know the Latin names of thousands of marine organisms (Isono 1988, 60). He was both a collector and a go-between who communicated with local fishermen to acquire specimens.

As at stations like Naples and Woods Hole, where a marine biological laboratory opened in 1888 under the direction of onetime Tokyo University zoology professor Charles Otis Whitman, Misaki was a place where animals could be acquired with the collecting net and the dredge. But the specimens that featured in some of the laboratory's most famous taxonomic publications came from the widespread application of local tuna fishermen's longlines (*haenawa*), composed of hundreds of hook-and-line rigs dangling from a much longer trunk line, to the task of sampling deepwater creatures (Iijima 1896). If one knew where to look, theretofore unnamed specimens, including a wide array of *Hexactinellid* glass sponges, could be found just outside the station in Sagami Bay. Director Yatsu Naohide would later boast: "The Misaki Station is far superior to the two world famous biological centres, Naples and Woods Hole, in respect to the richness of marine forms" (Yatsu 1926, 5). Following the move to Aburatsubo, a handful of researchers came from abroad each year, including Columbia University's Bashford Dean and Stanford University's William Starr Jordan (Dean 1904; Jordan 1914). Misaki's leaders enticed visitors to experience the region's aquatic fauna up close, something that was possible through the efforts of collectors associated with Misaki's fishing community.

People came to the Misaki Station at different times of the year, but it was busiest during July and August, when Tokyo Imperial University students and faculty made the summer trek (via variable combinations of rail, ship, bicycle, and foot) from the capital to the Miura Peninsula.[13] They were sometimes joined by middle and higher school teachers who applied to take special courses on the kinds of embryological, morphological, and systematic problems that formed the basis of university lectures and practicums under the laboratory's first two directors, Mitsukuri (whose formal directorship extended from 1898 to 1904) and Iijima Isao (1904–21) (Isono 1988, 31). Every zoology student at Tokyo Imperial University passed through Misaki. Their number was not large in absolute terms: only eighty-seven between 1881 and 1918 (Ueno 1976, 47). But some of these early graduates, steeped in pure and applied matters, went on to become leaders in the Japanese Empire's fisheries bureaucracy. They would take central positions in the construction of fisheries science as a field of its own.

Surveying the Science of Fisheries

In late nineteenth-century Japan, the word *fishery* could be translated as one of two distinct terms, *gyogyō* or *suisan*. Following the 1868 installation of the Meiji emperor and subsequent dismantling of the hundreds of ancien régime domain governments that had once regulated rivers and coasts, Meiji officials began to see *gyogyō*, "livelihoods of catching" water-dwelling plants and animals, as activities that had spiraled out of control. For the small ranks of fisheries bureaucrats who occupied government posts from the late 1870s, the catching of marine creatures, the people who caught them, and the fishing villages in which catchers lived became activities, roles, and places in need of regulation through mechanisms like fishing rights (*gyogyōken*) and fishing associations (*gyogyō kumiai*) (Takahashi 2007).

To be a fisheries booster, however, was to be a *suisan* booster. Mitsukuri Kakichi was a member of the Dai Nihon Suisankai, a group that came to be known after its 1882 establishment as the Fisheries Society of Japan. Mitsukuri and other Fisheries Society members spoke on behalf of *suisan* or "aquatic production." The term implied the application of scientific techniques in order to harness the productive powers of watery, and especially marine, spaces. In a country bordered "on all four sides by the ocean," as the Fisheries Society's supporters repeated over and over, *suisan* encapsulated a late-nineteenth-century vision of an industrial marine economy propelled by knowledge of the sea. "The word *suisan* existed before, but it is not long since it has come into widespread use," Mitsukuri wrote. "It began to appear on people's lips around ten or fifteen years ago. Ever since the Fisheries Society (*Suisankai*) and the Ministry of Agriculture and Commerce's Fisheries Bureau (*Suisankyoku*) started up, the word *suisan* has naturally come to be spoken" (Mitsukuri 1892, 232). As such, the fishery as *suisan* evoked a sense of novelty and imperial legitimacy in Meiji Japan.

State officials across the late nineteenth-century world sought knowledge of where aquatic creatures lived, what they ate, when they spawned, and how quickly they grew. In Meiji Japan, the methods, tools, and institutions for organizing this knowledge fell under the fisheries survey, or *suisan chōsa*. During the mid-1890s, amid a funding crunch at the Ministry of Agriculture and Commerce, a new Fisheries Survey Office (*suisan chōsajo*) took the place of the Fisheries Bureau (Nōshōmushō suisankyoku 1910, 15–16). Mitsukuri, at one point tapped to lead the office, served as a member of its advisory committee. Using an extended martial analogy, Mitsukuri wrote: "The fisheries survey is like the military planning office of the fishing industry [*gyogyō*]. If traders and industrialists are foot soldiers, then fisheries surveys serve as their rations" (Mitsukuri 1892, 239).

The survey included Kishinoue Kamakichi, an 1889 graduate of the Imperial University zoology program who began to refer to his work not in terms of zoology but rather in terms of fisheries. "I know something of zoological investigation but I have had absolutely no introduction to fisheries as a whole," Kishinoue wrote in a report on a survey of Tokyo Bay, his first job for the Ministry of Agriculture and Commerce (Kishinoue 1892, 139). But, by the early twentieth century, he had taken up a position as head of the Fisheries Survey Department (*suisan chōsaka*) in the reconstituted Fisheries Bureau. He was the first Japanese representative to attend sessions of a foundational northern European group in the development of oceanography, the International Council for the Exploration of the Sea (ICES) (Uda 1966, 2; Mills 1989; Rozwadowski 2002). He returned from a 1901 ICES planning meeting in Kristiania (present-day Oslo) calling for Japanese-led studies in the new field of oceanography (*oseanogurafi, umi no shirabe,* or, in the neologism taken up across Japan and much of East Asia, *kaiyōgaku*) (Kishinoue 1903). Robert-Jan Wille (2016) has underscored the difficulty of separating "national" fisheries science from "transnational" ocean biology or biological oceanography. Indeed, oceanographic methods could be reappropriated in a variety of ways. For Kishinoue, national and imperial interests provided a ready source of appeals for expansive marine study. One million people—"one in fifty residents of the archipelago"—made their livings from fishing. If one counted those who fished in part for their livelihoods, the ratio grew to one in twenty (Kishinoue 1905, 1). Kishinoue argued that Japan could contribute to an international project of mapping seas "that belong to no country" (Kishinoue 1903, 163) and direct fishermen away from the archipelago's crowded coastlines by engaging in oceanographic fisheries surveys.

The Fisheries Survey Department became an early twentieth-century conduit for adapting North Atlantic styles of biological oceanography to Pacific finfish and their marine environments. In Japan, surveyors' tool kits included the Kitahara net, devised by department member Kitahara Tasaku (an 1894 zoology graduate) as a variant of the plankton nets used by University of Kiel professor Victor Hensen. Alongside Kitahara nets were Akanuma hydrometers, developed by Akanuma Tokurō from a Kiel hydrometer design (Uda 1954, 139). Surveyors like Kitahara recorded some of the first quantitative data on plankton levels in Japan, data that were soon applied to investigations of Japan's most important coastal finfisheries. Kishinoue used Kitahara's calculation of 1 gram of plankton per 1,000 liters of Seto Inland Sea water to estimate that a sardine would have to swim nine miles in order to consume half a gram of food per day (Kishinoue 1908, 81; Herdman 1923, 316).

The tools, methods, and practitioners of fisheries survey work became part of fisheries science pedagogy. Fisheries science (*suisangaku*) was a new educational field, having appeared first at the Suisan denshūjo, a fisheries school founded by the Imperial Fisheries Association in 1888. After 1898, the Bureau of Fisheries took over the Suisan denshūjo, subsequently named the Suisan kōshūjo or Japanese Imperial Fisheries Institute. The Imperial Fisheries Institute was a model of practical fisheries instruction, notably for the University of Washington's College of Fisheries, the first American fisheries program when it opened in 1919 (Stickney 1989, 3). The institute's division of knowledge followed the organization of the Fisheries Bureau and its survey unit, which classified fisheries work into the triumvirate of catching, cultivating, and processing aquatic organisms. Its faculty, virtually all of whom had experience in fisheries surveys, engaged in the boundary-making task of determining what constituted fisheries science as an academic field.

Tokyo Imperial University zoology alumnus Fujita Tsunenobu believed that catching and cultivating were the most important elements of fisheries science. Fujita was among the faculty who looked at fisheries science through the lens of aquaculture. Perhaps as a result, he hesitated to demarcate fisheries science from agriculture (Fujita 1918, 8). Elsewhere, fisheries science instructors turned their focus to ocean space. Kitahara Tasaku wrote a biologically inclined oceanography textbook with Okamura Kintarō, a Tokyo Imperial University botany graduate and former fisheries survey member who joined Kitahara on the Imperial Fisheries Institute faculty as a plankton expert. The duo suggested that regular, careful measurement of ocean conditions might reveal marine processes not visible from the surface. Their formulation of hydrobiology (*suiri seibutsugaku*) was one of the subfields taking shape in and around fisheries science (Kitahara and Okamura 1910, 1).[14]

The first generation of Japanese fisheries science faculty exuded the managerial visions of men who saw themselves as a hybrid of imperial official and higher school educator. What resulted was a field that encompassed saltwater and freshwater environments, the flora and fauna that lived in aquatic spaces, and human communities that sought ways to know them better.

Fisheries Science Comes to Misaki

Fisheries science arrived at Japan's imperial research universities amid the Meiji state's post-1897 policy of subsidizing long-distance fishing vessel owners and the extension of Japan's colonial empire across the Pacific (Ninohei 1981). The first decade of the twentieth century witnessed the formation of a fisheries science department in Sapporo, affiliated in short order with first the Sapporo Agricultural College (1907), next Tōhoku Imperial University (1907–18), and then Hokkaido Imperial University (1918–). The Fisheries Bureau's and Imperial Fisheries Institute's Fujita Tsunenobu headed the new program, which arose amid long-standing prefectural support for American-style fish propagation experiments and surveys of the surrounding northern seas (*Hokudai suisan gakubu nanajūgo nenshi* 1982). At Tokyo Imperial University, College of Agriculture chair Kozai Yoshinao claimed that his field was a better place for fisheries science than the College of Science.[15] As one agriculture school student (and future professor of fisheries chemistry at Kyushu University) later recalled Kozai's argument: "Under agriculture's purview are the sciences that make biological use of the entire surface of the earth" (Okuda 1960, 120). Kozai spoke about the global scope of fisheries science, but he rolled out the new program over the course of several years, citing difficulties in finding (and paying) professors (Shisuikai 1960, 151). Starting in 1907, the Fishery Survey Department's Kishinoue Kamakichi lectured on catch fisheries, while Ishikawa Chiyomatsu focused on aquaculture. Others included Hara Jūta, an 1895 Imperial University zoology graduate who joined the fisheries science faculty in 1911. Hara, who had taught at the Sapporo Agricultural School and studied in Europe, offered some of Japan's first university courses in oceanography (Tezuka 1992, 234).

When it gained the status of a full academic degree program in 1910, Tokyo Imperial University's fisheries science course was institutionally affiliated not only with the College of Agriculture but also with the College of Science's Misaki Station. Misaki's physical appearance changed with the arrival of fisheries science. The Aburatsubo site expanded to accommodate the new program; a laboratory building, an aquarium room, and motorized water pumps were part of the construction that was completed in 1910 (Nishimura 2008, 39–52). Thereafter, two institutionally differentiated groups of professors and students began to arrive at Misaki. In short order, Kishinoue organized a sumo wrestling tournament between his fisheries science students and their College of Science counterparts. Fisheries science lost ("Misaki dayori" 1911, 288).

Zoology and fisheries science diverged and overlapped by the shore. Students in the zoology program dissected sea urchins, sea cucumbers, and squid. Nearly every day they identified plankton species under the microscope. Meanwhile, Hara Jūta's fisheries science students received training in oceanographic survey techniques. They learned to gather samples with quantitative plankton nets, used centrifuges to gather suspended matter, and took barometric readings ("Misaki dayori" 1912,

117). Even so, the two programs did not always follow separate tracks at Misaki. In the spring of 1912, around ten fisheries science students joined eight zoologists to collect specimens during one exceptionally low tide, after which "research was heatedly performed on the gathered creatures" (Kawamura 1912, 318). Yet, by the mid-1910s, the fisheries science program was using Misaki's collecting ship *Dōsun Maru* not only for specimen gathering but also for shipboard observations during "oceanography and marine zoology training" ("Fuyu no Misaki" 1918, 97). Fisheries science students were, moreover, as likely to visit fish markets and sushi restaurants as they were to examine organisms at the bench. As one fisheries graduate put it: "The animal dissections (*dōbutsu jikken*) of fisheries science required us to investigate how to eat [our specimens] and their flavor. This was a different approach from the College of Science's zoology" (Irie 1960, 208).

Along with fisheries science students came professors whose research ranged beyond Misaki. Take, for example, Ishikawa Chiyomatsu and Kishinoue Kamakichi. Ishikawa's long career is difficult to pigeonhole. He was involved in the publication of the first book on Darwinian evolutionary theory in Japanese (*Dōbutsu shinkaron*), an 1883 text adapted from Edward S. Morse's zoology lectures; he later studied at the Stazione Zoologica in Naples; he pursued research into mechanisms of heredity; he weighed in frequently on matters relating to the social formations of humans and other living creatures; and he directed the Tokyo Imperial Zoo until the untimely deaths of a pair of giraffes in 1907 (Morse 1883; Miller 2013, 49–50; Onaga 2015, 418; Godard 2017). Ishikawa left the zoo amid controversy but quickly found a place in fisheries science, remaining until retirement a faculty member of Tokyo Imperial University's College of Agriculture.

A variety of sweetfish known as "little *ayu*" or *koayu* exemplifies the convergence of Ishikawa's interests in evolutionary morphology with the fisheries science program. *Koayu* were smaller than migrating anadromous *ayu* and, unlike *ayu*, did not spend any part of their lives in the ocean. Earlier observers had proposed that *koayu* and *ayu* were different kinds of sweetfish, but Ishikawa countered that the two were one and the same species: *koayu* populations had at some point in the past become landlocked in bodies of freshwater, including Japan's largest lake, Biwa, which prevented them from reaching the sea. Ishikawa showed that lake-gathered *koayu* could grow into *ayu* through a series of trials that included attempts, based on existing village practices, of raising them in ponds until they grew into "big *ayu*" or *ōayu*. In 1913 and 1915, he made further efforts to reintroduce *koayu* to the dammed

Tama River near Tokyo (Ishikawa 1926, 436–37). He continued to investigate sweetfish—and reshape transpacific riparian environments—well after establishing a link between *ayu* and *koayu* in the 1910s (Ishikawa 1930). By the following decade, he was involved in attempts to bring sweetfish not only to rivers in other parts of the archipelago but also to California and the territory of Hawaii. All the while, sweetfish research took Ishikawa and his students to regional fisheries experiment stations and fish raisers across Japan (Ishikawa 1921, 1930).

While Ishikawa exhibited an affinity for investigating freshwater environments, Kishinoue Kamakichi followed finfish throughout Japan's "pelagic empire" (Tsutsui 2013). The importance of deepwater fisheries to Japan became clearer in the aftermath of the country's expansion to the north into Sakhalin Island after the 1904–5 Russo-Japanese War and to the south during the First World War, when the Japanese navy occupied German colonies in Micronesia. The Pacific—from Arctic to Antarctic and from the Asian continent to coastal North America—became a focus of the university fisheries science community (Muscolino 2013; Tsutsui 2013). A major strand of Kishinoue's research at Tokyo Imperial University examined far-flung Pacific tuna and bonito. More than a decade of work resulted in "the single most comprehensive treatment of scombroid biology ever written" (Graham and Dickson 2001, 136). Part of Kishinoue's research tried to explain a phenomenon that had long puzzled observers: tuna and bonito often maintained a body temperature higher than the surrounding water. His argument turned on anatomical observation: tuna and bonito exhibited a prominent network of subcutaneous blood vessels distinct from that of mackerel, which was grouped together with the other two in the Scombroidae family. He proposed that this vascular anatomy was unique and that tuna and bonito should therefore be placed in their own order, the Plecostei. Others in Japan disagreed, but his publications became prominent examples of efforts to integrate systematic and morphological methods under the rubric of fisheries science (Kishinoue 1921; Kishinouye 1923; Graham and Dickson 2001, 137).

Kishinoue approached fisheries science as the study not only of marine organisms but also of the human societies they sustained. In 1907, a village fisheries association at the eastern edge of Tokyo Bay asked him to survey its members' archives. Historical study revealed a pattern of fishermen moving farther and farther out to sea from the late nineteenth century on (Kishinoue 1914). Kishinoue interpreted fisheries expansion not as an imperial inevitability but rather as a sign of strained class structures in existing fishing communities. "Why has 'mid-level' [*chūkibo*] fishing (this is a neologism, although you surely understand its meaning) declined while very small fishing has persisted?" he asked (Kishinoue 1915, 43–44). His answer was that the distribution of the catch had itself become polarized. Just as some "small fishermen" tried in vain to go offshore using near-shore boats, others began to sell their labor to some of the world's largest fisheries companies. Research into tuna and bonito went alongside investigations of the effects that long-distance fisheries were having on communities like Misaki. As in other parts of the Japanese Empire, Misaki was home to fishermen who were going deeper into the sea to find tuna of their own, whether they had planned to or not (Aono 1933).

Fisheries science held the promise of showing fishers where to continue the hunt or how to raise aquatic creatures intensively. Describing marine spaces in terms of "societies in the sea" (*kaichū no shakai*), as Kishinoue did in his *Fish and Human Life* (*Uo to jinsei*), was both an imaginative exercise and a suggestion that knowledge could extend—along with control—from the shore to the pelagic depths (Kishinoue 1915, 63). By the early 1920s, Kishinoue was proposing a vision of fisheries science headlined by a central fisheries research station and equipped with dedicated survey vessels that could synthesize observations gathered across the Japanese Empire (Kishinoue 1920).[16] Over time, a new generation of faculty and students would reposition fisheries science outside the orbit of zoology and, in certain ways, outside Misaki.

Places of Experiment

From its founding until the present day, the Misaki Marine Biological Station's Japanese moniker has identified it as a "place of *jikken*" (*jikkensho*). But the meanings of *jikken*, now regularly translated as *experiment*, and *jikkensho*, a possible gloss of *laboratory*, have never been stable.[17] They certainly were not when Yatsu Naohide became the third director of Misaki in 1922, following the death of his predecessor, Iijima Isao. Yatsu reoriented zoology at Tokyo Imperial University away from morphology and systematics and toward a physiologically inspired experimental zoology (*jikken dōbutsugaku*).[18] For Yatsu, the path toward experimental zoology required a transformation of Misaki Marine Biological Station—the laboratory—into a place defined primarily by indoor laboratory work. The convergence of Misaki's *jikken* with the laboratory—rather than *jikken*'s earlier career as the practical experience of anatomical dissection of creatures gathered by the sea—reflected Yatsu's interests on his return to Japan in 1907, having completed a PhD (and subsequent investigations) at Columbia under the guidance of leaders in cytological, physiological, and genetic research: Edmund Beecher Wilson and Thomas Hunt Morgan (Ueno 1976, 52; Moriwaki 1979, 369).

The push for experimental zoology under Yatsu's directorship (1922–38) contributed to the construction of zoology as a field apart from applied matters. Yatsu sought to dispel conceptions of *jikken* as a form of practical experience and instead specify it as indoor, laboratory-based experiment. For Yatsu, this formulation of experiment was the end point of zoology's historical development, which he charted through a "pre-Darwinian period" of morphology, a "Darwinian period" of evolutionary systematics, a "period of cellular biology," and a "period of experimental zoology" (Yatsu 1920, 211). Elsewhere, he argued that an experimental age was coming to Japan: "For the past two or three years, biology has for the most part been morphological research. But recent research has slowly moved from morphological to physiological matters. Now we are coming to a point where morphologists and physiologists are difficult to distinguish" (Yatsu 1919, 8).

Support for physiologically focused experimental zoology went hand in hand with a move away from research about the shapes and particularly the names of animals. As noted by other historians, the irony of this is that Yatsu had specialized in morphological and taxonomic matters, both as a student at Misaki and for a time as a professor on his return from Columbia (Uchida 1957). His writings on systematics would become standard-issue reference volumes for Japanese biologists (Yatsu 1914; Yatsu and Uchida 1972). Even so, zoology students of the 1920s recalled him telling them to "toss away systematics," at least as a subject of their own research (Uchida 1957, 34).

Unlike previous Misaki directors—that is, Mitsukuri Kakichi and Iijima Isao—Yatsu saw applied zoology as the province of other fields, especially agriculture, fisheries, and medicine (Yatsu 1919, 3). He did not deny the importance of applied science per se, having, for example, argued on behalf of state funding for North Pacific oceanographic surveys (Yatsu 1915). He also pushed zoologists in training to gather their own specimens whenever possible. But he did not see collecting in itself as a goal of the zoology curriculum or research agenda. Students were instead encouraged to intervene in the internal workings of living organisms—whether sea urchins, starfish, or other creatures that could be found in abundance around Misaki (Yatsu 1920, 173–75).

The arrival of Yatsu's experimental zoology altered Misaki's daily rhythms. Electric lighting came in 1923 to the station's laboratories, making evening work possible (Isono 1988, 48). With Yatsu's promotion of laboratory-based research and teaching also came new demands for year-round access to living specimens, in particular sea urchin eggs. Pressure to provide specimens spurred new kinds of work in the water. The station's second chief collector, Deguchi Shigerō, developed a way to delay the spawning of red sea urchins by packing them inside submerged baskets or cages (Isono 1988, 129–30, 140–41). Although supported by littoral research like Deguchi's, experimental zoology moved further indoors.

Zoologists in interwar Japan, as elsewhere (e.g., the United Kingdom), remarked that their discipline was changing in fundamental ways (Kraft 2004; Erlingsson 2009). Some began to debate the merits of old and new methods of zoological research. Ōshima Hiroshi, a Kyushu Imperial University professor (and former Tokyo Imperial University zoology graduate), was skeptical of what he saw as the US-led "new zoology," lamenting "the thought of American students chasing the novelty of 'experiment, experiment'" (Ōshima 1921, 476). After all, budding zoologists were forgetting that renowned experimentalists like Edmund Beecher Wilson, Edwin Grant Conklin, and Thomas Hunt Morgan had started their careers as systematically inclined morphologists.[19] Their Japanese counterparts were also attempting to follow an experimental direction (*jikken hōmen*), at least to the extent that they could obtain the latest overseas publications. Being left behind were old-fashioned practices of systematics, anatomy, and embryology. Ōshima argued that experimental approaches should be attempted, if at all, only after gaining several years of experience in the old zoology (Ōshima 1921, 478).[20]

The rise of experimental zoology brought some old-school zoologists closer to fisheries science. Tanaka Shigeho, for one, turned toward what he dubbed *applied ichthyology* (*ōyō gyogaku*). Tanaka, perhaps best known for his early twentieth-century efforts to catalog the fishes of Japan with the Stanford ichthyologists David Starr Jordan and John Otterbein Snyder, emphasized the value of taxonomic research (Jordan, Tanaka, and Snyder 1913). In time, this orientation moved him into the fisheries science camp. His sketch of applied ichthyology resembled formulations of fisheries science—including the familiar triumvirate of catching, cultivating, and processing—although he kept his focus more on wild finfish and less on aquaculture or oceanographic plankton studies (Tanaka 1926, 250). Taxonomy was for him the fundamental problem. Were there one thousand species of fish along the Japanese archipelago's coasts or ten thousand? Not far behind was morphology. Tanaka's students started by examining specimens on the bench rather than watching fish in a tank or in the sea, but his goal was to get them closer to living creatures in the water. He pushed young zoologists to make use of fishermen's knowledge, including the names by which they referred to fish in local dialects (Tanaka 1934, 254–59). As he put it, this was a zoology based on experience (*keiken*) rather than experiment (*jikken*) (Tanaka 1942, 219–21). Yet, although he would briefly lead Misaki after Yatsu (1938–39), his vision of zoological practice did not alter zoological pedagogy so much as it reinforced trends in the fisheries science milieu with which he was increasingly familiar.

Interwar changes at Tokyo Imperial University's zoology program took place amid the appearance of new marine research facilities and aquariums at other Japanese imperial universities. Kyoto Imperial University opened a marine station by the Seto Inland Sea (and reorganized its Lake Biwa freshwater experimental station under the College of Science) in 1922, Tōhoku Imperial University unveiled a station in Aomori's Asamushi in 1924, and Kyushu Imperial University started a station in Kumamoto's Amakusa in 1928. The new stations engaged in experimental zoology, but they also focused on fisheries and taxonomic research. Kyoto and Kyushu had strong emphases on ecology (*seitaigaku*), spearheaded by the former Tokyo Imperial University zoology graduates Kawamura Tamiji and Ōshima Hiroshi (Endō 2007). A round of 1930s construction transformed Misaki as well, with a partly Asamushi-inspired aquarium in 1932 and, next door, a laboratory building in 1936. Whereas the earlier 1908–10 expansion had received justification from the arrival of the fisheries science program, the new buildings were monuments to indoor experimental research and the public spectacle of the expanded aquarium (Nishimura 2008, 73–152).

Fisheries science also looked different in the 1930s. While the first generation of fisheries science professors like Kishinoue, Ishikawa, and Hara came overwhelmingly from zoology and botany backgrounds, their replacements were more likely to be graduates of fisheries science programs. This was the case for Kishinoue's successor Amemiya Ikusaku, one of the first generation of fisheries science faculty with degrees in the subject. Amemiya joined colleagues from Hokkaido Imperial University and the Imperial Fisheries Institute to found Japan's first fisheries science organization (the Nihon suisan gakkai) in 1932 (*Hokudai suisan gakubu nanajūgo nenshi* 1982, 91). By this time, the Japanese state had also begun to sponsor the large-scale surveys for which

Kishinoue had called prior to his death in 1929. Along with a ministry-level fisheries experiment station came multiship surveys throughout the 1930s, including a series of forty-odd-vessel synoptic surveys from the East China Sea to Japan's Karafuto colony on southern Sakhalin Island (Uda 1966, 1–2). It was in this context that in 1936 the Tokyo Imperial University fisheries science program unveiled its own place of experiment, a stand-alone fisheries science station over 250 kilometers to the west of Misaki. The station was initially located in Aichi Prefecture's Shin Maiko, on a spit of land donated by the Nagoya Tetsudō railway company in return for a promise that the fisheries science station would operate an aquarium of its own (Nakamura 1960, 102). Amemiya Ikusaku served as the station's first director; he remarked at the opening ceremony that construction had started only after "two or three" previous efforts to gain support for a fisheries science station (Chiba et al. 1986, 51). Once it was up and running, Amemiya oriented the station toward fisheries survey and aquaculture research, including his own interests in sardine evolutionary morphology and edible oyster cultivation (Nakamura 1960; Tamura 1985).

The creation of a separate Tokyo Imperial University fisheries science station contributed to subsequent historiographic divergences. Institutional histories chronicled either the zoology or the fisheries science programs, just as they spotlighted the zoology station at Misaki or the fisheries science station in Shin Maiko (which would move again before arriving in 1970 at its current location near Lake Hamana in Shizuoka Prefecture) (Chiba et al. 1986; Isono 1988). The two stations were rarely mentioned together in print. Hints of this could be seen at the end of World War II, when a two-page letter written before the September 1945 arrival of American-led occupying forces to the Miura Peninsula became the most widely publicized document ever produced at Misaki. Signed "The Last One to Go," the letter asked that Misaki be left open for research (fig. 4.2). "This is a marine biological station with her history of over 60 years," it began, which placed Misaki alongside a list of American institutions including Woods Hole. *Time* magazine soon published the letter's contents; the document itself ended up in Woods Hole Marine Biological Laboratory ("Appeal to the Goths" 1945; Inoué 2016, 4; see fig. 4.2). "The Last One to Go" was Dan Katsuma, a product of Yatsu Naohide's focus on experimental zoology who had received an undergraduate degree from Tokyo Imperial University in 1929. Dan went on to study biology with Lewis Victor Heilbrunn at the University of Pennsylvania, during which time he made several research visits to Woods Hole. The Misaki of the 1930s and 1940s was

This is a marine biological
station with her history of
over sixty years.
If you are from the Eastern Coast,
some of you might know Woods Hole or
Mt Desert or Tortugas.
If you are from the West Coast,
you may know Pacific Grove or
Puget Sound Biological Station.
This place is a place like one of these,
Take care of this place and protect
the possibility for the continuation
of our peaceful research.

You can destroy
the weapons and
the war instruments
But save the civil equipments
for Japanese students
When you are through
with your job here
notify to the University and
let us come back to our
scientific home

The last one to go

FIGURE 4.2. "The Last One to Go" hanging on the wall of the MBL Library. Courtesy of the Marine Biological Laboratory, Woods Hole.

where he had continued his cytological and embryological work, much of his time being spent at the microscope observing cell division in fertilized sea urchin eggs (Dan 1987). In an immediate sense, his letter connected Miskai with American marine research while dissociating the station from the Japanese imperial state and military, including the naval forces that had been stationed on the Miura Peninsula since early 1944. But, in another sense, it was part of a rewriting of Misaki as a marine biological station devoted to a laboratory-focused strand of experimental research.

Conclusion

It is hard to avoid the impression that indoor, experimental zoology has come to characterize marine biological laboratories like Misaki and Woods Hole, in contrast to the outdoor, applied practices of fisheries stations and their agricultural or forestry cousins.[21] At Misaki, interwar changes in Tokyo Imperial University's zoology program redefined the scope of zoological research in ways that seemed to make the observation of marine conditions and organisms secondary to laboratory study (Fukui 2006). Yet, throughout the early twentieth century, marine stations such as Woods Hole permitted "a wide range of individuals interested in marine organisms to congregate and interact in the same location" (Muka 2016, 138–39). In the Japanese empire, too, historiographically overlooked opportunities were emerging in agriculture-affiliated programs and at other institutions for work in ecology, genetics, and natural history (Endō 2007; Setoguchi 2009; Iida 2015; Lee 2015; Onaga 2015).[22] One place of opportunity was Misaki itself.

Misaki may have fostered the birth of zoology in Japan, but it also attended the early years of imperial fisheries surveying and university fisheries science. The station began with Mitsukuri Kakichi's support for turning experimental embryological work on marine organisms into the applied zoology of aquaculture. It later became a center of experimental zoology, yet it remained the first shoreline among many for fisheries scientists looking to sample, collect, and cultivate. Even after the zoology and fisheries science programs came to operate separate stations, both still made Misaki a home.[23] Misaki is a marine biological station, but it has always been a disciplinarily unsettled space. Treating it as such allows us to consider the prospect of life sciences taking shape beside the water and in the water.

Notes

1. Authors have used both 1886 (the end of construction at the initial location) and 1887 (the start of academic work) to mark the beginnings of the Misaki Marine Biological Station. Isono (1988) and the station's own publications date its opening to 1886.
2. The present-day University of Tokyo has gone through several institutional reorganizations and attendant name changes, including periods as Tokyo University (1877–86), Imperial University (1886–97), and, following the establishment of a second imperial university in Kyoto until the promulgation of post–World War II reforms, Tokyo Imperial University (1897–1947).
3. More comparative work needs to be done in order to situate Japanese fisheries science vis-à-vis European and American styles of fisheries biology. Institutionally, ideologically, and in practical terms, fisheries science did not cleanly map onto the kinds of population modeling that came to characterize early twentieth-century fisheries biology. To some extent, this was because the rise of ideas like maximum sustainable yield were themselves in part a historical response—with one of its academic centers at the University of Washington's fisheries program, no less—to the massive early twentieth-century expansion of Japanese fishing activities in the Pacific (Finley 2011). Suffice it to say for now that Japanese fisheries science can be considered a political science in the broad sense that Jennifer Hubbard has described fisheries biology but that the specific political calculus that shaped fisheries science differed from that of fisheries biology in Europe and the United States (Hubbard 2014).

4. What *has* received attention is the work of Nishikawa Tōkichi, a Tokyo Imperial University zoology graduate who took leave from a government fisheries post in 1905 to work on methods of pearl cultivation at Misaki. With university funding, Nishikawa and his zoology program collaborators developed a still-used commercial process for inducing the formation of pearls in living shellfish. In recent decades, Nishikawa's project has come to serve as the premier example of Mitsukuri's approach to applied zoology at Misaki. At the same time, however, historians in Japan have also tended to view the project as a last stand of fisheries-focused zoological research at Misaki as both Mitsukuri and Nishikawa passed away in the middle of the project in 1909 (Isono 1988; Kuru 1987).
5. Mitsukuri was a member of one of the most prominent scholarly families in nineteenth-century Japan. Three generations of his family taught at the Tokugawa shogunate's Institute for the Investigation of Barbarian Books (Genpo, his grandfather, and Shūhei, his father) and at the institute's second act as the core of the imperially supported Tokyo University and its subsequent reorganizations (Kikuchi Dairoku, his elder brother, and, later, Mitsukuri Genpachi, his younger brother) (Tamaki 1998).
6. The garden began as one of the early modern shogunate's authorized medicinal herb gardens (relocated in 1684 to the estate of Lord Koishikawa) and became attached to the university in 1877 (Marcon 2015, xi, 121).
7. As Mitsukuri later described his Tokyo Imperial University–centered view of post-1868 science: "At least up to the present, the [university's] chronology is itself the history of science in Japan" (Mitsukuri 1907, 989).
8. By the 1880s, the animal was a familiar concept in Japan; the study of animals was not (Miller 2013, 28). Then head of Tokyo Imperial University Hamao Arata would, on Mitsukuri's passing in 1909, note: "Among Japan's scholars of *hakubutsugaku*, there were many people who did *honzōgaku* and botany (*shokubutsugaku*), but no one who did zoology (*dōbutsugaku*). . . . I think it's fair to say that Mitsukuri was the first person in Japan to devote himself to zoology" (cited in Tamaki 1998, 35).
9. Mitsukuri saw zoology as a field that could be newly applied to problems of state building, but in so doing he downplayed a longer nineteenth-century history of the instrumentalization of knowledge in formulations such as *jitsugaku* (Marcon 2015).
10. Mitsukuri pursued some of the first investigations into reptile embryology with turtles from the Tokyo aquaculturist Hattori Kurajirō, "collecting, with [Hattori's] kind consent," as he noted, "ample materials for my studies on the development of Chelonia" (Mitsukuri 1904, 260; see also Mitsukuri 1891, 105).
11. Mitsukuri's letters to Dohrn mention that Mitsukuri's brother Kikuchi Dairoku, who had spent time at the Naples Station, also contacted Dohrn about the new project (Isono 1988, 36–37).

12. Mitsukuri noted the funding appropriation in a July 24, 1885, letter to Anton Dohrn (Isono 1988, 39).
13. Most traveled by sea until the 1910s (Isono 1988, 47).
14. Raf de Bont has argued that, in late nineteenth- and early twentieth-century Europe, the field of study known as hydrobiology was among the "proto-ecological practices" taking place at zoological stations (de Bont 2014, 200). Kitahara and Okamura were well aware of such work in choosing to self-translate the subject of their monograph as *hydrobiology*.
15. Kozai was a central figure in Japan's agricultural chemistry community, known for his attempts to bring craft practices of fermentation, including the mold cultures used in soy sauce and sake production, into the realm of university microbial research (Lee 2015).
16. Like Ishikawa, Kishinoue often visited the prefectural fisheries experiment stations (*suisan shikenjō*) that had proliferated since the first was built in Aichi Prefecture in 1894. But Kishinoue remained unsatisfied with prefectural fisheries research, which he criticized as parochial and repetitive. Better, he proposed, to put regional stations under the control of an imperial center (Kishinoue 1915, 202–9). An institution resembling Kishinoue's proposals, the Ministry of Agriculture and Forestry's fisheries experiment station, would later open in 1929.
17. To be sure, the laboratory has not been a historically stable institution in other contexts either (Kohler 2008). Eugenia Lean has explored the history of experiment (*shiyan*, written with the same characters as *jikken*) in early twentieth-century China, arguing that "investigating through practice" did not map directly onto Euro-American conceptions of experimental science (Lean 2014).
18. Yatsu was not the first person in Japan to use the term *jikken dōbutsugaku*, but he appears to have been the first to use it as a gloss for the field as practiced in the early twentieth-century United States. An earlier set of works on *jikken dōbutsugaku* by the zoologist Gotō Seitarō, e.g., focused on the experiment of macroscopic and microscopic anatomy (Gotō 1900).
19. All three, like Mitsukuri, were among the dozens of biologists (over forty of them American) who received training from William Keith Brooks in the late nineteenth-century Johns Hopkins Zoology Department (Benson 1985, 164). Tokyo Imperial University zoology faculty members Gotō Seitarō and Watase Shōzaburō also went to Baltimore to study with Brooks (Uchida 1957).
20. Elsewhere in Asia, too, biologists did not uniformly embrace experiment, as Jiang (2016) has shown for early twentieth-century China.
21. Robert-Jan Wille has argued persuasively that compartmentalized accounts of biological and agricultural research overshadowed their coproduction at places like the Dutch colonial Buitenzorg Botanical Garden in Java (Wille 2015).

22. It should be noted too that research in fields like genomics has also pushed fisheries science and other applied agricultural sciences toward new meanings of *experimental* bioscientific laboratory work since the late twentieth century. At Tokyo University, e.g., agriculture has been paired since 1994 with life sciences (*seimei kagaku*) in the Faculty of Agriculture's graduate school, where one can find the successor to fisheries science, the Department of Aquatic Bioscience (*suiken seibutsu kagaku*). Agriculture also houses applied fields like animal resource science (*ōyō dōbutsu kagaku*).
23. The station's continued importance to fisheries science can be seen in the zoologist Mōri Hideo's diaries of activities at Misaki in the 1950s. As Mōri later recounted: "Affiliates of the Faculty of Science were not the only ones to make use [of Misaki]; people connected to fisheries in the Faculty of Agriculture did as well. There was research involving the cultivation of sea urchins, shrimp, oysters, *nori*, and finfish" (Mōri 2011, ii).

References

Allard, Dean C. 1967. "Spencer Fullerton Baird and the U.S. Fish Commission." PhD diss., George Washington University.

Aono Hisao. 1933. "En'yō gyogyō no konkyokō to shite no Misaki." *Chiri zasshi* 45, no. 3:149–50.

"Appeal to the Goths." 1945. *Time* 46 no. 24 (December 10): 92.

Benson, Keith R. 1985. "American Morphology in the Late Nineteenth Century: The Biology Department at Johns Hopkins University." *Journal of the History of Biology* 18, no. 2 (Summer): 163–205.

Carr, Michael Edward. 1979. "A Linguistic Study of the Flora and Fauna Sections of the 'Erh-ya.'" PhD diss., University of Arizona.

Chiba Kenji et al. 1986. *Tōkyō daigaku nōgakubu suisan jikkensho no gojūnen.* Maisaka: Tōkyō daigaku nōgakubu fuzoku suisan jikkensho.

Dan Katsuma. 1987. *Uni to kataru: Gekidō no jidai shizen o tomo to shita aru seibutsu gakusha no shōgai.* Tokyo: Gakkai shuppan sentā.

Dean, Bashford. 1904. "A Visit to the Japanese Zoological Station at Misaki." *Popular Science Monthly* 65 (July): 195–204.

———. 1909. "Memoir of Kakichi Mitsukuri." *Annals of the New York Academy of Sciences* 19 (March): 352–53.

de Bont, Raf. 2014. *Stations in the Field: A History of Place-Based Animal Research, 1870–1930.* Chicago: University of Chicago Press.

Endō Hideki. 2007. "Dōbutsugaku to nōgaku no kankeishi." In *Nihon no dōbutsugaku no rekishi*, ed. Mōri Hideo and Yasugi Sadao, 198–220. Tokyo: Baifūkan.

Erlingsson, Steindór J. 2009. "The Plymouth Laboratory and the Institutionalization of Experimental Zoology in Britain in the 1920s." *Journal of the History of Biology* 42:151–83.

Finley, Carmel. 2011. *All the Fish in the Sea: Maximum Sustainable Yield and the Failure of Fisheries Management*. Chicago: University of Chicago Press.

Fujita Tsunenobu. 1918. *Suisangaku tsūron*. Tokyo: Shōkabō.

Fukui Yuriko. 2006. "Yatsu Naohide to rinkai jikkensho: Kansatsu kara jikken e." *Seibutsugakushi kenkyū*, no. 76 (June): 61–64.

"Fuyu no Misaki." 1918. *Dōbutsugaku zasshi* 30, no. 352 (February 22): 97.

Godard, G. Clinton. 2017. *Darwin, Dharma, and the Divine: Evolutionary Theory and Religion in Modern Japan*. Honolulu: University of Hawaii Press.

Gotō Seitarō. 1900. *Jikken dōbutsugaku dai ichi kan*. Tokyo: Kinkōdō.

Graham, Jeffrey B., and Kathryn A. Dickson. 2001. "Anatomical and Physiological Specializations for Endothermy." In *Tuna: Physiology, Ecology, and Evolution*, ed. Barbara Ann Block and Ernest Donald Stevens, 121–66. San Diego: Academic.

Herdman, William Abbott. 1923. *Founders of Oceanography and Their Work: An Introduction to the Science of the Sea*. London: E. Arnold.

Hibiya Takashi. 1981. "Suisangaku no kenkyū (2)." In *Nihon no nōgaku kenkyū: Kindai 100 nen no ayumi to shuyō bunkenshū*, ed. Nihon nōgakkai, 146–55. Tokyo: Nōsangyoson bunka kyōkai.

Hokudai suisan gakubu nanajūgo nenshi. 1982. Sapporo: Hokkaidō daigaku suisan gakubu.

Hubbard, Jennifer. 2014. "In the Wake of Politics: The Political and Economic Construction of Fisheries Biology, 1860–1970." *Isis* 105, no. 2 (June): 364–78.

Iida, Kaori. 2015. "Genetics and 'Breeding as a Science': Kihara Hitoshi and the Development of Genetics in Japan in the First Half of the Twentieth Century." In *New Perspectives on the History of Life Sciences and Agriculture*, ed. Denise Phillips and Sharon Kingsland, 439–58. New York: Springer.

Iijima, Isao. 1896. "Long-Lines as Zoological Collecting Apparatus." *Dōbutsugaku zasshi* 8, no. 89 (March 15): 13–17.

Inoué, Shinya. 2016. *Pathways of a Cell Biologist: Through yet Another Eye*. Singapore: Springer.

Irie Haruhiko. 1960. "Omou kabu [*sic*] mama ni." In *Tōkyō Daigaku nōgakubu suisan gakka no 50-nen*, ed. Shisuikai, 207–9. Tokyo: Tōkyō Daigaku nōgakubu suisan gakka sōritsu gojū shūnen kinenkai.

Ishikawa Chiyomatsu. 1921. "Denran [Ayu-tamago] no Beikoku yusō shiken." *Dōbutsugaku zasshi* 33, no. 390 (April 15): 120–22.

———. 1926. "The Story of the Ayu." *Mid-Pacific Magazine* 32, no. 5 (November): 433–42.

———. 1930. "Ayu no hanashi." *Ōhara nōgyō kenkyūjo* 14 (February): 61–76.

Isono Naohide. 1987. *Mōsu sono hi sono hi: Aru oyatoi kyōshi to kindai Nihon*. Yokohama: Yūrindō.

———. 1988. *Misaki rinkai jikkensho o kyorai shita hitotachi: Nihon ni okeru dōbutsugaku no tanjō*. Tokyo: Gakkai shuppan sentā.

———. 1999. "Mitsukuri Kakichi to Anton Dohrn: Misaki rinkai jikkensho no setsuritsu." In *Napori rinkai jikkensho: Kyorai shita Nihon no kagakushatachi*, 23–41. Tokyo: Tōkai daigaku shuppankai.

Jiang, Lijing. 2016. "Retouching the Past with Living Things: Indigenous Species, Tradition, and Biological Research in Republican China, 1918–1937." *Historical Studies in the Natural Sciences* 46, no. 2:154–206.

Jordan, David Starr. 1914. "Record the Fishes Obtained in Japan in 1911." *Memoirs of the Carnegie Museum* 6, no. 4:205–313.

Jordan, David Starr, Shigeho Tanaka, and John Otterbein Snyder. 1913. "A Catalogue of the Fishes of Japan." *Journal of the College of Science, Tokyo Imperial University* 33:1–497.

Kawamura Tamiji. 1912. "Misaki dayori." *Dōbutsugaku zasshi* 24, no. 283 (May 20): 318.

Keiner, Christine. 2009. *The Oyster Question: Scientists, Watermen, and the Maryland Chesapeake Bay since 1880*. Athens: University of Georgia Press.

Kishinoue Kamakichi. 1892. "Tōkyōwan no suisan." *Dai Nihon suisankai hōkoku*, no. 119:139–46.

———. 1903. "Kaiyō no chōsa ni tsuite." *Chiri zasshi* 15, no. 2:158–67.

———. 1905. *Suisan genron*. Tokyo: Seibidō.

———. 1908. "Iwashi gyogyō chōsa." *Suisan chōsa hōkoku* 14:71–110.

———. 1914. *Awagun suisan enkakushi*. Hōjōmachi: Awagun suisan kumiai.

———. 1915. *Uo to jinsei*. Tokyo: Sanseidō shoten.

———. 1920. "Risōteki suisan kenkyūjo narabi ni suisan hakubutsukan kensetsu no kyūmu." *Suisan* 8, no. 5 (March): 1–3.

———. 1921. "Maguro katsuo no kenkyū." *Suisankai*, no. 466 (July): 22–25.

Kishinouye, Kamakichi. 1923. "Contributions to the Comparative Study of the So-Called Scombroid Fishes." *Journal of the College of Agriculture, Tokyo Imperial University* 8:293–475.

Kitahara Tasaku and Okamura Kintarō. 1910. *Suiri seibutsugaku yōkō: Gyogyō kihon*. Tokyo: Kondō shōten, 1910.

Kohler, Robert E. 2008. "Laboratory History: Reflections." *Isis* 99:761–68.

Kraft, Alison. 2004. "Pragmatism, Patronage and Politics in English Biology: The Rise and Fall of Economic Biology, 1904–1920." *Journal of the History of Biology* 37:213–58.

Kuru Tarō. 1987. *Shinju no hatsumeisha wa dare ka? Nishikawa Tōkichi to Tōdai purojekuto*. Tokyo: Keisō shobō.

Lean, Eugenia. 2014. "Proofreading Science: Editing and Experimentation in Manuals by a 1930s Industrialist." In *Science and Technology in Modern China, 1880s–1940s*, ed. Jing Tsu and Benjamin A. Elman, 185–208. Leiden: Brill.

Lee, Victoria. 2015. "Mold Cultures: Traditional Industry and Microbial Studies in Early Twentieth-Century Japan." In *New Perspectives on the History of Life Sciences and Agriculture*, ed. Denise Phillips and Sharon Kingsland, 231–52. New York: Springer.

Maienschein, Jane. 1989. *100 Years Exploring Life, 1988–1988: The Marine Biological Laboratory at Woods Hole*. Boston: Jones & Bartlett.

Marcon, Federico. 2015. *The Knowledge of Nature and the Nature of Knowledge in Early Modern Japan*. Chicago: University of Chicago Press.

Miller, Ian Jared. 2013. *The Nature of the Beasts: Empire and Exhibition and the Tokyo Imperial Zoo*. Berkeley and Los Angeles: University of California Press.

Mills, Eric L. 1989. *Biological Oceanography: An Early History, 1870–1960*. Ithaca, NY: Cornell University Press.

"Misaki dayori." 1911. *Dōbutsugaku zasshi* 23, no. 271 (May 15): 288–89.

"Misaki dayori." 1912. *Dōbutsugaku zasshi* 24, no. 280 (March 15): 117.

Mitsukuri Kakichi. 1883. "Kaki no hassei." *Dai Nihon suisankai hōkoku*, no. 13 (March): 25–30.

———. 1887. "The Marine Biological Station of the Imperial University at Misaki." *Journal of the College of Science, Imperial University, Japan* 1:381–84.

———. 1891. "Suisan jigyōjō gakujutsu no ōyō." *Nihon taika ronshū* 3, no. 1: 101–11.

———. 1892. "Suisan chōsa ni tsuite." *Dōbutsugaku zasshi* 4, no. 44 (June 15): 232–40.

———. 1895. *Tsūzoku dōbutsu shinron*. Tokyo: Keigyōsha.

———. 1904. "The Cultivation of Marine and Fresh-Water Animals in Japan." *Bulletin of the United States Bureau of Fisheries* 24, no. 1:259–89.

———. 1907. "Hakubutsugaku." In *Kaikoku gojūnenshi jōkan*, ed. Ōkuma Shigenobu, 983–1014. Tokyo: Kaikoku gojūnenshi hakkōjo.

Mitsukuri Kakichi and Ishikawa Chiyomatsu. 1884. "Dōbutsu saishū hōkoku." *Gakugei shirin* 14, no. 80 (March): 246–68.

Mōri Hideo. 2011. *Tōkyō daigaku Misaki rinkai jikkensho zakki*. Tokyo: Seibutsu kenkyūsha.

Moriwaki Daigorō. 1979. "Gotō sensei to Yatsu sensei." *Dōbutsugaku zasshi* 88, no. 4 (December): 368–70.

Morse, Edward. 1883. *Dōbutsu shinkaron*. Translated by Ishikawa Chiyomatsu. Tokyo: Higashinari Kamejirō.

Muka, Samantha. 2016. "Marine Biology." In *A Companion to the History of American Science*, ed. Georgina M. Montgomery and Mark A. Largent, 134–46. Malden, MA: Wiley.

Muscolino, Micah. 2013. "Fisheries Build Up the Nation: Maritime Environmental Encounters between Japan and China." In *Japan at Nature's Edge: The Environmental Context of a Global Power*, ed. Ian Jared Miller, Julia Adeney, and Brett L. Walker, 56–72. Honolulu: University of Hawaii Press.

Nakamura Nakaroku. 1960. "Suisan jikkensho." In *Tōkyō Daigaku nōgakubu suisan gakka no 50-nen*, ed. Shisuikai, 102–5. Tokyo: Tōkyō Daigaku nōgakubu suisan gakka sōritsu gojū shūnen kinenkai.

Ninohei Tokuo. 1981. *Meiji gyogyō kaitakushi*. Tokyo: Heibonsha.

Nishimura Kimihiro. 2008. *Daigaku fuzoku rinkai jikkensho suizokukan: Kindai Nihon daigaku fuzoku hakubutsukan no ichi chōryū*. Sendai: Tōhoku daigaku shuppankai.

Nōshōmushō suisankyoku. 1910. *Suisan ni kan suru shisetsu jikō yōroku*. Tokyo: Nōshōmushō suisankyoku.

Okuda Yuzuru. 1960. "Sōritsu no koro." In *Tōkyō Daigaku nōgakubu suisan gakka no 50-nen*, ed. Shisuikai, 119–22. Tokyo: Tōkyō Daigaku nōgakubu suisan gakka sōritsu gojū shūnen kinenkai.

Onaga, Lisa. 2015. "More Than Metamorphosis: The Silkworm Experiments of Toyama Kametaro and His Cultivation of Genetic Thought in Japan's Sericultural Practices, 1894–1918." In *New Perspectives on the History of Life Sciences and Agriculture*, ed. Denise Phillips and Sharon Kingsland, 415–38. New York: Springer.

Ōshima Hiroshi. 1921. "'Kyūshiki' dōbutsugaku to 'shinshiki' dōbutsugaku." *Dōbutsugaku zasshi* 33, no. 398 (December 15): 475–78.

Phillips, Denise, and Sharon Kingsland. 2015. *New Perspectives on the History of Life Sciences and Agriculture*. New York: Springer.

Rozwadowski, Helen M. 2002. *The Sea Knows No Boundaries: A Century of Marine Science under ICES*. Seattle: University of Washington Press/International Council for the Exploration of the Sea.

Setoguchi Akihisa. 2009. *Gaichū no tanjō: Mushi kara mita Nihonshi*. Tokyo: Chikuma shobō.

Shisuikai, ed. 1960. *Tōkyō Daigaku nōgakubu suisan gakka no 50-nen*. Tokyo: Tōkyō Daigaku nōgakubu suisan gakka sōritsu gojū shūnen kinenkai.

Stickney, Robert R. 1989. *Flagship: A History of Fisheries at the University of Washington*. Dubuque, IA: Kendall/Hunt.

Takahashi Yoshitaka. 2007. *"Shigen hanshoku no jidai" to Nihon no gyogyō*. Tokyo: Yamakawa shuppansha.

Tamaki Tamotsu. 1998. *Dōbutsu gakusha Mitsukuri Kakichi to sono jidai*. Tokyo: San'ichi shobō.

Tamura Osamu. 1985. "Sensei no ma-iwashi kenkyū." In *Amemiya sensei o shinobite*, ed. Amemiya Ikusaku sensei kinen jigyō jikkō iinkai, 113–17. Tokyo: Kōei bunkasha.

Tanaka Shigeho. 1926. *Sakana: Rigakuteki oyobi kagakuteki kōsatsu*. Tokyo: Kōgaku shuppanbu.

———. 1934. *Uo to jinsei*. Tokyo: Rakurō shoin.

———. 1942. *Shinkaigyo: Kenkyūshitsu shūhen*. Tokyo: Sangabō.

Tanaka Yoshio. 1874. *Dōbutsugaku jō*. Tokyo: Hakubutsukan.

Tezuka Akira. 1992. *Bakumatsu Meiji kaigai tokōsha sōran dai 2 kan*. Tokyo: Kashiwa shobō.

Tsutsui, William. 2013. "Pelagic Empire: Reconsidering Japanese Expansion." In *Japan at Nature's Edge: The Environmental Context of a Global Power*, ed. Ian Jared Miller, Julia Adeney, and Brett L. Walker, 21–38. Honolulu: University of Hawaii Press.

Uchida Tōru. 1957. *Seibutsugaku no namikimichi*. Tokyo: Uchida rōkakuho.
Uda Michitaka. 1954. "Nihon kaiyōgaku no shinpo no ashiato." *Chigaku zasshi* 63, no. 3:139–44.
———. 1966. "Gyokyō to kaikyō." *Showa 41 nendo gyokyō kaikyō yohō jigyō zenkoku kaigi kōen naiyō*, ed. Suisanchō chōsa kenkyūbu, 2–8. Tokyo: Suisanchō chōsa kenkyūbu.
Ueno Masuzō. 1976. "Nihon kindai dōbutsugaku no ichidai tenkanki (1)." *Iden* 30, no. 9 (September): 47–52.
Wille, Robert-Jan. 2015. "The Coproduction of Station Morphology and Agricultural Management in the Tropics: Transformations in Botany at the Botanical Garden at Buitenzorg, Java, 1880–1904." In *New Perspectives on the History of Life Sciences and Agriculture*, ed. Denise Phillips and Sharon Kingsland, 253–75. New York: Springer.
———. 2016. "Stations and Statistics: Paulus Hoek and the Transnational Discipline of Ocean Biology." In *Soundings and Crossings: Doing Science at Sea, 1800–1970*, ed. Katharine Anderson and Helen M. Rozwadowski, 179–211. Sagamore Beach, MA: Science History Publications.
Wilson, Roderick I. 2015–16. "Placing Edomae: The Changing Environmental Relations of Tokyo's Early Modern Fishery." *Resilience: A Journal of Environmental Humanities* 3 (Winter/Spring/Fall): 242–89.
Yatsu Naohide. 1914. *Dōbutsu bunruihyō*. Tokyo: Maruzen.
———. 1915. "Kita taiheiyō no tanken." *Gendai no kagaku* 3:821–23.
———. 1919. *Seibutsugaku kōgi*. Tokyo: Shōkabō.
———. 1920. *Dōbutsugaku kōwa*. Tokyo: Tōadō.
———. 1926. *The Misaki Marine Biological Station, Guide-Book Excursion C-10, Pacific Science Congress, 1926, Japan*. Tokyo: Tōkyō chigaku kyōkai.
Yatsu Naohide and Uchida Tōru. 1972. *Yatsu Uchida dōbutsu bunruimei jiten*. Tokyo: Nakayama shoten.

PART TWO

Marine Practice

FIVE

Illuminating Animal Behavior: The Impact of Laboratory Structure on Tropism Research at Marine Stations

SAMANTHA MUKA

At the turn of the twentieth century, experimentalists interested in studying the reactions of animals to light stimuli did much of their work at marine stations throughout the United States and the Caribbean. Working at stations in Massachusetts, Bermuda, Florida, and California, they extended research on plants using everything from planktonic crustaceans nearly invisible to the naked eye to starfish, mollusks, and fish. Over the course of twenty years, two groups emerged: those who believed that light-sensing behavior was an automatic (or involuntary) response in proportion to the power of the stimuli presented and those who believed that response to stimuli was based on a whole host of variables in each individual, only one of which was the power of stimuli presented.

Jacques Loeb and Herbert Spencer Jennings became the faces of a larger debate about the nature of behavior. A mechanistic outlook on physiology led Loeb to perform experiments with large colonies of invertebrates, which he believed to be a basic evolutionary form from which behavior could be extrapolated to higher organisms. Over time, he sought to quantitatively account for all the variables in his experiments, especially the amount and type of light entering his enclosures, in order to understand the impact of light on behavior. He argued that tropism behaviors were ingrained, involuntary, and universal. In contrast, Jennings believed that individual research subjects' responses to light stimuli were based on a combination of life history and general disposition. This combination of internal and external variables made it impossible to extrapolate behaviors from a single experimental organism. Jennings used a variety of species and paid close attention to where and when each subject was collected. His work was largely qualitative, focusing on the movements and behaviors of individual subjects to understand the reaction of each to light (Gussin 1963; Pauly 1981, 1987; Kingsland 1987).

Both quantitative and qualitative tropism research was centered at marine stations during this period. Loeb did the majority of his work at the Marine Biological Laboratory in Woods Hole, Massachusetts. Jennings also worked at Woods Hole, as did most of the researchers discussed in this essay. This space was the primary location for most tropism research. However, Jennings also worked at the Stanford Hopkins Marine Station, the Naples Zoological Station, and the Carnegie Laboratory in the Dry Tortugas. As tropism research matured, experimentalists worked at additional stations far and wide. For example, William John Crozier, a student of Loeb's, worked primarily at the joint Harvard–New York University research station on Bermuda. Experimentalists such as Jennings and Crozier worked on similar experiments at several of these locations, sometimes in the same summer (Muka 2014). In this essay, I discuss marine stations as a single entity, not because small differences between individual stations were absent, but because we can view them as similar types of institutions during this period and it is the institutional impact on research design in which I am particularly interested.

Both Loeb and Jennings chose to work at marine stations because of the physical and social aspects of these spaces. Many historians of science have called attention to these aspects, highlighting the role of location, community, and scientific resources available during this period (Benson 1988a, 1988b, 2001; Maienschein 1988; Pauly 1988; Muka 2014; de Bont 2015). Marine stations occupy an interesting position: they exist within a space that allows extensive field observations and exploration while also facilitating experimental research in the laboratory.

This essay focuses on a different angle by highlighting the importance of the malleability of experimental materials and of the observational spaces and technologies at marine laboratories to explain how they became integral to the work involved in the Loeb-Jennings debate during this period. Thus, these stations provided researchers the basic spaces, access to specimens, and basic technological components needed to build a variety of experimental and observational research programs, often with fundamentally different underlying theoretical and experimental assumptions (Muka 2014, 2016).

This essay is divided into two sections. The first looks at the importance of organisms to tropism debates during this period and highlights the importance of marine stations to this endeavor; it would be impossible to have the Loeb-Jennings debate over tropism and related contributions in such a short period without the availability in large quantities of the organisms on which the debate relied, together with the ability to track detailed features of these organisms because of the proximity to the marine environment from which they were sourced. The second section looks at the role of the laboratory spaces and especially the plasticity of basic technologies and architecture contained therein, which in turn made it possible for studies of both quantitative and qualitative tropism behavior. Tropism research required a wide range of organisms, access to a variety of glassware and lighting technologies, and spaces that facilitated the ability to manipulate natural and artificial light quickly.

By paying close attention to the experimental choices made by tropism researchers at marine stations during this period, I seek to call attention not only to the experimental work but also to the myriad choices available to those pursuing opposing research agendas at these stations. Attention to this plethora of choices that permitted diverse and even competing research programs to flourish can help historians and contemporary researchers understand the requirements for experimental research, particularly when diverse approaches to similar or the same problems are in use.

Organism Choice and Use in Tropism Research

Animal behaviorists utilized a wide range of organisms at the turn of the twentieth century. American-based behavioral studies most commonly used rats and pigeons during this period, but experimentalists often stepped outside this diptych for examinations of specific behavioral questions (Dewsbury 1989; Petit 2010). One such group of questions involved tropism studies. Tropism researchers commonly worked with single-celled aquatic organisms, such as the planktonic crustacean *Euglena*. However, a large portion of the Loeb-Jennings debate revolved around the variety of species chosen and the condition of the individual organism at the time of experimentation. This section explores the aspects of the Loeb-Jennings debate relating to organismal choice and use, the reasons why aquatic organisms took a prominent place in these studies, and the changes made regarding experimentation with and reporting about the organisms utilized.

Jennings and Loeb had differing views on the importance of the experimental organism in tropism studies. Loeb's organismal choices were driven by his belief that tropic responses were mechanistic and relied primarily on "forced movements" of body orientation based on light stimulus. Simply put, the movement toward or away from light stimulus of a plant or lower organism is based on what part of the body the subject instinctually orients toward light. Loeb stated that similar reactions to light were found in plants, lower organisms, and mammals. This similarity in levels of organisms was important because it allowed Loeb to draw conclusions about the nature of heliotropic behavior by working with any of these levels of organisms. To that end, he most often worked with colonies of invertebrates, such as tube worms (Loeb 1918, 15).

Most often Loeb focused his research on data drawn from group reactions to light. In one set of experiments, he and John Northrup tested the Bunsen-Roscoe Law, which states that biological reaction to light is directly proportional to the energy in the dose of light. They placed thousands of barnacle larvae in a single tank and exposed them to different light intensities. They did not collect these organisms themselves, nor were they concerned with their individual (or colony) life histories. They chose these organisms because "they [were] small and [could] be obtained in large numbers" and, when released in a tank with a dark bottom, the white larvae "form a straight line towards the source of light": "It is possible to measure the angle of this trail. In this way in each observation the trail of thousands of individuals is measured" (Loeb and Northrup 1917, 541). Loeb and Northrup believed that the aggregation of these individual trails proved their hypothesis that the Bunsen-Roscoe Law held true for barnacle larvae and therefore all organisms. This focus on the aggregate allowed Loeb to draw sweeping conclusions from his data in line with his desire to produce a generalized theory of heliotropic behavior and to fit thousands of individual reactions statistically into what he believed to be universal numbers attached to behavior.

Jennings believed that the best way to understand tropic behavior was to understand more thoroughly the internal conditions of the experimental subject. In the first pages of his work on the behavior of starfish, he states:

> It is of the utmost importance, if we are to understand the behavior of organisms, that we think of them as dynamic—as processes, rather than as structures. The animal is something happening. In connection with these internal processes, we find that most organisms have a system of movements, of the body as a whole or of its external parts. This system of movements we call behavior. It is closely bound up with the internal processes; indeed, the two sets of activities are really one, and we shall be led far astray if we try to think of the behavior separately from the internal processes.

In a more succinct statement of his stance on the next page, he asserts: "The general problem of physiology is: How are the bodily processes kept going? The general problem of behavior is: How are the bodily processes kept going by the aid of movements?" (Jennings 1907, 56, 57).

This focus on the internal condition of the organism influencing behavior led Jennings to report that behavior was variable and determined by the amount of time a specimen had lived in captivity, its feeding schedule, and, sometimes, its natural disposition (Jennings 1907, 70). Thus, researchers needed to pay close attention to these variables when recording behaviors. Jennings thought that an individual's behavior could tell the experimenter more about tropic reactions than could aggregation of behaviors of a group of organisms. Where Loeb sought universal laws with aggregate data from large groups, Jennings focused on individual paramecium (*Stentor*), positing a theory that light reactions were largely "trial and error" in organisms (Jennings 1907, 22). To study this, he chose a few subjects and kept them in the laboratory tanks for extended periods to habituate them to captivity. He recorded behaviors before, during, and after experimentation to understand how reactions to light fit into larger patterns of behavior. This focus on the individual reaction, as opposed to watching the reaction of thousands of minute organisms and grouping the data together, opened the door to using multiple species for research as well as to more extensive experiments with individual specimens.

Investigators on both sides of the Loeb-Jennings debate congregated at marine stations because doing so allowed a wide range of easily accessible experimental subjects. Protozoa and infusoria are prevalent in water environments; thousands of experimental organisms can be collected in a small amount of water (Yerkes 1900). The collected specimens are easy to maintain in the laboratory owing to their elastic diet and the plasticity of the environment in which they flourish. In the "material and apparatus" section of Samuel Ottmar Mast's work with the fresh water protozoa *Stentor coeruleus*, he states: "The animals used in the following experiments were obtained by letting aquatic plants collected in a pond known to contain *Stentor*, decay in battery jars nearly filled with water" (1906, 360). Researchers also used readily available "higher" organisms. Lower metazoa, including echinoderma (sea urchins), anthropoda (lobsters and crabs), and cnidaria (sea slugs and jellyfish), are easily accessed in the shallow, littoral zone near the shore. These organisms require more care in captivity, including a varied diet and larger tanks. However, there was such ready availability that subjects that fared poorly in captivity could be easily replaced daily by newly collected specimens.

Loeb and experimentalists interested in studying aggregate movements of subjects chose colony dwellers to be analyzed as a unit. Loeb was particularly fond of both protozoa and hydroids. His work on the hydroid *Eudendrium* displays his typical process (see fig. 5.1). In a paper seeking to establish the efficiency of different light spectrums on curvature production in hydroids, he and Hardolph Wasteneys placed a colony of "newly regenerated polyps" of *Eudendrium* in a small tank. They exposed the colony to varied intensity and duration of light and measured the curvatures of the colony toward the light. The colony is considered a single experimental subject in this experimental setup, and little beyond the newly regenerated form of the organisms is mentioned, although the authors highlight the difficulty of working with young hydroids that is caused by their delicate forms (Loeb and Wasteneys 1915, 45).

Other researchers used groups to make statistical observations about behavior. Charles Davenport's tropism research used *Daphnia*. Davenport and his collaborators sought to test the hypothesis that light receptors on specific aspects of the body affected behavior (Davenport and Lewis 1899). Davenport used *Daphnia*, a freshwater crustacean. He does not mention where he obtained them or anything about the health of any individual *Daphnia*; the experiment suggests that each *daphnia* is interchangeable and that the outcomes are generalizable. In one experiment, Davenport and Walter Bradford Cannon draw conclusions based on the behavior of the majority of the group: "58

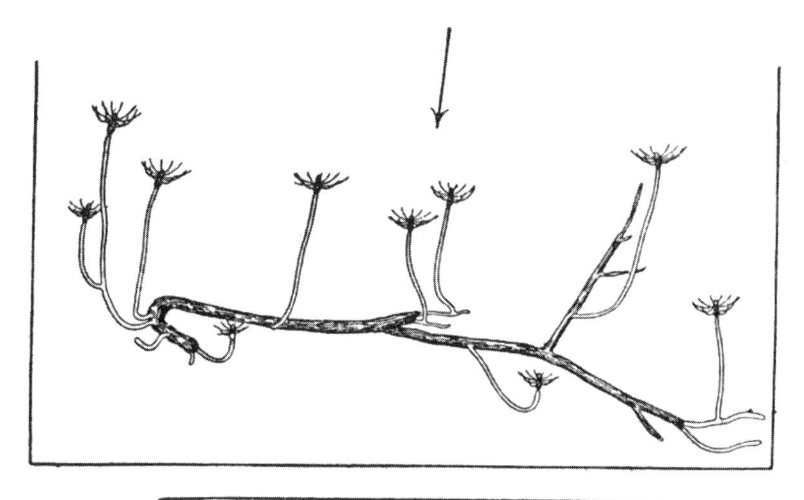

FIGURE 5.1. Loeb recorded the movement of a large number of subjects at the same time in an attempt to produce a generalized theory of heliotropic behavior. He worked with colony-dwelling organisms, such as the polyps of a *Eudendrium*, and recorded general movements of the colony, not the individual specimen. From Loeb (1918, 66).

were introduced, 33 of these (57%) went to *B* in intervals of one to two minutes. 6 more have moved towards *B*, so that in all 67% had passed from a region of greater intensity to a region of less intensity, but in the direction of incoming rays." The remaining 23 percent are explained as "being prevented by a bubble of gas from entering the water and consequently did not act normal." Of the 10 percent that moved in the opposite direction, the authors explain that "the presence of so many accidents of an internal and external nature" makes these outliers acceptable but does not negate their findings (Davenport and Cannon 1897, 27). On the basis of this statistical analysis, the authors assert that light intensity is not the only important factor influencing movement.

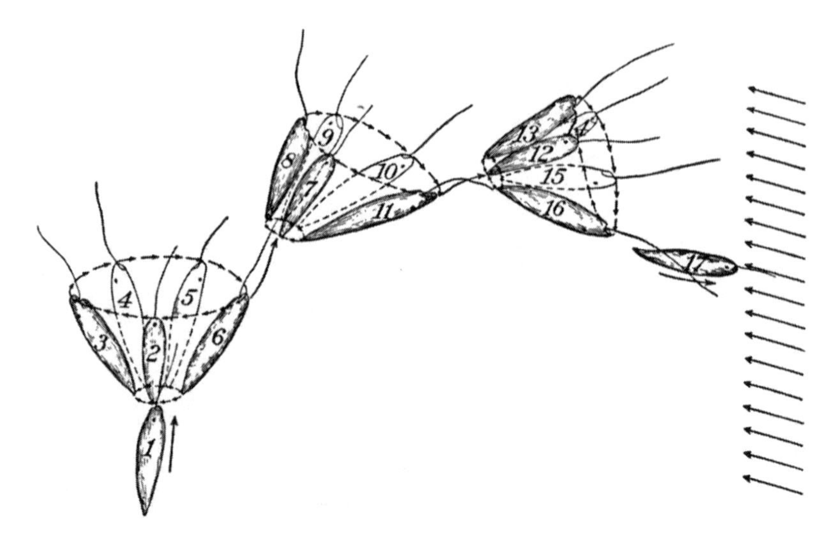

FIGURE 5.2. Jennings followed individual subjects' movements over long periods in an attempt to understand their heliotropic responses. He used a wide variety of subjects, starting with lower-level *Euglena* and *Stentor* and eventually progressing to starfish and other higher invertebrates. From Jennings (1906, 139).

Jennings's focus on the internal basis of behavior and the importance of comparative studies resulted in an emphasis on the individual subject and a wider range of species. In *Behavior of Lower Organisms* (1906), Jennings dedicates over half his work to protozoa but, unlike Loeb and Davenport, concentrates on the movements of individuals as opposed to the group (see fig. 5.2). He used the same species of protozoa as did Loeb to analyze individual courses of movement. In his *Contributions to the Study of the Behavior of Lower Organisms*, he states: "The light reaction is thus somewhat inconstant, and varies among different individuals. It varies considerably with *Stentors* of different cultures; from some cultures almost all the individuals show it, while from others it is barely noticeable. This variability and inconstancy run through all manifestations of the light reaction in *Stentor*" (Jennings 1904, 32).

But Jennings did not study only individual protozoa; he also worked with a variety of subjects, including starfish, on which he conducted extensive research. During a summer of research at the Marine Biological Association of San Diego's marine station (now known as the Scripps Institute of Oceanography), he used *Asterias forreri* starfish for his light studies. He did extend his observations to starfish both in the shallows of tidepools near the marine station and in different tanks in the laboratory. He stated that "when the starfish is undisturbed in the natural conditions that are throughout favorable to its life processes . . . it keeps rather quiet" but that if it is disturbed in any way, "as for example when it is removed from the large tank to a small one that serves for observation," it begins to explore its surroundings and behave differently (Jennings 1907, 95). After a short time, during which he felt that the starfish had acclimated to the smaller tank, he performed a variety of experiments testing starfishes' reactions to light. By moving the tank into direct sunlight under a window, he found that the subject moved to the darker side of the tank. In addition, he found that, if a tank was fully illuminated, the starfish often took a defensive position and became agitated. But he was always careful to state that reactions were "extremely variable" and could be "modified or quite prevented by various internal conditions as well as by external influences" (Jennings 1907, 102). Other researchers were attentive to these individual differences and worked with a wide range of subjects and species. Robert Yerkes worked with clapper jellyfish, a species found at this time almost exclusively at the Marine Biological Station in Woods Hole (Yerkes 1902, 1903, 1904; Muka 2016). Also working in Woods Hole, Edmund Beecher Wilson compared two subspecies of hydra: *h. Fusca* and *h. viridis* (Wilson 1891). Mast also chose to vary his subspecies, comparing light reactions in the larvae of the ascidians *Amaroucium Constellatum* and *Amaroucium Pellucidum* (Mast 1921).

One important variable for choosing organisms on which to test tropism was the ability to follow their movements closely. The size and speed of organisms influenced investigators' use of certain species. In his work on *Ranatra* (water stick insects), Samuel Jackson Holmes explained his choice of organism by comparing difficulties found working with a variety of other organisms, noting: "Animals vary greatly as regards both the definiteness of their reactions to light and the ease with which their movements can be followed." According to Holmes, copepods, cladocera, and ostrapoda react noticeably to light but are difficult to follow individually, while larger invertebrates are easier to follow but have sometimes indefinite reactions to lighting changes, making it difficult to draw robust conclusions (Holmes 1905, 305–6). William John Crozier and Leslie B. Arey chose to work with *chiton* (a mollusk) because the "behavior of very slowly moving animals in an illuminated field" proved easy to follow and map for clearer results. In their publication, the authors bemoan the inability to utilize multiple organisms at the same time because of their large size but state that the ability to follow a slow-moving organism over a superimposed grid drawn onto the aquarium far outweighed these concerns (Crozier and Arey 1918, 488). Wilson also expressed his delight at working with a slow organism: hydras were easy to follow and record "on account of their slowness" (Wilson 1891, 413).

Experimentalists seeking to test previous findings regarding tropism behavior greatly expanded the range of organisms preferred for this type of research; the number of organisms employed in tropism research continued to expand as a result of reactions to the Loeb-Jennings debate. Holmes tested body axis orientation conclusions in Loeb's tropism theory by experimenting with fiddler crabs, which orient "sidewise" (Holmes 1908). Philip B. Hadley included a paper on heliotropic reactions in American lobster larvae within his larger work on the species' behavior (Hadley 1908). Louis Murbach's extensive research on the behavior of clapper jellies also included research on phototactic responses (Murbach 1909). When testing Loeb and Wasteneys's conclusions regarding the Bunsen-Roscoe theory, Crozier utilized available organisms while teaching at the Bermuda Biological Station for Research. He looked at light reactions in *Balanoglossus* (sea worms) and *Chiton* (Crozier 1917a, 1917b), and his research on nudibranch behavior included a section on phototropic responses (Crozier and Arey 1919). All these organisms were available at marine stations and provided the ability to test the claims about body orientation made by Loeb and the impact of life history made by Jennings.

The Loeb-Jennings debate influenced not merely the choice of organisms but also the variables that researchers recorded when handling and maintaining organisms in the laboratory. Researchers testing the theory that behavior stemmed from internal conditions paid close attention to and in turn reported in official papers a multitude of variables, including age, time in captivity, and amount of rest time before experimentation, all of which they theorized influenced the behaviors of individual organisms. One of the most common variables reported was age. Both Loeb and Jennings came to the conclusion that the age of the organism mattered when studying phototropism. Young or newly hatched specimens commonly proved to be positively phototropic, whereas the mature form was negatively phototropic (photonegative). Crozier and Arey found that young *chitons* were the opposite: the young were photonegative. Analyzing the age difference in phototropic responses required researchers to include information about the collection and maintenance of these organisms in the laboratory. These experimentalists often utilized two specimens collected in separate locations to compare light reactions (Crozier and Arey 1918).

In addition, Jennings and other researchers observed that certain organisms were positively phototropic only if they were underfed. When fed sufficiently, they were continuously negatively phototropic, suggesting that light reactions might be tied to feeding reactions. Testing these hypotheses involved the recording of maintenance procedures in the laboratory. In his research on the tropic responses of hydra, Wilson starved the organisms and then recorded their tropic responses. He came to the conclusion that hydra become positively phototropic (or more so than "normally") in order to place themselves "in the position of maximum food supply." He noted that the common food sources of the hydra are all positively tropic and that the organism may therefore have evolved a response mechanism of positive tropism (Wilson 1891, 417).

When reporting results, researchers also included handling details to explain the organisms' condition at the beginning of an experiment. Those interested in the impact of internal conditions on behavior recorded feeding times, lighting conditions during rest, water temperature, and even the number of organisms placed in a tank at any time. Cora Reeves emphasized the need to maintain "natural" and "normal" conditions for the fishes: "Otherwise response is often inhibited by unnatural conditions or by manipulations that induce fright" (Reeves 1919, 3). Crozier emphasized the importance of tracking the conditions of his research subjects before experiments. In his work on *balanglossus,* he noted that there was marked light sensitivity differences between specimens in "bad condition" from handling and those in physiologically "good condition" (Crozier 1917a, 242).

Working with protozoa was often simpler, but experimentalists still thought they needed to record the amount of time that organisms were allowed to rest after being transferred from the general aquarium to the experimental apparatus. Charles Frank Phipps kept his amphipods for "some weeks" before beginning his experiments (Phipps 1915, 216). William Tower allowed his organisms to have twelve to eighteen hours of rest after transfer from the main aquarium to the experimental aquarium (Tower 1899). Wilson used only hydras that had been maintained in the aquarium for two months before the experiments began. The time allotment for settling ranged from thirty minutes for protozoa to hours or days for organisms such as medusa and echinoderma (Wilson 1891).

In summary, marine stations facilitated experimental programs that supported both sides of the Loeb-Jennings debate because they allowed researchers to have two major forms of access to experimental organisms. First, the marine station gave researchers a relatively wide choice of materials with which to work, whether they were interested in aggregating results or studying individuals. Those interested in working with a large number of the same species or an entire colony could easily access these just outside the laboratory doors. This easily accessible set of organisms also made it possible to focus on individual behavior and allowed those interested in doing so the ability to work with a wider variety of species.

Second, marine stations gave researchers the necessary data relating to their experimental subjects, such as age, natural environment, and factors relating to collection. Unlike laboratories situated in urban areas that relied on organisms provided by specimen warehouses, researchers could view their subjects in their natural habitats before collecting them. Marine stations provided researchers with the ability to track their subjects from the moment they were collected to the moment they were placed in an experimental tank. The location of marine laboratories near water sources with a wide variety of fresh specimens and collecting areas made them integral to the scientists studying phototropism during this period.

Technologies, Spaces, and Enclosures for Studying Tropism

In a report that Jennings submitted on March 25, 1904, to the Carnegie Institution following his stay at the Naples Zoological Station on a Carnegie research grant (no. 83), he outlined the future of his animal behavior research: "This line of work does not primarily require extensive or novel apparatus, nor great laboratories. While new apparatus may be needed from time to time as the work develops, ordinary well-equipped laboratories, such as are found in the zoological and physiological departments of many of our universities, amply suffice for most of the work."[1] Instead of technological requirements, he highlighted time requirements; animals must be watched extensively, and this type of watching requires uninterrupted stretches of time. His report makes tropism studies appear nontechnologically bound as he stated that little *new* technology was needed for animal behavior studies but that an "ordinary well-equipped laboratory" at a university would suffice. His tropism experiments used available lighting and basic glassware already available in most laboratories and marine stations.

In contrast, Loeb's work was technologically intensive. Loeb wanted to replace the "anthropomorphic method" of animal behavior with what he termed the "objective and quantitative methods of the physicist" (Loeb and Northrup 1917, 539). He sought to do this by stringently controlling and reporting the exact stimuli to which his subjects reacted. He came to use cutting-edge artificial lighting technologies along with experimental instrumentation to specify and quantify his experimental processes and results. In his earliest experiments, he employed a very simplistic experimental setup: protozoa were placed in a beaker and exposed to sunlight through a window in the laboratory (see fig. 5.3). Each successive experiment brought more technological interventions, including multiple lighting sources, enclosures built to test reactions to differing colors of light and minimize reflection, and cloths meant to dampen the amount of heat given off by artificial light.

At the turn of the twentieth century, marine stations contained almost no specialized equipment beyond aquariums and circulating

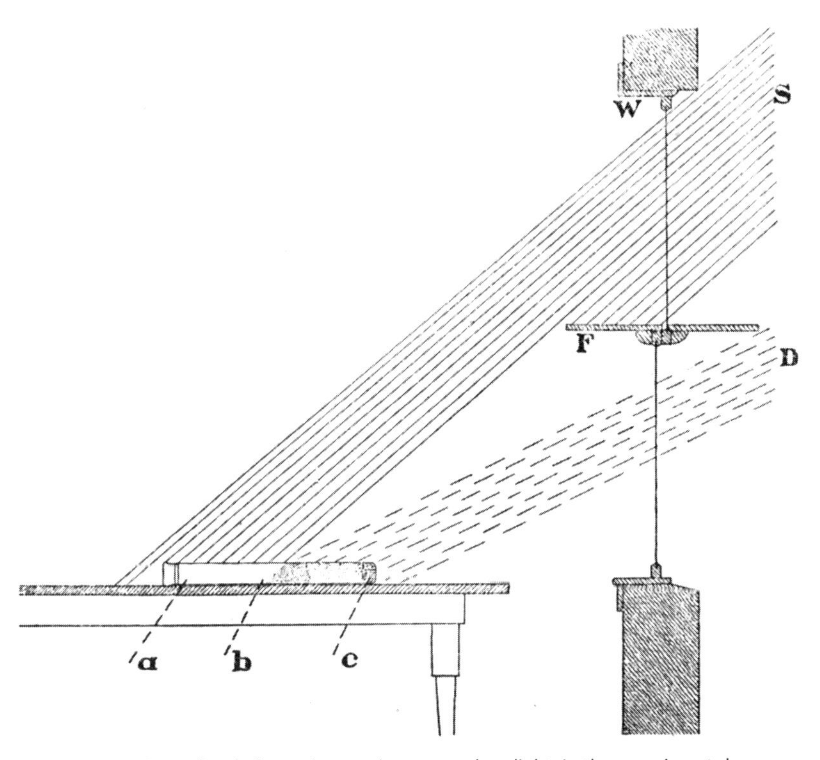

FIGURE 5.3. The earliest heliotropic experiments used sunlight. In the experiment shown here, Loeb placed newly hatched barnacle larvae in a test tube on a table and exposed them to direct sunlight (S) and diffused sunlight (D) from a window (W) with a film over half of it (F). From Loeb (1918, 50).

water to keep organisms alive indoors, plus high-quality microscopic equipment; lighting consisted of overhead lights in laboratories and natural light from windows and outside working space. Any specialized equipment required for experiments had to be either brought by the researcher or created out of the simpler glassware already available. Tropism researchers used simple, found objects to create specialized enclosures to test their theories and eventually brought various types of artificial lighting technologies into these spaces. The addition of these technologies facilitated comparative work in photochemistry and photobiology with marine organisms and made the laboratory spaces capable of accommodating the types of research pursued by both qualitative and quantitative phototropic researchers.

This section investigates the development of enclosures, lighting technology, and the shifting architecture of marine stations during this period to highlight how tropism researchers used simple spaces to create complex experimental programs. Marine laboratories facilitated a wide range of research not only by providing experimental materials but also by providing simple spaces that could be easily transformed depending on the temporary requirements of seasonal researchers.

Four-sided glass aquariums were the basic structure for studying phototropism in marine organisms. Behaviorists working with aquatic species could view the organism from a multitude of angles and easily introduce stimuli from multiple directions. Marine stations contained hundreds of glass containers that could be utilized for experiments, although tropism researchers eventually settled on several prominent variations of the glass-sided aquarium. Quantitative and qualitative researchers created aquariums de novo with found materials at stations, and almost every marine station was guaranteed to have the same simple components required for each apparatus. Often included in publications were the schematics or directions for creation of the apparatus so that the result could be re-created by other researchers.

Researchers studying the impact of light stimuli on small aquatic protozoa often built an aquarium that fit onto the stage of their microscope, termed the *stage* or *slide aquarium*. In 1893, Carl J. Cori first described a modification to the microscopic slide that would allow researchers to view minute organisms in a liquid solution. He called this modification a stage aquarium. The design was made from a strip of glass bent in a U formation to serve as the sides and bottom of the structure. The object holder (5 × 10 centimeters) formed the back wall, and a small cover (30 × 40 millimeters) formed the front. While easy to construct in the laboratory, this design allowed limited visibility and was difficult to load into the stage of the microscope. In 1894, Cori modified his invention to increase the visibility of the organism within the enclosure and allow the removal of the enclosure from the microscope stage (Cori 1894).

William Tower mentions the use of the stage aquarium in tropism studies in his treatise on *hydra viridis* (freshwater green hydra). He placed the hydra in a slide aquarium for 12 to 16 hours and then "carefully placed [the aquarium] upon the stage of the projection microscope" (Tower 1899, 505). Mast made his own slide aquarium, so named because it was made out of microscope slides glued together with balsam boiled in linseed oil. He employed the slide aquarium when analyzing the movements of amoeba, but the setup would have worked equally well for any organism that could fit in the relatively small enclosure (Mast 1911, 93–152). Raymond Pearl and Leon Cole also placed their specimens in slide aquariums (Pearl and Cole 1902, 77). The combination of a stage aquarium and a projection microscope allowed researchers interested in the individual movements of specific specimens to use a wide variety of species, not limited to those large enough to be tracked easily by the human eye.

Tropism researchers were interested in reactions to both light and dark, and their experimental equipment required strict control over these variables. Thus, their experimental setups required enclosures that permitted light to be systematically introduced to subjects and minimized unwanted light or shadow. The structure settled on was rectangular in shape to reduce the possibility of light distortion, commonly in the form of a traditional aquarium (which was all glass). Researchers modified the standard glass aquariums that were available at any marine station in order to fit their experimental needs. For example, Davenport and Cannon (1897, 25) built a glass enclosure painted "dead black" inside and out for their 1897 study. Loeb also fashioned a glass enclosure, but, instead of painting the aquarium completely black, he surrounded it with black paper (Loeb 1918, 107). The common modification involved using black paper or fabric to reduce reflection, thus allowing researchers to use readily available aquariums instead of building specialty pieces for short-term experimentation. This setup proved important when studying reaction to light directionality: light could be directed into the enclosure through a given set of openings, and then a new set of openings could be cut into the paper. In addition, organismal reaction to specific colors of light was a major research agenda during this period. If researchers utilized an all-glass aquarium, they could easily replace the dark paper with a plate of colored glass. For instance, Wilson divided one side of an aquarium into four quadrants, covering one area with yellow glass, one with opaque glass, and one with blue glass, and leaving the fourth uncovered (thus permitting the entrance of white light); the top and other three sides of the aquarium he covered with black paper (Wilson 1891, 421–22). The use of opaque screens with a traditional aquarium allowed experimenters to modify their enclosures easily, thus permitting maintenance of a consistent environment while supporting quick changes to lighting variables. The ability of researchers to make any needed apparatus out of found equipment helped both quantitative and qualitative researchers shift experimental setups quickly depending on interest, organismal availability, and findings from peers.

Many experiments included bringing in specialized lighting technology. Tropism researchers incorporated in their research new lighting technologies that allowed them to vary light intensity, more tightly control lighting conditions, and work any time of day. The use of artificial lighting caused many investigators to worry that light sources were not being reported correctly and that experiments could not be replicated. But new tools that measured UV spectrums, heat, and light intensity helped with standardization and experimental replication. With the introduction of these new lighting technologies, researchers were also forced to question the supposedly natural reactions of their subjects under experimental conditions. Concerns about calibrating stimuli for experimental purposes and the naturalness of technology became built into the experimental processes.

The three most common lighting technologies available at marine stations were gas burners, carbon-arc lamps, and Nernst glowers. The gas burner—specifically the Welsbach burner—was one of the first artificial lighting technologies used in tropism studies. Baron Carl von Auer Welsbach invented the Welsbach burner—a lamp consisting of a gas burner and gauze soaked in cerium and thorium that became incandescent when heated—in 1885 and over the course of the next four years slowly perfected the design to emit brighter, whiter light ("The Welsbach Light" 1900). Yerkes explained that different intensities of light could be obtained by either reducing the gas or moving the instrument; he tended to move the burner (Yerkes 1900, 408). The Welsbach was a convenient light source, but it did not produce a high-intensity light, nor was it similar to sunlight. The carbon arc lamp was a popular choice for researchers investigating the effect of intense light on behavior. While the carbon arc lamp was invented long before the Welsbach burner, it was not until 1890 that a more efficient and cost-effective design was produced in the United States. Arc lamps produce light through the application of an electric arc between two carbon electrodes, resulting in a large amount of UV light, especially useful for the study of spectrum differences in phototropism. Loeb and Wasteneys (1915, 45) used them to test the effects of different spectrums on *Eudendrium*. Pearl and Cole (1902, 77). studied the effects of intense light on phototropic responses in a multitude of organisms with arc lamps. But these lighting sources also had drawbacks: they were so bright that they could cause eye problems among the people operating them, and extremely intense arc lamps could give users sunburns (Hockberger 2002). In addition, they proved costly because of a short life span (only 8–16 hours).

Nernst glowers were not introduced to the United States until 1898, and production stabilized in only in 1901. The light source worked similarly to an arc lamp or incandescent bulb does, but it did not require a vacuum to produce light. Instead, electricity was conducted through a ceramic mixture of zirconium oxide heated to incandescence. The light produced was softer than that produced by the arc lamp and thus was thought to be closer to natural sunlight. Loeb chose Nernst glowers for several experiments, and Mast stated that, of all artificial light sources, he found that "the Nernst single glower lamp was the . . . most satisfactory source of light for all experiments, both quantitative and qualitative, providing the intensity required was not great" (Mast 1911, 362). The ceramic glower was long lasting and provided softer light, but it had a downside: it required a separate heating filament to prepare the ceramic to conduct electricity on its own. Even with this downside, the Nernst glower became very popular in tropism studies.

Artificial lighting technologies did not immediately produce lighting conditions that were ideal for studying tropic behaviors; they often required additional technological fixes to focus and temper their rays. To lessen the impact of the heat given off by artificial lights, researchers placed a thin aquarium filled with alum between the light source and the experimental enclosure to absorb electrical heat. A maze of mirrors could be used to fix a single point of light into a tank. And, for an experiment that called for graded light intensities in a single tank, Yerkes developed a *light grader.*

Yerkes first describes his light grader, built to provide "a band of light regularly graded," in his work on light and heat reactions of *Daphnia pulex* (Yerkes 1905, 363; see fig. 5.4). The grader was composed of two light sources passed through alum aquariums, contained by black fabric, and reflected by mirrors to produce a graded light band focused

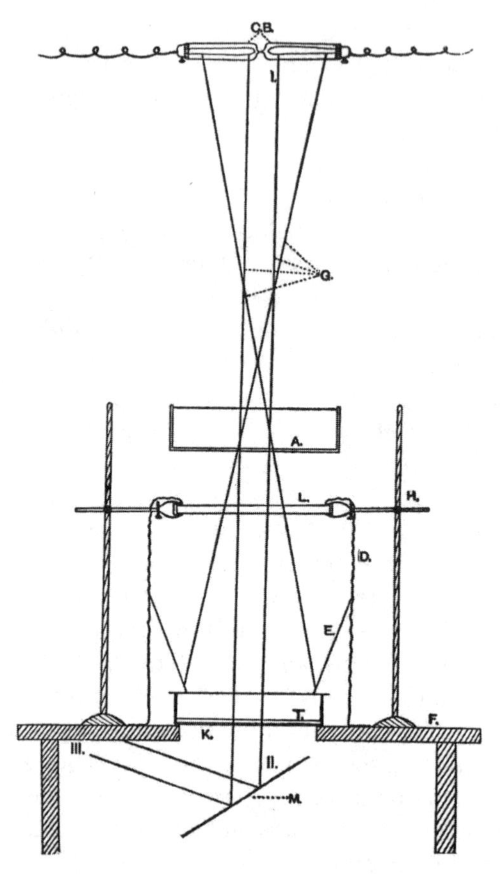

FIGURE 5.4. Controlling artificial light in the laboratory required a variety of new tools. Yerkes's original light grader allowed two light sources (at the top of the diagram) to be altered via mirrors (M), alum (A), and black cloth (D). From Yerkes (1905, 363).

directly onto the experimental enclosure, usually located on a stage in a stage aquarium.

Mast (1911) utilized a light grader modified to "suit the conditions of the experiments" in his experiment with plants. Victor Shelford (1914) and C. F. Phipps (1915) built light graders for their work, both consulting Yerkes's and Mast's designs for inspiration and blueprints.

As Loeb and other researchers focused on external stimuli analysis and increasingly relied on lighting technologies, their results reflected the increased use of quantifiable stimuli. Loeb's results pages became increasingly graph based. Large tables outlined light intensity, water temperature, chemical makeup of the water, and duration of stimulus exposure. The desire to quantify variables led to the introduction of photometers to measure light intensity and thermopiles to measure heat output (Laurens and Hooker 1920).

The use of natural sunlight in tropism research initially declined with the introduction of artificial lighting technologies. Researchers quantifying stimuli found it difficult to work with direct sunlight for more than a few hours a day. The shifting angles of the sun's rays, the inability to work on overcast days, and the possibility that other factors were creating uncontrollable variables made natural sunlight an unreliable light source. Davenport and Cannon criticized previous researchers on the subject because they did not record the angle of the sun properly. They claimed that these findings were insufficient because, "so far as the data go, there might well have been, in this case, a movement in the direction of the sun's rays" (Davenport and Cannon 1897, 24). After Loeb began experimenting with artificial light sources in the laboratory, he rarely used sunlight again in his research, instead choosing to work with quantifiable and universally reproducible light sources. But the introduction of artificial lighting in the laboratory did not make natural light obsolete. In fact, it became a control for testing organismal behaviors in the laboratory.

The use of lighting technologies caused some researchers to question the results as unreflective of natural behaviors. In 1918, Crozier and his research partner Leslie B. Arey, Loeb's student and a researcher at the Bermuda Biological Station, enumerated arguments against Loeb's universal understanding of tropisms. They state that one argument against Loeb was that his results were "laboratory product[s]" but immediately counter this argument by stating that it is "quite beside the point" (Crozier and Arey 1919, 487). For Loeb, this criticism was indeed beside the point; he was interested in understanding and controlling the tropic responses of organisms, but he was not invested in the finding the deep-seated causes of these reactions. But other researchers, including Crozier, took these concerns seriously. In addition to testing results by following the original experimental procedure closely, Crozier also tested them with direct and diffused sunlight (Crozier 1914). Mast also exposed organisms to multiple sources of light, including sunlight. In his research on the paramecium *Stentor*, he worked with a gas burner, incandescent bulbs, a carbon arc lamp, sunlight (direct and diffused), and a Nernst glower (Mast 1921). Elizabeth Towle (1900) compared reactions of organisms to light from a Welsbach burner and both diffused and direct sunlight. Crozier and Arey (1919, 489) placed organisms in "diffused light from a north window" at the Bermuda Biological Station.

To accommodate the use of both natural and artificial light, researchers sought out new spaces in marine stations. The earliest marine stations were built with open, shared laboratories and large windows meant to allow as much natural light as possible indoors; while some senior scholars worked in smaller spaces, most researchers worked in the main laboratory. But, as a wider variety of scholars visited, the large, multiuse laboratories were joined by specialized spaces, such as rooms for chemical use containing hoods and ventilation. One such specialized space that phototropism researchers used was the darkroom. Photographic documentation of experimentation rose in the early years of the twentieth century, and darkrooms were added to marine stations during this period. Between 1917 and 1929, the Hopkins Marine Laboratory at Stanford University expanded to include six specialty darkrooms (having originally been built with none): one photographic darkroom, three for spectroscopy, one reserved for polarimetry and photometry, and one with a heliostat that drew its light directly from the roof of the building (Fisher 1918, 301).

Photographic darkrooms allowed researchers to diminish the effects of ambient light while using mirrors and minimal black fabric scrims to direct light stimuli. Crozier utilized darkrooms in his research with sea cucumbers, an organism with such intense light responses that he kept them in the darkroom between experiments to prevent ill effects from low-level ambient light. He exposed the cucumbers to light "by admitting sunlight, or light from a 40 c.p. tungsten filament, through a diaphram into a blackened box containing the holothurians in a flat-sided glass aquarium" (Crozier 1914, 10). In addition, researchers did not have to abandon the use of sunlight once they moved into darkrooms. Many had heliostats to direct sunlight into the room. If researchers chose to work with sunlight in a comparative study, they could easily direct the light into the required area without losing effectiveness because of light saturation.

In summary, the marine station environment made it possible for researchers interested in controlling a wide range of lighting variables to do so easily using found equipment and basic architectural spaces. Because marine stations were built to accommodate a wide range of research goals, it was possible for individuals wishing to do comparative work throughout a single summer season to use natural and several different types of artificial lighting and to build several different apparatuses that allowed them to pursue different types of research. The plasticity of the spaces available at marine stations was a perfect fit for highly technical tropism research.

Conclusion

Marine stations of the early twentieth century served an important role in tropism research; the mutual interdependence of field and lab made them capable of hosting a wide array of experimental programs simultaneously. While most animal behavior experimentalists were at inland university laboratories working with vertebrates such as rats, tropism researchers congregated at marine stations throughout the United States to perform experiments that would help them engage in the ongoing Loeb-Jennings debate.

The first major component of the marine station that was an ideal fit for the study of tropism was the field environment. Marine stations gave researchers the ability not only to work with a wide variety of organisms permitting pursuit of different experimental theories and practices; they also allowed them to access the field in which those organisms where found. Loeb, and others interested in mechanical explanations of behavior, did not necessarily need personal access to the field environment. But they did need the healthy and abundant colony-dwelling subjects that were plentiful in marine spaces. In fact, they were too delicate to work with inland, a good reason to travel to marine stations.

Jennings, and others interested in the impact of internal and external conditions on behavior, required extensive field collection notes to provide the necessary information to support their experimental parameters. As the debate intensified, researchers such as Crozier expanded the number of species with which they worked and spent more time collecting these animals and studying their individual development and life histories for answers to behavioral questions. Both large numbers of fresh experimental subjects and intimacy with the field environment were important to tropism studies: marine stations offered both of these in abundance.

The second major reason why marine stations were an excellent fit for tropism research was the easily manipulated nature of these laboratory spaces. Each marine station contained basic glassware and open-floor laboratory space or lab benches and tables to be adjusted as needed. Jennings's low-tech experimentation required little more than modifying available aquariums and finding the nearest window in which to place them. A focus on quantifying external variables would seem to have had very different requirements, but the same marine laboratory space also accommodated Loeb: he modified available glassware, brought in new lighting technologies, and utilized the architecture of various types of darkrooms to perform highly quantifiable experiments at the same time that others were still using sunlight.

Because marine stations were built as multipurpose spaces for anyone hoping to study at the shore, they were capable of supporting both quantitative and qualitative research at the same time. The traits of these spaces, including direct access to the field with abundant levels of organismal life, simple glassware, and spaces easily changed and manipulated with multiple lighting capabilities, made marine stations the perfect location for both quantitative and qualitative phototropism studies in the first decades of the twentieth century.

Note

1. Folder 3, "Carnegie Institution of Washington," Herbert Spencer Jennings Papers, American Philosophical Society, Philadelphia.

References

Benson, Keith R. 1988a. "Laboratories on the New England Shore: The 'Somewhat Different Direction' of American Marine Biology." *New England Quarterly* 56:53–78.

———. 1988b. "Why Marine Stations? The Teaching Argument." *American Zoologist* 28:7–14.

———. 2001. "Summer Camp, Seaside Station, and Marine Laboratory: Marine Biology and Its Institutional Identity." *Historical Studies in the Physical and Biological Sciences* 32, no. 1:11–18.

Cori, C. J. 1894. "Stage-Aquarium." *Journal of the Royal Microscopical Society*, February, 121–22.

Crozier, W. J. 1914. "The Illumination of a *Holothurian* by Light." *American Journal of Physiology* 36, no. 1:8–21.

———. 1917. "The Photic Sensitivity of *Balanoglossus*." *Journal of Experimental Zoology* 24, no. 2:211–17.

———. 1917. "The Behavior of Holothurians in Balanced Illumination." *American Journal of Physiology* 43, no. 4:510–13.

Crozier, W. J., and Leslie B. Arey. 1918. "On the Significance of the Reaction to Shading in *Chiton*." *American Journal of Physiology* 46, no. 5:487–92.

———. 1919. "Sensory Reactions to *Chromodoris Zebra.*" *Journal of Experimental Zoology* 29, no. 2:261–310.

Davenport, C. B., and W. B. Cannon. 1897. "On the Determination of the Direction and Rate of Movement of Organisms by Light." *Journal of Physiology* 21, no. 1:22–32.

Davenport, C. B., and F. T. Lewis. 1899. "Phototaxis of Daphnia." *Science* 9, no. 219:368.

de Bont, Raf. 2015. *Stations in the Field: A History of Place-Based Animal Research, 1870–1930.* Chicago: University of Chicago Press.

Dewsbury, Donald A. 1989. "A Brief History of the Study of Animal Behavior in North America." *Perspectives in Ethology* 8:85–122.

Fisher, W. K. 1918. "The New Hopkins Marine Station of Stanford University." *Science* 47, no. 1217:410–12.

Gussin, Arnold E. S. 1963. "Jacques Loeb: The Man and His Tropism Theory of Animal Conduct." *Journal of the History of Medicine and Allied Sciences* 18, no. 4:321–36.

Hadley, Philip B. 1908. "The Behavior of the Larval and Adolescent Stages of the American Lobster (*Homarus Americanus*)." *Journal of Comparative Neurology and Psychology* 18, no. 3:199–301.

Hockberger, Philip E. 2002. "A History of Ultraviolent Photobiology for Humans, Animals and Microorganisms." *Photochemistry and Photobiology* 76, no. 6:561–79.

Holmes, S. J. 1905. "The Reactions of Ranatra to Light." *Journal of Comparative Neurology and Psychology* 15, no. 4:305–49.

———. 1908. "Phototaxis in Fiddler Crabs and Its Relation to Theories of Orientation." *Journal of Comparative Neurology and Psychology* 18, no. 5: 493–97.

Jennings, Herbert Spencer. 1904. *Contributions to the Study of the Behavior of Lower Organisms.* Washington, DC: Carnegie Institution of Washington.

———. 1906. *Behavior of Lower Organisms.* New York: Columbia University Press.

———. 1907. "Behavior of the Starfish, *Asterias Forreri* De Loriol." *Contributions from the Laboratory of the Marine Biological Association of San Diego* 4, no. 2:53–185.

Kingsland, Sharon. 1987. "A Man out of Place: Herbert Spencer Jennings at Johns Hopkins, 1906–1938." *American Zoologist* 27, no. 3:807–17.

Laurens, Henry, and Henry D. Hooker Jr. 1920. "Studies on the Relative Physiological Value of Spectral Lights." *Journal of Experimental Zoology* 30:345–68.

Loeb, Jacques. 1918. *Forced Movements, Tropisms, and Animal Conduct.* Philadelphia: Lippincott.

Loeb, Jacques, and John H. Northrup. 1917. "Heliotropic Animals as Photometers on the Basis of the Validity of the Bunsen-Roscoe Law for Tropic Reactions." *Proceedings of the National Academy of Sciences* 3, no. 9:539–44.

Loeb, Jacques, and Hardolph Wasteneys. 1915. "On the Identity of Heliotropism in Animals and Plants." *Proceedings of the National Academy of Science* 1, no. 1:44–47.

Maienschein, Jane. 1988. "History of American Marine Laboratories: Why Do History at the Seashore?" *American Zoologist* 28, no. 1: 15–25.

Mast, Samuel Ottmar. 1906. *Rhythmical Pulsation in Scyphomedusae*. Washington, DC: Carnegie Institution of Washington.

———. 1911. *Light and the Behavior of Lower Organisms*. New York: Wiley.

———. 1921. "Reactions to Light in the Larvae of the Ascidians, Amaroucium Constellatum and Amaroucium Pellucidum with Special Reference to Photic Orientation." *Journal of Experimental Zoology* 34, no. 2:148–87.

Muka, Samantha. 2014. *Working at Water's Edge: Life Sciences at American Marine Stations, 1880–1930*. Philadelphia: University of Pennsylvania Press.

———. 2016. "The Right Tool and the Right Place for the Job: The Importance of the Field in Experimental Neurophysiology, 1880–1945." *History and Philosophy of the Life Sciences* 38, no. 3:1–28.

Murbach, Louis. 1909. "Some Light Reactions of the Medusa *Gonionemus*." *Biological Bulletin* 17, no. 5:354–68.

Pauly, Philip. 1981. "The Loeb-Jennings Debate and the Science of Animal Behavior." *Journal of the History of Behavioral Sciences* 17, no. 4:504–15.

———. 1987. *Controlling Life: Jacques Loeb and the Engineering Ideal in Biology*. Oxford: Oxford University Press.

———. 1988. "Summer Resort and Scientific Discipline: Woods Hole and the Structure of American Biology, 1882–1925." In *The American Development of Biology*, ed. Ronald Rainger, Keith R. Benson, and Jane Maienschein, 121–50. Philadelphia: University of Pennsylvania Press.

Pearl, Raymond, and Leon J. Cole. 1902. "The Effect of Very Intense Light on Organisms." In *Third Report of the Michigan Academy of Science*, 77–78. Lansing, MI: Wynkoop Hallenbeck Crawford.

Pettit, Michael. 2010. "The Problem of Raccoon Intelligence in Behaviorist America." *British Journal of the History of Science* 43, no. 3:391–421.

Phipps, C. F. 1915. "An Experimental Study of the Behavior of Amphipods with Respect to Light Intensity, Direction of Rays and Metabolism." *Biological Bulletin* 28, no. 4:210–23.

Reeves, Cora Daisy. 1919. *Discrimination of Light of Different Wave-Lengths by Fish*. Behavior Monographs. Cambridge, MA: Henry Holt.

Shelford, Victor E. 1914. "An Experimental Study of the Behavior Agreement among the Animals of an Animal Community." *Biological Bulletin* 26: 294–315.

Tower, William. 1899. "Loss of the Ectoderm of *Hydra Viridis* in the Light of a Projection Microscope." *American Naturalist* 33, no. 390:505–9.

Towle, Elizabeth. 1900. "A Study of Heliotropism in Cypridopsis." *American Journal of Physiology* 3, no. 8:345–65.

"The Welsbach Light." 1900. *Science* 12:951–56.

Wilson, Edmund B. 1891. "The Heliotropism of *Hydra*." *American Naturalist* 25, no. 293:413–33.

Yerkes, Robert. 1900. "Reactions of Entomostraca to Stimulation by Light, II: Reactions of *Daphnia* and *Cypris*." *American Journal of Physiology* 4, no. 8: 405–22.

———. 1902. "A Contribution to the Physiology of the Nervous System of the Medusae Gonionemus Murbachii, Pt. I: The Sensory Reactions of Gonionemus." *American Journal of Physiology* 7, no. 2:434–49.

———. 1903. "A Study of the Reaction and Reaction-Time of the Medusa *Gonionema murbachii* to Photic Stimuli." *American Journal of Physiology* 9: 279–307.

———. 1904. "The Reaction-Time of *Gonionemus Murbachii* to Electric and Photic Stimuli." *Biological Bulletin* 6, no. 2:84–95.

———. 1905. "Reactions of Daphnia Pulex to Light and Heat." In *Mark Anniversary Volume to Edward Laurens Mark, Hersey Professor of Anatomy and Director of the Zoological Laboratories at Harvard University, in Celebration of Twenty-Seven Years of Successful Work for the Advancement of Zoology, from His Former Students, 1877–1902*, 359–78. New York: Henry Holt.

SIX

The Scientific Fishery: Sampling, Dissecting, and Drawing in the Gulf of Naples

KATHARINA STEINER

This discussion focuses on the research program of the Naples Zoological Station (the Stazione Zoologica di Napoli), which the station's founder, Anton Dohrn, termed the "Scientific Fishery."[1] That program centered on a historically young and at the time relatively unstudied research object, the marine invertebrate. Within this research framework, a range of interacting individuals, practices, and artifacts came together in a pathbreaking monographic series, the Fauna und Flora des Golfes von Neapel und seiner angrenzenden Meeresgebiete (Fauna and flora of the Gulf of Naples and adjacent marine areas).[2] While the series was the outcome of systematic study, this was not what Dohrn referred to as the *old systematics*; rather, the work reported on was carried out from a biological perspective (Nyhart 2009) taking account of information gained from both ecological fieldwork and experimental laboratory investigation. In his reflections on Fauna und Flora, Dohrn did not refer to a *new systematics*; rather, a central feature of the station's program was its demarcation from its old version through the study of living organisms in their natural environment and in comparison with other zoogeographic regions. In this manner, Dohrn moved beyond taxonomic traditions of embryology, anatomy, and physiology that had been decisive in defining and classifying a species. In this new approach, realized in the framework of Fauna und Flora, zoogeography and behavior received equal consideration. This meant that the condition of the specimen was also central: it was now to be studied in its various biological life stages, in the process of dying, and when dead.

The Scientific Fishery combined work in the laboratory and field in three ways. First, specimens were not simply caught in the field and studied in the lab; rather, scientifically trained fishermen and researchers working for the station identified, preserved, and studied specimens in the field and the laboratory. Second, many researchers physically moved between the field and the lab, conducting both sampling trips and laboratory investigation. Third, the Fauna und Flora series, including its artifacts—its illustrations, prepared specimens, data tables, plankton nets, etc.—was the joint product of field and laboratory research that, among other things, integrated experimental interventions on embryos and zoogeographic data. In this essay, I offer an overview of the Scientific Fishery's institutional and research foundations. As we will see in some detail, the pioneering research of Wilhelm Giesbrecht, a scientist at the station specializing in the study of plankton, encapsulates all three of these key dimensions of work at the Scientific Fishery.

The Fauna und Flora Series: Implementing a Research Program

Anton Dohrn never offered a formal definition of what he meant by the term *Scientific Fishery*. In 1880–81, in one of the first annual reports detailing the activities and finances of the station,[3] the basis for its annual publications,[4] he gave an overview of material- and personnel-connected aspects of its program. The program's research activities were closely tied to the station's founding ideals and to further developments in the course of that institute's extension of research units. I am here referring to the development of an infrastructure consisting of laboratories and a fleet whose furnishings and instruments were meant to offer a methodological-technical foundation, followed by a research library and specimen collections serving as crucial aids in examining new species. Alongside the technical infrastructure, the program was also based on careful organization of personnel.

From 1876 on, putting in place the institute's own staff of officials for executing its research program was a central administrative goal.

Required here was a highly heterogeneous team whose qualifications and areas of deployment were in line with the wide range of tasks and huge work expenditure tied to the field of marine zoology. The bulk of the institute's employees consisted of various technical staff members, including draftsmen, laboratory assistants, zoologically trained fishermen, technicians, engineers, and so-called scientific assistants. The assistants usually stayed at the station for a number of years, a transitional stage that often led to an academic appointment. Both these groups were accorded official status, with a distinction in classification corresponding to degree of education: the former group was classified as "lower," the latter as "higher," officials. Both groups received a monthly income and enjoyed German state medical insurance and pensions (Steiner 2018).

Together with external researchers, both lower and higher personnel developed instruments and methods for examining marine invertebrates. The station assistants Paul Mayer and Salvatore Lo Bianco, for example, contributed pioneering work to support this effort. Such developments have for the most part been discussed in relation to services offered the station's guest researchers, not as part of its internal research program. With a shift of perspective from the guest researchers to the station's permanent employees (or from services offered contractually to the research program), a spectrum of instruments from sampling net to microscope comes into focus, instruments that were used in a context of scientific production and remained of enduring value for research in plankton.

Although the technical and personnel infrastructure was in part conceived as a facility providing fresh marine material for guest researchers doing work in Naples (Müller 1976; Groeben 2002; de Bont 2015, 51–69; see also Groeben, in this volume), it was also the operative foundation for detailed exploration of the Tyrrhenian Sea in the framework of the Fauna und Flora project, representing the program in its published form. The series amounted to the practical application of the institute's "productive strengths" (Dohrn 1880, v), which is to say technical innovation in both laboratory and field together with the shared production of knowledge by all those working at the station. "Prompted by knowledgeable people to demonstrate the importance and productive energy of the Zoological Station through independent publications," Dohrn wrote, "I had to plan a specific program equally taking account of and articulating scientific needs and the [institute's] specific strengths. A faunistic exploration of the gulf seemed to me the most appropriate framework for this" (Dohrn 1880, v).

When the first volume of Fauna und Flora appeared in 1880, the station had already assembled a team consisting of twenty-five to thirty individuals[5] putting into practice the institute's multidimensional program. When Dohrn retired as director in 1907, handing the position over to his son Reinhard, Wilhelm Giesbrecht took over as the series' chief editor. The series' focus was on the systematic examination of forms of zooplankton, with additional volumes on fish, coral, and algae.[6]

Dohrn understood the series in its entirety as a kind of reference work compiling data on the ecosystem of the Tyrrhenian Sea, especially the Gulfs of Naples, Salerno, and Gaeta and the Pontic Islands; at the same time, each volume was meant to function as a microstudy of a species or family. The pool of authors was made up not only of officials at the station such as Giesbrecht, Paul Mayer, Raffaele Valiante, Gottfried Berthold, Angelo Andres, Hugo Eisig, and Karl Brandt but also guest researchers such as Carl Chun, who authored the first volume, and Gottlieb von Koch. The study of a species was here laid out in a framework of individual research.

With a first viewing, our attention is naturally drawn to the many richly colored and finely drafted illustrations in the volumes' incorporated pictorial atlases. As a result, the series might seem to be one of those taxonomically oriented works that a younger generation of zoologists doing experimental research would have wished to move past. But an evaluation of the series as a whole[7] reveals a more complex picture than does one centered on morphological description alone. In the series, marine species are discussed in their ecological, behavioral, and geographic dimensions together with the anatomical, physiological, and embryological factors—all these facets of the research process being relevant to taxonomy.

In the foreword to the series' first volume, Dohrn outlined the programmatic basis for establishing an interdisciplinary research program for the station. The Scientific Fishery was designed to solve scientific problems with methodological principles, the aim being to expand knowledge of marine invertebrates and improve the instruments and techniques used for their investigation. In the course of this process, Dohrn drew on the Italian government's economic and industrial interests to benefit the fishery's program, financing the physiological laboratory exclusively through collaboration with regional and national Italian institutions (Steiner 2018, 93–94).[8] Dohrn described the Scientific Fishery as *systematic* in two senses. First, the marine areas around the Gulf of Naples were to be surveyed in a planned and structured way, allowing the long-term "statistical study of the mode of life of marine animals" (Dohrn 1886, 102). This involved sampling according to predetermined methods, not fishing in a haphazard manner. This was a central concern in Dohrn's founding of the station.

The possibility of obtaining fresh material for laboratory work has until now been considered only from the aforementioned perspective of service. Logbooks, fishing records, and charts from the station serve as historical documentation of the sampling procedure, furnishing information about date and time, meteorological and oceanographic conditions, loci, techniques, and instruments. A key step in this process was a vertical sampling of depths extending to 4,000 meters in the vicinity of the Pontic Islands.

Second, the term *systematic* referred to taxonomy. Systematics is where all the biological subdisciplines practiced at the station came together. Dohrn underscored the distinction between an approach resting on an old-style examination of dead and preserved animals in order to draw conclusions about their "natural condition"[9] and the approach made into a program at the Naples institute. The new approach was marked by an observational and an experimental spectrum covering both the specimen's living state and its natural and artificial environments, the sea, on the one hand, and laboratory and aquariums, on the other. In placing taxonomy at the center of its epistemological interest, the Scientific Fishery integrated all productive scientific methods into its program; recognizing field research as a central element in the station's innovative scientific activities widens the perspective of a historiography tending to focus on the experimental laboratory work carried out as it were behind closed walls and to view the station's fishing activities as basically isolated from that work. As Dohrn intended, then, the precise function of the Scientific Fishery was to unite these two realms.

Data generated in both field and lab formed the basis for investigating the marine organisms. Among the data collections, we find not only measurements of sampling depth and water temperature along with oxygen and light levels beneath the water surface but also records of the multiple stages of specimen examination. When called for, identification and appropriate preservation of the sampled organic materials—often referred to at the station as *materia prima*—took place on board the research boat, a procedure central for the success of the continuously undertaken sampling trips. The individual steps involved here—fishing, sorting, identifying, conserving, and preserving—were carried out by fishermen and laboratory assistants who had received zoological training at the station to this end, together with engineers and the authors of the series contributions, the academically trained zoologists.

In this way, laboratory practices were already deployed in the field, just as the knowledge that the fishermen and the laboratory assistants had accrued through training in preservation methods—the anatomical and histological study (examination of tissue structure) of species with the help of illustrated volumes and direct observation of organic objects—flowed back into the field during the sorting and preserving process, directly after the nets had been pulled up on board. Because of the fragile organic constitution of lower marine animals (many of their anatomical components were transparent and composed of thin but tough tissue), preservation methods developed on dead animals in the laboratory could be brought into the field. Conversely, the living, injured, dead, and preserved organisms, together with ecological notes collected at the places they were found, were brought back to the laboratory and the authors' workplaces. The Scientific Fishery here connected field and lab in a very concrete manner, with the continuity and interaction between tasks, accumulated knowledge, objects, and data in each realm both given an institutional foundation and placed in a programmatic framework. In the process, the practice of random collecting was replaced by systematic-analytic sampling. The program thus counteracted an idea and practice of taxonomy approaching systematics as a merely descriptive morphological undertaking.

In order to meet their goals, those working at the Naples Zoological Station had to overcome certain problems. When the program got under way, research on plankton had been being undertaken for two decades,[10] but the necessary instruments—for example, plankton nets—had not been developed to the degree required for the sort of biologically oriented research initiated in the project's framework. Another problem was posed by a lack of adequate preservation methods. The incapacity to produce first-rate specimens, Dohrn indicated, together with the limited possibilities for examining living specimens, would diminish the researchers' ability to carry out the systematics he championed.

Biological Experiments and Observation

Against the backdrop of transformative processes unfolding in the life sciences since the 1850s, the program's relevance for the establishment of a biological approach becomes clear. Taxonomy in this period had central status. The Fauna und Flora series appeared at a time when the development of systematized nomenclature had become extremely important. This was especially the case not only because it took account of new nineteenth-century theories and methodological approaches but also because prevailing and often simultaneously existing nomenclature codes were not unified through international regulation. Starting in 1890 with the foundation of the German Zoological Society, a scientific board consisting of Karl Möbius, Ludwig Döderlein, and Victor Carus took up the unification of disparate classification schemes, offering recommendations during its annual meeting. Until the institution of international guidelines in the *International Rules of Zoological Nomenclature* (International Commission on Zoological Nomenclature 1905), regional nomenclatures were present in debates about the classification of organisms and species, genera, and families (Melville 1995). In the course of that process of transforming taxonomic practice, Giesbrecht's first volume in the Fauna und Flora series appeared under the title *Systematik und Faunistik der pelagischen Copepoden* (Giesbrecht 1892); this was followed in 1899 by *Asterocheriden* and in 1910 by *Stomatopoden*.

In our context, *Systematik und Faunistik* and *Stomatopoden* can serve as examples of the ways in which practices and products were combined in the Scientific Fishery within the perspective of the new systematics and then integrated into Fauna und Flora. The two volumes, both of which have retained their scientific importance, approach systematics in different ways. In his book on copepods, Giesbrecht undertook a revision of extant systematics with the species in the Bay of Naples as his starting point. In the *Stomatopoden* volume, however, focusing on physiology and behavior, he accepted the systematics that were already in use. His practice in *Systematik und Faunistik* itself is grounded in the approach to nomenclature embraced by the German Zoological Society from 1890 on: "When the 'Entwurf von Regeln für die zoologische Nomenclatur im Auftrage der deutschen zoologischen Gesellschaft' [Outline of rules for zoological nomenclature commissioned by the German Zoological Society] . . . appeared, I saw to my satisfaction that in my volume on the Pelagian copepods of the Gulf of Naples the synonyms were processed in harmony with these rules in all essential points" (Giesbrecht 1894, 87–88).

Giesbrecht's research at the Naples Station invites historians of science to follow the paths of marine specimens through an investigative web laid out by the station's scientific program. This included description of *materia prima* captured at sea as well as its identification and preservation, its manipulation, staining, and magnification, and, finally, its description and illustration. Furthermore, Giesbrecht's volumes in the Fauna und Flora series offer historiographic insight into the series' importance for marine biology in the context of a set of concrete epistemological interests emerging at the Zoological Station around 1900.

The early stages of this German Wilhelmian scientist's career help illuminate his scientific work in Naples. Following his doctoral work with Möbius in Kiel and his PhD dissertation on sea urchins' incisors (Giesbrecht 1880), in 1881, at age twenty-seven, he received on Möbius's recommendation a one-year fellowship to conduct research on copepods at the station's "Prussian workplace,"[11] in connection with a project situated at the University of Kiel on "exploration of the German seas" (Giesbrecht 1884). During this fellowship period, which he devoted to studying genera of parasitic copepods, he was approached by Dohrn concerning the possibility of contributing the above-mentioned Fauna und Flora volume on pelagic copepods, this being tied to receipt of a permanent position at the station. In 1882, Giesbrecht thus began official work there, serving as a scientific assistant until his sudden death in 1913. Over those years, he edited the station's annual report together with Paul Mayer and also functioned as director of the specimen collection and, later, chief editor of Fauna und Flora. In addition, he had produced thirty-five scientific publications at the station, along with many more popular articles treating the institution's activities and southern Italian life. While most of his work was devoted to copepods, in his later years he returned to his academic beginnings through research on the regeneration of sea urchin larvae. All these activities were closely intertwined with his role as one of Dohrn's closest advisers. In this manner, he was deeply engaged in both operative business and conceptual development, including institutionalization of the Scientific Fishery in the context of a projected institute of oceanography tied to the station.

All told, Giesbrecht's research offers a panorama of biological methods in use in the waning nineteenth century and the early twentieth. Much of that research was carried out as part of the Fauna und Flora project, as was mandatory for research associates at the station. The research took place within a field developed by prominent zoologists, hence defined by relevant late nineteenth-century debates. Among these was the controversy about quantitative-qualitative research on plankton (Torma 2014). This was best reflected in a debate between Ernst Haeckel and Victor Hensen concerning fishing nets, part of a wider controversy over sampling methods and a discussion of the objective status of scientific drawing (Haeckel 1891, 857–64; Hensen 1891). We should note that, while Giesbrecht did not participate theoretically in these debates, his empirical research amounted to a significant contribution to them.

In this context, it is notable that use of the net as a metaphor is particularly suited for capturing both the operative and the methodological dimensions of scientific fishing. For we should understand the station's research program not as a primarily linear process but rather in terms of the interconnections of individual elements. While centering on their object of investigation, participants in the research process wove a net consisting of various tools applied to the living, dead, preserved, microscopically observed, and drawn specimen. The products that emerged were epistemic objects, starting with the plankton nets and colored microscope slides and extending to preserved plankton separated into body parts then reassembled as scientifically precise but aesthetically refined drawings (so-called habitus pictures), scientific photographs, and mature animals manipulated with needles. Such products were both the tools for studying and the objects of study.[12]

In order to understand the nature of the practical research dynamic unfolding at the Scientific Fishery better, let us now look at Giesbrecht's work and approach in greater detail. *Systematik und Faunistik* represented the first general revision of copepod systematics since the 1860s. Giesbrecht here applied standards for identifying and defining species that were being discussed in recommendations for nomenclature accompanying the revised, international regulation of zoological systematics. In *Systematik und Faunistik,* he identified species and genus. In doing so, he took a consistently comparative approach, with the result that his analysis also offered an overview of the mutual relationship between species. In his conception of what constituted a species, he aligned his approach with that of his teacher Möbius, avoiding a "split of species into varieties" in favor of drawing demarcations between species when an adequate material basis was present or sources that could undermine the classification could be excluded. "Subconcepts," Giesbrecht indicated, had "practical value: to facilitate an overview of the range of forms"; nevertheless, the "lowest systematic concept" was that of species (Giesbrecht 1892, 86–88; see also Möbius 1886). The illustrations he produced in the course of classifying species remain important for systematics, being drawn on as a reference for new classifications and definitions of synonyms.[13] The German Zoological Society's recommendations for nomenclature specified that the publication of a so-called *Kennzeichnung*,[14] an illustration identifying a species, together with its anatomical description, represented a mandatory reference to new taxonomic identification. The coupling of the "rule of priority" to a visual taxonomic identification in turn determined possible synonyms. In this way, not only did pictorial reception emerge as a central step in taxonomic research, but zoological pictorial production as a whole also became subject to regulation.

Arguably, Giesbrecht made use of this normative function of illustration, as already recommended by Möbius, Carus, and Döderlein. In his Fauna und Flora volumes, we see the reception of extant pictorial material treated as part of an organism's analysis. Published drawings here served as a basis for scrutinizing his own observations, just as they could be deployed as an orientating plan for studying a specimen. Giesbrecht encapsulated this research method in a response to Carl Claus, a fellow zoologist who consistently raised objections to his methods. Referring to the "Entwurf von Regeln für die Zoologische Nomenclatur im Auftrage der deutschen Zoologischen Gesellschaft zusammengestellt von J. Victor Carus, L. Döderlein und Karl Möbius" (Outline of rules for zoological nomenclature, commissioned by the German Zoological Society, compiled by . . .), Giesbrecht indicated: "Since I consider it practical to refer back to old names as much as possible, in the authors' depictions I look for everything that can lead to recognition of the species." If this was unsuccessful, then he subjected an illustration's multistepped production procedure to critical assessment: "If some information of the authors . . . contradicts my interpretation of their species, then I examine whether these contradictions . . . perhaps stem from mistakes . . . caused by inadequate observation, or inadequate technical aids, or poor preservation of the material, etc." (Giesbrecht 1894, 89–90). By means of visual reception, he thus confronted constants of biological research that could have a negative impact on use of the principle of priority and imperil the classification of species and genus and the connected deployment of synonyms.

In this way, the station's research program took current developments in zoology into account and even undertook systematic-faunistic studies before nomenclature was regulated. A central goal was seeing to the successful execution of a range of necessary research steps—including the step of illustration—through the maintenance of a smoothly functioning and innovative technical infrastructure. The station had to develop ways of examining living organisms in order to carry out the new systematics. In line with the approach of Paul Mayer, such examination was meant to precede the specimen's killing, fixing, embedding, cutting, staining, and, finally, study. "This approach," explained Mayer and Arthur Bolles Lee in their textbook *Elements of Microscopic Technique* (originally published in 1898), "is the only correct one particularly in the case of lower animals, to the extent that they are transparent enough—and that applies to many marine animals. For the living animal immediately reveals a great deal that can be ascertained only very laboriously or not at all with specimens. In addition, this is the only approach offering certainty that changes in the tissue have not taken place through manipulation in preserving, cutting, etc.—changes whose dimensions cannot be calculated in advance. Even if a researcher has only a single example of an animal to work with, it is best to start with the living organism" (Mayer and Lee 1910, 1).

In a study that Giesbrecht began with Mayer in 1886 and then continued in 1896, the two zoologists made clear how important the choice was between studying live animals and studying dead specimens. The study, focused on three large living examples of *Euphasia*, involved an effort to verify the hypothesis of James Murray and George O. Sars that the organs identified by Carl Claus and Carl Semper in the *Euphasia* plankton species were not visual but rather luminescent. In an editor's note added to William Patten's article "Eyes of Molluscs and Arthropods," Giesbrecht and Mayer describe their earlier results as follows:

> Isolated thorax eyes, crushed beneath the cover glass during observation with the compound microscope, glowed strongly; very probably the light streamed from the rod bundles. An intact animal was stimulated with ammoniac solution while placed on its back; immediately all the so-called eyes protruded like so many luminous points, while the rest of the body remained dark. For circa half a minute the thorax and belly organs glowed intensely, the organs in the shafts of the compound eyes more weakly; even in the half dark the blue-green points of light were clearly visible with the naked eye. The information of Sars is here confirmed. (Patten 1886, 738)

Carl Chun pursued the assumption that the luminescence comes from the animals' rod bundles, arguing further that the phosphorescent organs known as photosphaerae at whose center the rod bundles are located were best designated as *Streifenkörper*, "striated bodies" (see fig. 6.1). Nevertheless, Giesbrecht observed that this did not constitute proof that the luminescence actually came from the rods and not, as Claus continued to assert, from the adjacent reflector. For, Giesbrecht emphasized, Chun had not "examined the question with the living

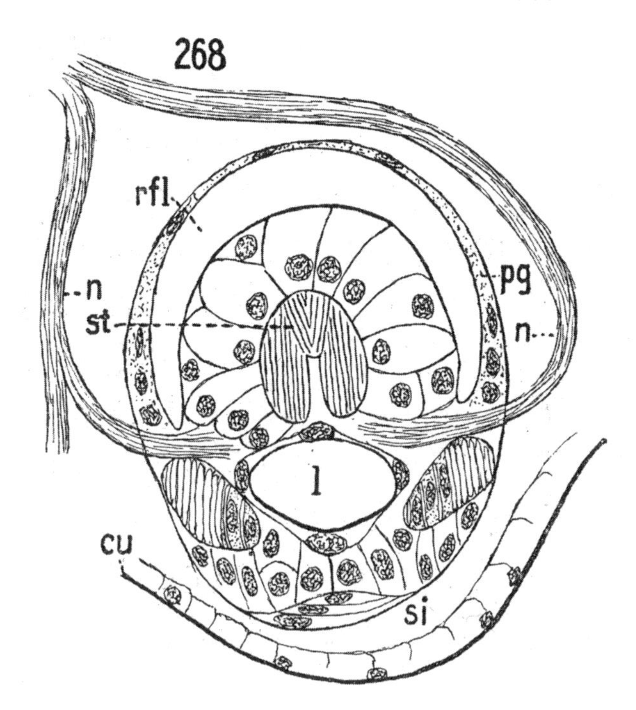

FIGURE 6.1. "Section of a luminescent organ from the thorax of *Nematoscelis* (taken from Chun)," from Giesbrecht (1921, 158). *Nematocelis* is a genus in the family Euphausiidae.

animal" (Giesbrecht 1896, 487). Although Giesbrecht and Mayer had worked with living specimens in 1886, they had merely shown that a *luminescent organ* was at work here, not that the animal had "eyes."

In 1896, Giesbrecht offered proof of the earlier hypothesis regarding the rods. By contrast with his first experiment, he now worked with individual animals in the larval state; here, he could not find anything "in the living or crushed animal . . . either in longitudinal or cross section . . . that could be interpreted as a reflector," while the rods were clearly manifest. But he did identify deep rod cell nuclei in the sections. These flattened cells appeared wrapped in a mantel. If one of the animals was pressed under a microscope or exposed to ammoniac, the rod bundles emitted a blue glow. From the lack (yet) of a reflector in the larval state, Giesbrecht concluded that light was emitted "simply from the rod bundle, the cells that it encloses, and perhaps the flat cells that are wrapped around it." Since the rod bundle was not reduced in size in the fully developed animal but to the contrary grew into a striated body, it was clear to Giesbrecht "that also in the developed bioluminescent assemblage light development is located in the striated body" (Giesbrecht 1896, 490).

Such biological experiments were closely related to a study entitled "Über das Leuchten der pelagischen Copepoden und das thierische Leuchten im Allgemeinen" ("On the luminescence of Pelagian copepods and animal luminescence in general") (Giesbrecht 1895, 648–89) that Giesbrecht carried out following his *Systematik und Faunistik*. In that volume, Giesbrecht had not yet pointed to the bioluminescence phenomenon; the start of its study was marked by his article on the "pigment head" of *Pleuromma*, sparked by a discussion with Friedrich Dahl (Giesbrecht 1893a, 212–13; Giesbrecht 1895, 690–94). The phenomenon of copepod luminescence was here explained as a biological function[15] used to threaten predators. Giesbrecht's investigation of luminescence incorporated anatomical and physiological factors as well as behavior, informed by physiological experiments and morphological observations.

Biological characteristics of a species raised taxonomic questions. As was also the case with *Euphasia*, evidence of luminescence and its source stood at the forefront, but, here, this was located in a framework of a systematics that reappraised the relation of a species' phosphorescent capacity to that of other species in the same genus. Giesbrecht thus called into question the assumption that "almost all crustaceans living on the ocean's surface glow," for, if that were the case, then "it would be very striking that the capacity for luminescence has until now been verified through observation of too few species" (Giesbrecht 1895, 648). In his study, from a broad sample of copepods Giesbrecht collected five species for which he could observe this phenomenon in the Gulf of Naples: *Pleuromma abdominale*, Lubbock; *Pleuromma gracile*, Claus; *Leuckartia flavicornis*, Claus; *Heterochaeta papilligera*, Claus; *Oncaea conifer*, Giesbrecht. "I identified these species," he indicated, "by pouring the daily sampling through a gauze filter in small portions . . . then using a needle to collect the animals that were already glowing as soon as the water had run out, holding the filter with the remainder in the dark box over ammoniac vapors" (Giesbrecht 1895, 653). As a result of the Scientific Fishery's thorough sampling of the Tyrrhenian Sea, he was able to show not only that not all copepod species were luminescent but also that most species were not. He conducted his sampling using different kinds of nets, two of which he designed and constructed together with an engineer and the zoologically trained fisherman and chief of sampling cruises Aniello Fortanosa. This provided the basis for his hypothesis and follow-up experiments. The Giesbrecht net, a type of closing net, made it possible to sample at different depths and provided the basis for Giesbrecht's hypothesis and follow-up experiments (Giesbrecht 1893b).

In this study, the darkroom (Müller 1876, 120)[16] served as a laboratory so that reflection from the copepod's chitin shell could be avoided. Giesbrecht identified the number and loci of the luminescent glands by squishing the living specimens under a cover glass plate and stimulating the organism with heat treatments. Later, in an article for Arnold Lang's *Crustacea: Handbuch der Morphologie der wirbellosen Tiere,* he defined the photosphaerae organs by contrasting them with skin glands. Both, he indicated, have the function of light organs. But, while the glands emitted luminous material, the photosphaerae were "nodula organs that glow in the interior" and emitted no such material (Giesbrecht 1921, 158). In addition, the photosphaerae were more complex than the skin glands and their ontogenesis was unclear. His explanation of the luminescence is still accepted and has been built on by recent biologists (Bannister and Herring 1989, 523).

The results of the "experimental study of the process of luminescence" offered empirical material for solutions to the following questions: "Is the glowing of the luminescent material a physiological process similar to contraction, assimilation, and other processes appearing exclusively in the living protoplasm, or does luminescence rather have no direct connection with the activity of living material and is only an effect accompanying a purely chemical or physical process: a process the conditions for whose existence have indeed been created by physiological processes but that is itself not physiological?" If the latter were the case, Giesbrecht then asked, "through what physiological processes does luminescent material develop," and "what is the physical or chemical process itself that announces itself to the eye through the development of light?" (Giesbrecht 1895, 653). He concluded that physiological processes provide the grounds for the production of luminescent substances; luminescence itself, however, was mainly part of a chemical process and in some cases perhaps also a physiological one. The study of such a biological process could be conducted only in an interdisciplinary manner; in this particular case, physiologists, chemists, and morphologists had to work together cooperatively (Giesbrecht 1895, 687–88). This example shows us the way in which experimentation was integrated both into the larger program of systematics and into the Scientific Fishery.

Just as Giesbrecht took account of illustrations made by colleagues when he was identifying a species, he also developed his drawings during a process of reception involving observation, assessment of published illustrations, and comparison of research objects (both living and dead). Although he acknowledged the subjective component of seeing and observing, it would thus be inaccurate to describe the drawings as subjective representations. In the illustrations, the different stages that the *materia prima* went through to form the object of study were together; at the same time, he made use of practices and methods bringing together draftsmanship with specimen study. For systematic biological study, sketches of living and dead organisms were indispensable, so the objects were drawn in a series of observational stages before the specimen was fixed on a microscope slide and its illustration published. The stages included the following: objects examined at sea after sampling, colored microscope slides, aquarium-bred larvae, preserved plankton separated into body parts then put together again as scientifically precise but aesthetically refined drawings (so-called habitus pictures), and mature animals manipulated with needles. This constituted the above-mentioned investigative web: the station's particular contribution to marine-biological practice.

Giesbrecht worked with two different pictorial genres: schematic drawings and habitus pictures. Sketches for producing these images emerged during the specimen's analytic dissection into individual parts and then its reintegration into a complete organism. Against the backdrop of the Scientific Fishery as a program grounding these individual research steps, the habitus picture took on a central position, for the intertwining of field and laboratory was not only a dimension of applied research but also immanent to that type of image itself. In this framework, we can consider the habitus picture as a model encapsulating specific characteristics of a species emerging in its various stages of study.

To illustrate the function of such images as models, it is useful to consider the individual stages, manifest as anatomical and histological features of an organism illuminated by detailed drawings before a final illustration brought the parts together. Living specimens were observed in the water to record, for example, their coloration but also, as mentioned, to take account of the differences between living, dying, dead, and preserved animals. Dissection took place with use of a microtome. Different individuals from a single species were killed in alcohol and sectioned in specific ways—depending on research interest and the species or genus being studied, use was made of ammoniac and other suitable substances (Giesbrecht 1881a, 1881b). By means of a camera lucida, this step (executed on both the nonpreserved and the preserved specimens) was recorded in schematic drawings prepared with the help of other published illustrations, the goal being not only eventual preparation of a final illustration but also identification of a species. Photographs served as an aid in this process,[17] but they could not do justice to the fine details of anatomical structures or their histological constitution. In the schematic drawing, individual body parts were depicted from a dorsal, lateral, and ventral perspective before being reconstructed as an entire organism in the habitus picture; by displaying bones, tissue, and pigment, this type of picture marked a contrast with the earlier, schematic drawings, which displayed an essentially flat surface. The habitus picture was not developed on the basis of an individual animal but reconstructed on the basis of sections of different individual animals.

At the same time, such pictures presented natural characteristics of a species in an artificial situation, and they were also accompanied by observations made in the field. Giesbrecht thus reported that species belonging to the *Sapphirina* genus were visible to attentive observers from the boat in calm seas because of the "color-play of the male" (Giesbrecht 1892, 621; see fig. 6.2). He conducted studies with males of both living and nearly dead specimens, working with an artificial external light source to catalyze the opalescence. He observed the living specimen from different angles in a lab glass but also under a microscope. When viewed statically from a dorsal perspective, *Sapphirina ovatolanceolata* appeared colorless. Its opalescent colors appeared only when the object was rotated around its long axis (drawing 7 in fig. 6.2). As the specimen was drying out and slowly dying, its colors dulled, and it became pale (drawing 8 in fig. 6.2), turning yellow when dead (Giesbrecht 1892, 632).[18] There was, however, no adequate drawing technique for rendering the luminescence of visualizing *Pleuromma gracile* and *Pleuromma abdominale.*

In comparing specimens to identify species and synonyms, Giesbrecht, it is worth emphasizing again, did not limit himself to, in his words, "simply the description and drawings of authors" but also took into account what they wrote about the organism's "habitat, appearance, way of life, etc. . . . for characterizing a species" (Giesbrecht 1894, 90). In proceeding this way, he was taking up another point that would be central for the *International Rules of Zoological Nomenclature.* Zoogeographic data were visible in the habitus picture in that knowledge about both horizontal and vertical habitat was crucial for determining both a species and a synonym. Giesbrecht argued that synonyms reflect that data. In establishing synonyms, authors not only accepted the validity

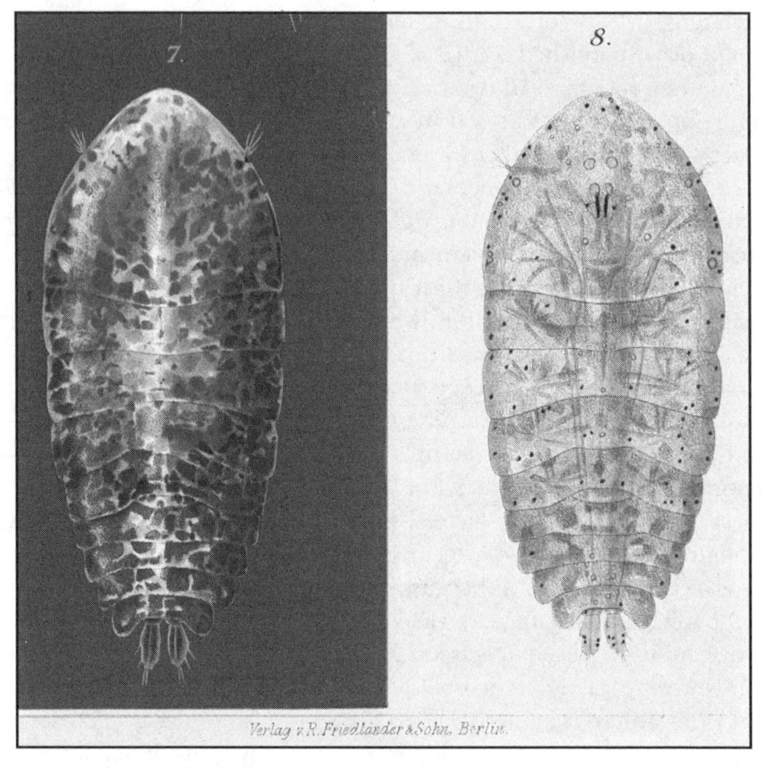

FIGURE 6.2. "*Sapphirina ovatolanceolata*, male, dorsal," from Giesbrecht (1892, plate 1, figs. 7 and 8).

of the oldest name but also acknowledged zoogeographic facts. Given the challenges involved in identifying specimens from different world regions, Giesbrecht saw the question of whether something was a new species or a synonym for an existing species as minor. The bigger, undecided question was whether geographic differences were relevant in deciding whether a specimen constitutes a new species or a synonym (Giesbrecht 1894, 90–91).

Conclusion

In the late nineteenth century, the Naples Zoological Station combined traditional systematics with physiological and experimental approaches in the framework of its Scientific Fishery research program. The program's scientific and organizational orientation allowed combined exploration of physiological and ecological questions under the rubric of classification of species. A team composed of permanently employed scientists, technicians, and zoologically trained fishermen and laboratory assistants together sampled and did research in the Gulf of Naples and adjacent marine areas; in doing so, they also contributed their own specific knowledge and experience to the Fauna und Flora series of scientific studies initiated in the framework of the Scientific Fishery program.

Appearing between 1880 and 1926, this series offered a unique window into the world of modern biology. It both took account of the unification of zoological nomenclature that had been under way since the 1880s and pragmatically realized it through the program. In this manner, the series documented a central late nineteenth-century development—one exemplified, I have argued, in the career of Wilhelm Giesbrecht. On the one hand, this development was manifest in pictorial practice: while for a long time illustration had served merely as something like a prop accompanying the verbal classification of species, from 1905 on it was granted normative status in the official *International Nomenclature* guidelines. On the other hand, the series embraced an interdisciplinary methodology, incorporating disparate material that took on significance—for instance, in the areas of physiology and ecology—in the emerging discussion of how species and genera were meant to be classified.

In the decades following the 1880s, other institutes of marine biology were established on the Neapolitan model along the coasts of Europe, North America, and Asia. Nevertheless, I would argue that the station remained unique in several ways. First, with the inception of the Fauna und Flora project, it developed organizational and technical structures to such an extent that repeatedly new developments were possible in the course of studying newly discovered species; Giesbrecht's sampling net and his methods of both preservation and drawing are examples of this innovative potential. Second, the station served over a very long period as the crystallization point for a network bringing together not only generations of scientists but also their different disciplines—zoology, botany, physiology, morphology—and schools. It thus possesses an extraordinary potential for multidimensional knowledge transfer, which could be made useful for the research structure pioneered by Anton Dohrn. Third, with the institute's focus on a new research object, the marine invertebrate, a new program entered the agenda of scientists pursuing a spectrum of disciplines. The rigorous study of the appearance and way of life of these tiny marine animals, their characteristics and physiology, work defining the interdisciplinary approach taken at the Naples Station, would come to exemplify a nexus of good scientific theory and practice in the years leading up to World War I and beyond.

Notes

This essay builds on two conference papers ("Depicting Species: The Role of the Image in Modern Biology" [History of Science Society, 2016] and "Scientific Fishery: Social Organization and Working Culture at the Naples Zoological Station 1880–1913" [History of Science Society, 2017]). Focusing on the intertwining of field and laboratory in the framework of a research program with interdisciplinary orientation, it was developed in parallel with a paper on that program's social organization and institutionalization.

1. *Reichsbericht* 1881, MAB Stazione Zoologica Anton Dohrn di Napoli—Archivio Storico (henceforth MAB SZN-AS), R 1881, p. 48.
2. The series was published in thirty-seven volumes between 1880 and 1926. The uninterrupted appearance of the volumes over those years was made possible by emigration during World War I of the station's entire administration, including those responsible for publications, to the University of Zurich's Zoological Institute. By 1885 at the latest, the institute had changed presses while retaining the firm responsible for the series's graphics and lithography throughout its years of publication. Production costs were covered by both the basic budget and outside funding.
3. *Reichsbericht* 1880/81/82, MAB SZN-AS, R 1880–1882, pp. 25–70.
4. Until 1893, Dohrn would report regularly on the program in the *Mittheilungen aus der Zoologischen Station zu Neapel*. Later, Giesbrecht's article on developments at the station (Giesbrecht 1907) would appear in the *Deutsche Rundschau*. But Wilhelm His, Victor Hensen, and others also publicized the Scientific Fishery program for both scientific and lay audiences.
5. Payment roll, MAB SZN-AS, M. XVII staff, a 6–7.
6. Although studies were also meant to focus on phytoplankton, only one saw print (Berthold 1882). The term *plankton* was coined in 1887 by the University of Kiel physiologist Victor Hensen; it runs across systematic-taxonomic categories. According to Hensen, plankton life-forms were organisms moving in water without any form of self-propulsion. This definition remains in force.
7. I began such an evaluation in the framework of my doctoral work on Wilhelm Giesbrecht and the Zoological Station and have continued it in a postdoctoral project titled "Depicting Species: The Role of the Image in Modern Biology."

8. The extent to which the exchange with the Italian ministry responsible for education and agriculture, Naples's provincial administration, and the consortium of southern Italian marine provinces took account of the scientific knowledge accumulated by the Scientific Fishery and made it useful for industry still needs researching.
9. Zoology students learned "to see, i.e., to observe and do research" from textbooks, for "starting with catching methods, from the culture and killing of the animals to the complicated forms of macroscopic and microscopic preparation . . . a true assessment of what is observed [is] . . . possible only when we are familiar with all the effects of what is undertaken, when observations are made in a lasting way . . . in order to judge what corresponds to the natural condition . . . and what is to be attributed to the undertaken preparation" (Schuberg 1910, 7).
10. This research began with Johannes Müller and remains a basic dimension of biological oceanography. Like the term *biological oceanography* itself, its techniques and practices point to an interdisciplinary research field focused on in Kiel starting in the late 1860s and carried forward in the framework of the Scientific Fishery in Naples. The station's connections with Kiel have not been sufficiently examined (see Mills 1989, 9–188; and Rozwadowski 2005).
11. On the organizational principle of these workplaces for guest researchers, see Groeben (in this volume).
12. On epistemic objects, see Rheinberger (1997).
13. See, e.g., the Species Card for *Pleuromma gracilis*, Claus 1863, on the Marine Planctonic Copepod Web site: https://copepodes.obs-banyuls.fr/en/fichesp.php?sp=1000.
14. In both the English and the French versions of the *Rules*, *Kennzeichnung* was rendered as *indication*, *definition*, or *description*. The association of *illustrations* with *Kennzeichnung* was specific to the regulations' German version, and this was not accidental. In the recommendations of the German Zoological Society, a species' author should ideally return to the oldest description that includes an illustration. At the same time, a species' new description was tied to the existence of an illustration.
15. Biological functions were relevant for taxonomy. In contrast to the volume on *Asterocheriden* and other volumes in the series, *Systematik und Faunistik* has no separate *Biologisches* rubric in the table of contents; instead, the theme is integrated into the diagnostic section. This is connected with Giesbrecht's intention to devote a second volume of *Pelagischen Copepoden* to *Biologisches*. He appears never to have written the volume.
16. Müller offers a list of instruments and laboratory units in use.
17. In his published scientific work, Giesbrecht used mainly lithographs, not photographs. But he intensely engaged in amateur streetscape photography alongside his zoological research (Steiner 2013).

18. With the exception of one plate and one figure in his study of stomatopodes, Giesbrecht drew all the illustrations in his scientific publications himself. In *Systematik und Faunistik*, fig. 7 on plate 1 (here, drawing 7 in fig. 6.2)—and only that figure—was colored by Comingio Merculiano, the station's employed craftsman and artist.

References

Bannister, Neil J., and Peter J. Herring. 1989. "Distribution and Structure of Luminous Cells in Four Marine Copepods." *Journal of the Marine Biological Association of the United Kingdom* 69, no. 3:523–33.

Berthold, Gottfried. 1882. *Die Bangiaceen des Golfes von Neapel und seiner angrenzenden Meeresabschnitte*. Fauna und Flora des Golfes von Neapel und seiner angrenzenden Meeresabschnitte, vol. 8. Leipzig: Wilhelm Engelmann.

de Bont, Raf. 2015. *Stations in the Field: A History of Place-Based Animal Research, 1870–1930*. Chicago: University of Chicago Press.

Dohrn, Anton. 1880. "Editor's Note." In *Die Ctenophoren des Golfes von Neapel und der angrenzenden Meeresabschnitte* (Fauna und Flora des Golfes von Neapel und seiner angrenzenden Meeresabschnitte, vol. 1), by Carl Chun, v–x. Leipzig: Wilhelm Engelmann.

———. 1886. "Bericht über die Zoologische Station während des Jahres 1882–1884." *Mittheilungen aus der Zoologischen Station zu Neapel zugleich ein Repertorium für Mittelmeerkunde* 6, no. 1:93–148.

Giesbrecht, Wilhelm. 1880. "Der feinere Bau der Seeigelzähne." *Morphologisches Jahrbuch: Eine Zeitschrift für Anatomie und Entwicklungsgeschichte* 6, no. 1: 79–105.

———. 1881a. "Methode zur Anfertigung von Serien-Präperaten." *Mittheilungen aus der Zoologischen Station zu Neapel zugleich ein Repertorium für Mittelmeerkunde*, nos. 1–2:184–86.

———. 1881b. "Zur Schneidetechnik." *Zoologischer Anzeiger* 4, no. 92:483–84.

———. 1884. "Die freilebenden Copepoden der Kieler Föhrde." In *Vierter Bericht der Commission zur wissenschaftlichen Untersuchung der deutschen Meere*, ed. Heinrich Adolf Meyer, Karl Möbius, Gustav Karsten, Victor Hensen, and Adolf Engler, 87–168. Berlin: Paul Parey.

———. 1892. *Systematik und Faunistik der pelagischen Copepoden des Golfes von Neapel und seiner angrenzenden Meeresgebiete*. Fauna und Flora des Golfes von Neapel und seiner angrenzenden Meeresabschnitte, vol. 19. Berlin: Friedländer.

———. 1893a. "Über den einseitigen Pigmentkopf von Pleuromma." *Zoologischer Anzeiger* 16, no. 421:212–13.

———. 1893b. "Zur Entwicklung des Schliessnetz." *Mittheilungen aus der Zoologischen Station zu Neapel zugleich Repertorium für Mittelmeerkunde* 11, nos. 1–2:306–24.

———. 1894. "3. Bemerkungen zu Claus' neueren Arbeiten über Copepoden-Familie der Pontelliden." *Zoologischer Anzeiger* 17, no. 442:87–95.

———. 1895. "Mittheilungen über Copepoden 7–9." *Mittheilungen aus der Zoologischen Station zu Neapel zugleich ein Repertorium für Mittelmeerkunde* 11, no. 4:631–94.

———. 1896. "Über den Sitz der Lichtentwicklung in den Photosphaerien der Euphausiiden." *Zoologischer Anzeiger* 19, no. 519:486–90.

———. 1899. *Die Asterocheriden*. Fauna und Flora des Golfes von Neapel und seiner angrenzenden Meeresabschnitte, vol. 25. Berlin: Friedländer.

———. 1907. "Altes und Neues von der Zoologischen Station in Neapel." *Deutsche Rundschau* 33:76–92.

———. 1910. *Stomatopoden*. Fauna und Flora des Golfes von Neapel und seiner angrenzenden Meeresabschnitte, vol. 33. Berlin: Friedländer.

———. 1921. "Arthropoda." In *Crustacea: Handbuch der Morphologie der wirbellosen Tiere 4*, ed. Arnold Lang, 9–239. Jena: Gustav Fischer.

Groeben, Christiane. 2002. "The Stazione Zoologica: A Clearing House for Marine Organisms." In *Oceanographic History: The Pacific and Beyond*, ed. Keith Benson and Philip F. Rehbock, 537–48. Seattle: University of Washington Press.

Haeckel, Ernst. 1891. *Anthropogenie und Entwicklungsgeschichte des Menschen: Keimes- und Stammesgeschichte*. Leipzig: Wilhelm Engelmann.

Hensen, Victor. 1891. *Die Plankton-Expedition und Haeckels Darwinismus: Über einige Aufgaben und Ziele der Naturwissenschaften*. Kiel: Lipsius & Tischer.

International Commission on Zoological Nomenclature. 1905. *International Rules of Zoological Nomenclature*. Paris: Rudeval.

Mayer, Paul, and Arthur B. Lee. 1910. *Grundzüge der mikroskopischen Technik für Zoologen und Anatomen*. Berlin: Friedländer.

Melville, Richard V. 1995. *Towards Stability in the Names of Animals: A History of the International Commission on Zoological Nomenclature, 1895–1995*. London: ICZN.

Mills, Eric. 1989. *Biological Oceanography: An Early History*. Ithaca, NY: Cornell University Press.

Möbius, Karl A. 1886. *Die Bildung, Geltung und Bezeichnung der Artenbegriffe und ihr Verhältnis zur Abstammungslehre*. Jena: Gustav Fischer.

Müller, Irmgard. 1976. "Die Geschichte der Zoologischen Station in Neapel von der Gründung durch Anton Dohrn (1872) bis zum Ersten Weltkrieg und ihre Bedeutung für die Entwicklung der modernen biologischen Wissenschaften." Habilitation diss., University of Düsseldorf.

Nyhart, Lynn K. 2009. *Modern Nature: The Rise of the Biological Perspective*. Chicago: University of Chicago Press.

Patten, William. 1886. "Eyes of Molluscs and Arthropods." *Mittheilungen aus der Zoologischen Station zu Neapel zugleich ein Repertorium für Mittelmeerkunde* 6, no. 4:542–756.

Rheinberger, Hans-Jörg. 1997. *Toward a History of Epistemic Things: Synthesizing Proteins in the Test Tube*. Stanford, CA: Stanford University Press.

Rozwadowski, Helen M. 2005. *Fathoming the Ocean: The Discovery and Exploration of the Deep Sea*. Cambridge, MA: Harvard University Press.

Schuberg, August. 1910. *Zoologisches Praktikum*. Leipzig: Wilhelm Engelmann.

Steiner, Katharina. 2013. "Die Alben von Wilhelm Giesbrecht: Private Fotosammlung und Institutionsarchiv—Ein Blinder Fleck im neapolitanischen Bilderkanon." In *Fotografie und Film im Archiv: Sammeln, Bewahren und Erforschen*, ed. Ulrich Hägele and Irene Ziehe, 163–78. Münster: Waxmann.

———. 2018. "Copepods and Fisherboys: Advanced Biological Research and Street Poverty in the City of Naples." In *Urban History of Science: Making Knowledge in the City, 1820–1940*, ed. Oliver Hochadel and Agustí Nieto-Galan, 80–101. London: Routledge.

Torma, Franziska. 2014. "Wissenschaft, Wirtschaft und Vorstellungskraft: Die 'Entdeckung' der Meeresökologie im Deutschen Kaiserreich." In *Weltmeere: Wissen und Wahrnehmung im langen 19. Jahrhundert*, ed. Alexander Kraus and Martina Winkler, 25–45. Göttingen: Vandenhoeck & Ruprecht.

SEVEN

A Dual Mission: Research and Education as Critical Factors for the Scientific Integrity of the Marine Biological Laboratory

KATE MACCORD

The Marine Biological Laboratory (MBL) in Woods Hole, Massachusetts, was founded during a wave of educational reform in American biology (Pauly 1984). Loose university curricula and lack of scientific training facilities and courses for American undergraduates and teachers led to the establishment during the 1870s of a number of seaside schools (Pauly 1984). These schools, such as the Anderson School of Natural History on Penikese Island and the Annisquam School, both also in Massachusetts, were aimed at acquainting teachers and other interested members of the public with the natural world around them (see Maienschein, in this volume). While these natural history schools served to promote better understanding of natural science, they did not address the problem of science teaching at the university level (Benson 1988). In tandem with the wish to reform scientific education and understanding among the public arose the desire to create laboratories for research akin to the Stazione Zoologica in Naples, founded by Anton Dohrn in 1872 (Maienschein 1985b).

The MBL was founded in 1888 and followed by the establishment of a number of other marine laboratories in the United States, like Hopkins Station in California (founded 1892) and Friday Harbor in the state of Washington (founded 1903), that sought to promote seaside research and education. Many of the more well-known European marine stations, like the Stazione Zoologica in Italy, Roscoff in France (founded 1872), and Plymouth in England (buildings opened in 1888), had places for students and offered limited courses but did not treat education as an equal to research.[1] The newly minted American stations served dual purposes; however, only the MBL has been able to maintain both the research and the education components of its mission throughout the many years since it first opened its doors.

Courses have been continuously offered at the MBL from its inaugural year. Over the decades following its founding, the numbers and prestige of MBL investigators grew, and the MBL persevered in providing extensive courses for a national, and increasingly international, body of students. It promoted the inclusion and education of women and minorities within its courses, all while the dual missions of many other US marine stations fell out of balance or crumbled altogether.[2] Despite the continued success and continuous expansion of courses offered at the MBL throughout its history, it was not a given that the pedagogical commitments of the institution would succeed. How and why did the MBL find a way to continue to grow and enhance its educational efforts and research despite periods of financial instability and internal division and as other marine stations chose to bolster one or the other? The answer lies in the history of the founding of the MBL and the vision of its first director, Charles Otis Whitman, as well as in the nature of the courses that developed at the MBL. In particular, for Whitman, research and education were inextricably linked.

This essay begins with the origins of the MBL and Whitman's vision for a national marine laboratory wherein instruction and investigation were intertwined and mutually beneficial, focusing on some of the struggles that Whitman encountered while cementing his vision into the foundation of the MBL. Following discussion of Whitman, a brief review of the embryology course from its founding in 1893 serves as a case study in how Whitman's understanding of the connectedness of research and education played out and how education in conjunction with research has been central to the scientific integrity of the MBL.

C. O. Whitman, a National Laboratory, and a Mission "Twice Blest"

Woods Hole, located at the southwestern tip of Cape Cod, is a unique environment. The warm waters from the Gulf Stream to the south mingle with the cold waters coming from Newfoundland to the north in the oceans and bays surrounding the Cape. The mix of these currents produces a rich biodiversity of marine life. Additionally, despite the large number of freshwater ponds and streams throughout the Cape, Woods Hole maintains a very steady year-round level of water salinity, making it an ideal place to conduct research on marine species (Galtsoff 1962). These factors made it the prime location for a research station of the US Fish Commission in the 1870s (Galtsoff 1962; Allard 1990). Because of the proximity of the commission, access to a wide range of marine organisms, and the perfect conditions in which to conduct experiments, Boston-area philanthropists and scientists, including the Boston's Women's Educational Association, chose Woods Hole to build a marine laboratory after the demise of the Anderson and Annisquam schools (Benson 1988; MacCord and Maienschein 2018).

In the agreement of association that founded the institution, signed by the board of trustees, the MBL officially assumed a mission "to establish and maintain a laboratory or station for scientific study and investigation, and a school for instruction in biology and natural history" (MBL 1888a). The aim of the founders was to build a national laboratory for both research and education, one that relied on the cooperation of academic institutions from across the United States. After some deliberation, Charles Otis Whitman, the director of the Allis Lake Laboratory in Milwaukee, Wisconsin, and soon-to-be head of the Biology Department at Clark University, was chosen to serve as the first director.

The MBL opened its doors on July 17, 1888—late in the season by modern standards—with an address from Whitman that laid the foundation for his views of the importance, and even necessity, of the marriage of instruction and investigation. Previous US marine labs (such as Anderson and Annisquam), he noted, failed to advance these two concerns because they had privileged one, usually education, over the other. One distinguished scientist who had written to Whitman about the opening of the MBL relayed the problem thus: "I have no sympathy with anything merely devoted to elementary instruction; and unless the greater part of the energy is given to original work, it is of no interest to me" (MBL 1888b, 28). Whitman understood that there was only the desire to avoid an imbalance between instruction and investigation, not hostility to the educational aspects of the MBL. Because the most important thing to Whitman was an "organized corps of investigators," he realized that the key to maintaining balance was to recognize that instruction and investigation were both necessary and mutually enforcing (MBL 1888b, 27). Why were these two objectives, which could easily have torn the institution in very different directions, necessarily linked for Whitman? He tells us:

> Other things being equal, the investigator is always the best instructor. The highest grade of instruction in any science can only be furnished by one who is thoroughly imbued with the scientific spirit, and who is actually engaged in original work. Hence the propriety—and, I may say, the necessity—of linking the function of instruction with that of investigation. The advantages of so doing are not by any means confined to one side. Teaching is beneficial to the investigator, and the highest powers of acquisition are never reached where the faculty of imparting is neglected. Teaching is an act twice blest, it blesseth him that gives and him that takes. To limit the work of the Laboratory to teaching would be a most serious mistake; and to exclude teaching would shut out the possibilities of highest development. The combination of the two functions in mutually stimulating relations is a feature of the Laboratory to be strongly commended. (MBL 1888b, 16–17)

Over the brief course of the six weeks of the MBL's first season, the institution hosted seven investigators (including four women) and eight students (including four women) in its first course. These investigators and students traveled to Woods Hole from women's colleges, universities, and high schools throughout the Northeastern and Midwestern United States—an amazing feat given that the first circulars announcing the opening of the lab were not sent out to colleges until June of that year.[3] Independent investigators worked at private tables, while students learned in a shared lab space in the same single building that constituted the MBL in its opening year. The first course, zoology, was designed to carry on the tradition of the earlier Anderson and Annisquam schools, providing introductory lectures and lab work for students and high school teachers. Instruction in the course was carried out by Balfour H. Van Vleck and consisted mostly of the study of the structure and life history of a variety of marine organisms and histological techniques (MBL 1888b). While Van Vleck was the primary instructor, other investigators at the laboratory aided in teaching. Students during the first year paid $25.00 for their place in the course, and independent investigators paid $50.00 for a private table. The atmosphere of that first summer season reflected Whitman's desire to integrate education and research—the initial group of investigators worked in close quarters with students (MBL 1888b).

Reflecting on the first year of operation, Whitman, who served as director for over twenty years, was optimistic about the possibilities that the new laboratory had to offer the scientific community. His vision, which he remarked on at length in the first annual report, was to expand the laboratory, providing both more courses and more space for research to flourish (MBL 1888b). The working proximity of the attendees and the involvement of investigators within the courses acted, in Whitman's view, as a sort of crucible of science, refining the free flow of ideas between students and investigators toward answering the big questions in biology and training the next round of brilliant young minds. After all, as Whitman noted frequently over the years, one of the main objectives in founding the MBL was, in an era of increasing disciplinary specialization, to create a space where all researchers in the biological sciences (both students and established investigators alike) could convene their diverse perspectives around these big questions. Whitman noted: "I have in mind, then, not a station devoted exclusively to zoology, or exclusively to botany, or exclusively to physiology; not a station limited to the study of marine plants and animals; not a lacustral station dealing only with land and fresh-water faunas and floras; not a station limited to experimental work, but a genuine biological station, embracing all these important divisions, absolutely free of every artificial restriction" (Whitman 1898, 42).

Over the course of the next few seasons, attendance of investigators and students at the MBL grew, and new buildings were added to make way for the growing population of scientists. In 1893, the MBL became financially self-supporting for the first time, but this financial situation was short-lived.[4] Compelled by the constant increase in attendance, Whitman continued to order construction of new facilities and improvements to the old ones, leading the institution back into a deficit in 1896. Reacting to this reversal of fortunes and Whitman's unwillingness to listen to their cautions, the board of trustees demanded that the laboratory be closed for the 1897 season unless funds could be raised in excess of the deficit (for further details, see Lillie [1911]). Whitman and several trustees on his side raised the money, and, following some internal political machinations, the government of the MBL was revolutionized. The seat of power, which had formerly been Boston (where many of the trustees, who were not scientists, lived), moved to Woods

Hole in 1897, and the institution became a truly democratic one. Trustees were now elected largely from the body of MBL investigators, and the corporation (open to all MBL investigators) gained power over the future of the institution (Maienschein 1985a).[5]

During this time of uncertainty, in addition to expanding the MBL's facilities, Whitman also expanded its educational offerings. Courses in botany, physiology, and embryology were added, designed to teach largely university students both the fundamental details within the field and the methods necessary to conduct their own future investigations. In so doing, these courses addressed some of the concerns with scientific training that was not a part of earlier seaside schools (Benson 1988). To prepare students to enter these emerging fields in biology, the advanced courses drew heavily on the community of summer investigators at the MBL to serve as either instructors or lecturers, further connecting research and education within the daily operations of the lab.[6] As recompense for their teaching duties, instructors were given a stipend to cover their living expenses and access to research space (Lillie 1944).

In spite of the continued growth and success of the MBL on a limited budget[7] as well as the advent of true scientific democracy and intellectual freedom at the lab, tensions arose over the dual nature of the MBL's mission. The intimate and necessary connection between research and education that Whitman envisioned was, it seems, not the ideal of every member of the MBL community. Following further financial hardship at the turn of the twentieth century, the trustees sought outside financial support and found a possible marriage of convenience with the Carnegie Institution, located in Washington, DC. In addition to scuttling the freedom of the MBL to decide its own fate, there was one other major problem—the Carnegie Institution required that the courses be eliminated (Lillie 1944). Some of the MBL trustees, including Edmund Beecher Wilson and Thomas Hunt Morgan, were delighted at the prospect of a well-funded *research* institution and pushed for the merger to go through (Whitman 1902a). A well-publicized scuffle ensued in the pages of *Science* in October of 1902 between Whitman and Wilson, and Whitman rallied support for rejecting the Carnegie Institution's terms (Whitman 1902a, 1902b, 1902c). The initial proposal, which would see the MBL become a research-only branch of the Carnegie Institution, fell through. Whitman won, cementing a vision that has carried through to this day of the MBL as a research and educational institution.

Scientific Integrity: Research, Education, and the MBL Embryology Course

Whitman believed that the scientific integrity of the MBL rested on research and education. For him, these were not dual pillars; they were an integrated foundation. In order better to understand how research and education were both connected and integral to advancing science at the MBL, I now turn to the MBL's embryology course. This course was first offered in 1893 and continues to be taught every summer. In many ways, it represents the heart of the MBL; more scientific and institutional leaders of the MBL have been involved with the course throughout its history as students, lecturers, instructors, and even course directors than any other MBL course (table 7.1).[8] And, just as the MBL has acted as a *biological station* in Whitman's sense of embracing all fields and questions related to life, the field of embryology has

Table 7.1 Embryology course directors from 1893 to 2019

Years	Name(s)
1893–98	Charles O. Whitman*
1899–1903	Frank R. Lillie*
1908–14	Gilman A. Drew
1915–18	William E. Kellicott
1919	Gilman A. Drew
1920–21	David H. Tennent
1922–41	Hubert Goodrich
1942–45	Viktor Hamburger
1946–50	Donald Costello
1951–55	S. Meryl Rose
1956–61	Mac Edds Jr.
1962–66	James Ebert*
1967–71	Malcom Steinberg
1972–74	Eric Davidson
1975–77	David Epel, Tom Humphreys
1978–79	Joan Ruderman,* Tom Humphreys
1980–82	Rudy Raff
1983–87	Bruce Brandhorst, William Jeffrey
1988–91	Eric Davidon, J. Richard Whittaker
1992–96	Eric Davidson, Michael Levine, David McClay
1997–2001	Marianne Bronner-Fraser, Scott Fraser
2002–6	Richard Harland, Joel Rothman
2007–11	Lee Niswander, Nipam Patel*
2011–15	Richard Behringer, Alejandro Sánchez Alvarado
2015–19	Richard Schneider, David Sherwood

*Served as director of MBL

embraced diverse methods and perspectives, including morphology, physiology, cell biology, genetics, and even evolution, throughout its history. The embryology course has kept abreast of these changing perspectives on developmental research and the cutting-edge techniques and research problems of the day, all while maintaining the strong focus on biodiversity that has been crucial to marine research.

From its inception, the embryology course was a novel approach to seaside education; it was designed as an advanced course to teach students both the fundamentals of the field and the methods necessary to conduct their own future research. This ran alongside and eventually replaced the older model derived from the Anderson and Annisquam schools, represented at MBL by the zoology course, which provided general lectures designed to introduce new students to the wonders of natural history. Instead, embryology students were groomed to become the next generation of investigators, both by the instruction offered in the course and by the scientific contacts that they made during their time at Woods Hole (MBL 1937).

In a show of how "teaching is an act twice blest," Whitman served as the first director of the embryology course, setting the tone of advanced study that continues to make the course famous. He directed the six-week course from 1893 through 1898, overseeing the research of 135 students over those six years with only a small staff of instructors, and providing a yearly set of lectures. During these early years, in addition to offering the embryology course, the lab also provided tables for beginning instructors in the field; for a modest fee, a newcomer to research could rent space at a shared table that was supervised by an experienced investigator. This investigator oversaw the work of only a few students and provided instruction and support and often guided students toward a unified project.[9] Eligibility for such a space was contingent on completing either the zoology or the embryology course (MBL 1901). The embryology course thus became a conduit providing the MBL with future investigators.

For the first few years of the course, Whitman, assisted by Frank Lillie (his protégé and a future MBL director), focused the embryology course on vertebrates, confining study mainly to fish eggs (fig. 7.1).[10] Each student was provided with eggs and was expected to work out each step of development, beginning with fertilization, by using the various methods of preparing surface views—including embedding in paraffin wax and celloidin, staining and mounting, drawing, reconstruction, and modeling—that Whitman and Lillie taught them (MBL

FIGURE 7.1. Members of the 1894 embryology course. Frank R. Lillie (*middle row, second from right*) met his future wife, Frances Crane (*middle row, standing next to the woman with the bicycle*), when she came to the course as a student in the 1894 embryology course. Courtesy of the MBL History Project, the MBL Archives, and Arizona Board of Regents.

1893, 1894, 1895). By 1895, prospective students in the course had to have taken the MBL's zoology course, or its equivalent, to be eligible to apply, marking the embryology course as a truly advanced course of research (MBL 1895).

In 1899, Lillie, who was by then the assistant director of the MBL, assumed leadership of the embryology course. Under his guidance, and provoked by Whitman's desire to assess biological problems broadly, the course evolved into its more modern form. Whereas initially it was focused on students carefully following the development of fish embryos, under Lillie's guidance it revolved around the fundamental problems of development that consumed the field of embryology. In this and the use of live embryos it was special. According to the description of the course in the 1900 annual report: "It supplements the usual college course in embryology, by laying special weight on questions of general importance, that can best be approached through the study of living material, much of which is available only during the summer at the seashore" (MBL 1900, 155).

Problems like cleavage and cell lineage, artificial parthenogenesis, and fertilization were emphasized across broad swaths of marine organisms through labs and lectures. While fish were still used extensively, amphibians, echinoderms, tunicates, molluscs, annelids, and nematodes made their ways into the lectures and labs by 1900.[11] Further, Lillie added methods that brought students into contact with the new experimental approaches pioneered around this time by Wilhelm Roux (Roux 1894).

This was the time period during which MBL researchers scoured the local waterways and landscapes of the Woods Hole region for research materials, from Lillie's *Unio* (freshwater mussels) (Lillie 1893, 1895) to Thomas Hunt Morgan's pycnogonids (sea spiders) (Morgan 1891). Embryonic cleavage and the closely associated study of cell lineages were two of the research topics that established the MBL as the center of American biology in the late nineteenth century and the early twentieth (Maienschein 1981, 1991). Whitman had pioneered studies of cell lineage with his *The Embryology of Clepsine*, clepsine being a genus of freshwater leeches (Whitman 1878). When he assumed the position as the first director of the MBL, he attracted students and young researchers who were interested in utilizing the techniques he had deployed to follow the earliest cleavages in the embryo. Early career scientists such as Edmund Beecher Wilson and Edwin Grant Conklin, who had formerly conducted research at the nearby US Fish Commission, brought their cell lineage and embryonic cleavage research across the street, where they published massive monographs that contributed to the growing reputation of the MBL (Wilson 1892; Conklin 1897). By the time that Lillie assumed the mantle of embryology course director, the most famous and enduring of these cell lineage studies were in print.[12] Students under Lillie took full advantage of this local expertise in cell lineage work, learning about how cleavage works in fish, annelids, molluscs, and amphibians, and then comparing development in these very different organisms.

Other new and exciting research areas under discussion in Lillie's embryology course included artificial parthenogenesis (an induced form of asexual reproduction). Jacques Loeb introduced the term *artificial parthenogenesis* in a groundbreaking paper (Loeb 1899), demonstrating how unfertilized *Arbacia* (sea urchin) eggs could develop to the pluteus stage following immersion in a solution of magnesium chloride and saltwater. Loeb's work stirred recognition from the popular press and ignited study of this phenomenon for decades (Pauly 1987; Elliott 2009). Less than a year after this publication, Loeb gave a lecture to the embryology course about the subject and led the students in a lab devoted to "Parthenogenic development of Ova of *Arbacia* in salt solutions" (MBL 1900).

Students attended demonstrations of and labs about fertilization in a variety of species and learned how to section, mount, and otherwise prepare embryos for observations and experiments. They took dredging excursions on the US Fish Commission steamer, skimmed the local waterways to collect their own pelagic specimens, and spent days in the lab preparing their own serial sections of various embryos. Under the guidance of Whitman and Lillie, the embryology course was defined by embedding students at the intersection of education and research. The course became focused on addressing big biological questions using living embryos culled from an array of species—an innovation in teaching embryology at the time. By providing students with a thorough knowledge of development and the skills to conduct their own experiments, Whitman and Lillie trained the next generation of scientists.

Lectures from the brilliant scientists that swarmed to the MBL each year brought students into contact both with the newest ideas and techniques in the field and with a network of researchers who would become crucial to their careers as scientists. In these founding years, the embryology course flourished. During the first decade, it took in an average of 50 percent female students, with ratios of women in any given year ranging from a low of 36 percent in 1898 to a high of 71 percent in 1902 (fig. 7.2).[13] After Lillie's final year directing the embryology course in 1903, it was not offered again at the MBL until 1908.[14]

Continuing to Educate the Next Generation of Researchers

The embryology course resumed in the summer of 1908 under the guidance of the MBL's assistant director, Gilman Drew.[15] Since this time, the course has continued to operate under thirty course directors on the model that Whitman and Lillie laid out (table 7.1 above).[16] Students still crowd the course every year to get access to live embryos unavailable in their home institutions, to gather hands-on knowledge of how to observe and experiment on embryos, and to build their incipient networks of colleagues through their interactions in Woods Hole. All these factors have ensured that the embryology course has continued to train the next generation of researchers and has done so by integrating research and education.

FIGURE 7.2. MBL students and instructors on the dock before a collecting trip, 1893. Note that twelve of sixteen people in this photograph are women. Courtesy of the MBL History Project, the MBL Archives, and Arizona Board of Regents.

For decades following the tenures of Whitman and Lillie, the model of investigator-as-instructor remained unchanged; in return for their teaching efforts, investigators were given stipends and access to research space. Their work in teaching reinforced their knowledge gained at the bench, and they gained new collaborators as bright new minds rose through the ranks from student to beginning investigator to professional scientist. One excellent example of this is Ernest Everett Just, one of the foremost African American scientists of his time. Just took the zoology course in 1909 and the embryology course in 1910, and the following year he began a nearly twenty-year tenure as a summer investigator at the MBL. He became Lillie's PhD student at the University of Chicago and one of the leading authorities on the mechanics of fertilization (Manning 1983). In the 1920s, he gave back to the embryology course by providing lectures on aspects of fertilization based on his most recent research. His was a common path for students of the embryology course.[17]

New course directors have always changed course content to keep pace with trends in the field. As genetics grew into a field in its own right, for example, MBL course director Hubert Goodrich brought in pioneers in the field like the longtime MBL investigator (and student and collaborator of Thomas Hunt Morgan's) Alfred Sturtevant to give lectures like "Genetics and Embryology" and "Genes and Cytoplasm" in the 1930s. As the field of embryology was undergoing the change in the late 1950s that resulted in what is now called *developmental biology*, various course directors increasingly emphasized understanding the cellular, biochemical, and molecular components of developing embryos, which involved looking at these phenomena as problems in and of themselves rather than as phenomena within the greater scheme of the developing embryo. This change in perspective, toward the reduction of the embryo to its constituent components, reflected a larger trend in the field of embryology (Horder 2010; Crowe et al. 2015).[18]

The embryology course, as did the MBL as a whole, continued to be a place of inclusion, welcoming African American, Jewish, Southeast Asian, and female scientists in a time in which these groups faced systemic institutional barriers. Especially notable is the participation of women in the embryology course. Cornelia Clapp, one of the first people to get to Woods Hole in the MBL's inaugural summer, gave lectures during Lillie's tenure. Jane Oppenheimer, a noted embryologist and historian of science, became the embryology course's first official female instructor in 1945, during the directorship of Viktor Hamburger.[19] And, from the opening of the course through 1926, the student body welcomed at least 50 percent female students each year. Unfortunately, except for a sizable bump during World War II, the attendance of women in the embryology course averaged under 50 percent after this time.[20]

In the spirit that Whitman and Lillie had set and research at the MBL has always embraced, biodiversity has always been emphasized, to varying degrees, within the embryology course. In an article in *The Collecting Net* (a weekly publication at the MBL),[21] the embryology course director, Viktor Hamburger, explained the importance of utilizing the diversity of marine species in the Woods Hole region: "The main purpose of this course is to bring students of biology in contact with the living, developing organism. . . . The variety and wealth of invertebrate material, in particular, afford a unique opportunity for the study of fundamental processes of development such as fertilization, gastrulation, and morphogenesis on a broad comparative basis, and open new vistas to the student who has been trained in the traditional frog, chick, and pig embryology" (Hamburger 1942, 6).

Over the years, course directors have used a variety of taxa; fish (especially teleosts), molluscs, annelids, tunicates, and echinoderms have been employed consistently to demonstrate everything from the mechanics of fertilization, to cell cleavage, to regeneration, to the regulation of the cell cycle. Other taxa, like crustaceans and cephalopods, have come in and out of favor, depending on the interests of the course director and the availability of summer investigators to teach the topic (see fig. 7.3).

Beginning in the 1970s, the course began to change. Whereas the vast majority of instructors had been drawn from the corpus of summer investigators at the MBL throughout the first eighty-odd years, 45 percent of the staff listed for the course in 1973 were not MBL researchers.[22] By 1979, the percentage of non-MBL instructors jumped to 85 percent as course directors increasingly brought in specialists who would parachute into Woods Hole for one or two modules of lectures and labs.[23] This shift in emphasis away from utilizing the expertise contained within the MBL community did not affect the innovative nature of the course that Whitman and Lillie had set. Biodiversity continued to be a hallmark of lectures and lab work, and course directors continued to refocus the content of the course to meet the demands of teaching beginning investigators the latest problems and techniques in the field.

Conclusion

In 1888, Whitman set out with a vision. Armed with the dual mission chiseled into the founding of the MBL, through persistence and personal resolve he ensured that research and education became and remained inextricably intertwined at the MBL. As the first director of the MBL and as the first director of the embryology course, he led by example, building an innovative program of education that drew on the local expertise that he built around him in Woods Hole. Through their participation in courses, investigators gained access to laboratory space and equipment and reinforced their knowledge through instruction. Students in the courses gained experience in cutting-edge methods and assembled networks of colleagues to help them advance as burgeoning young researchers, often returning to the MBL as summer investigators. The biodiversity that Whitman embraced as crucial to investigating the big questions in biology became and remains a vital aspect of the embryology course.

With this in mind, let us return to the driving question brought up at the beginning of this essay: How and why did the MBL find a way

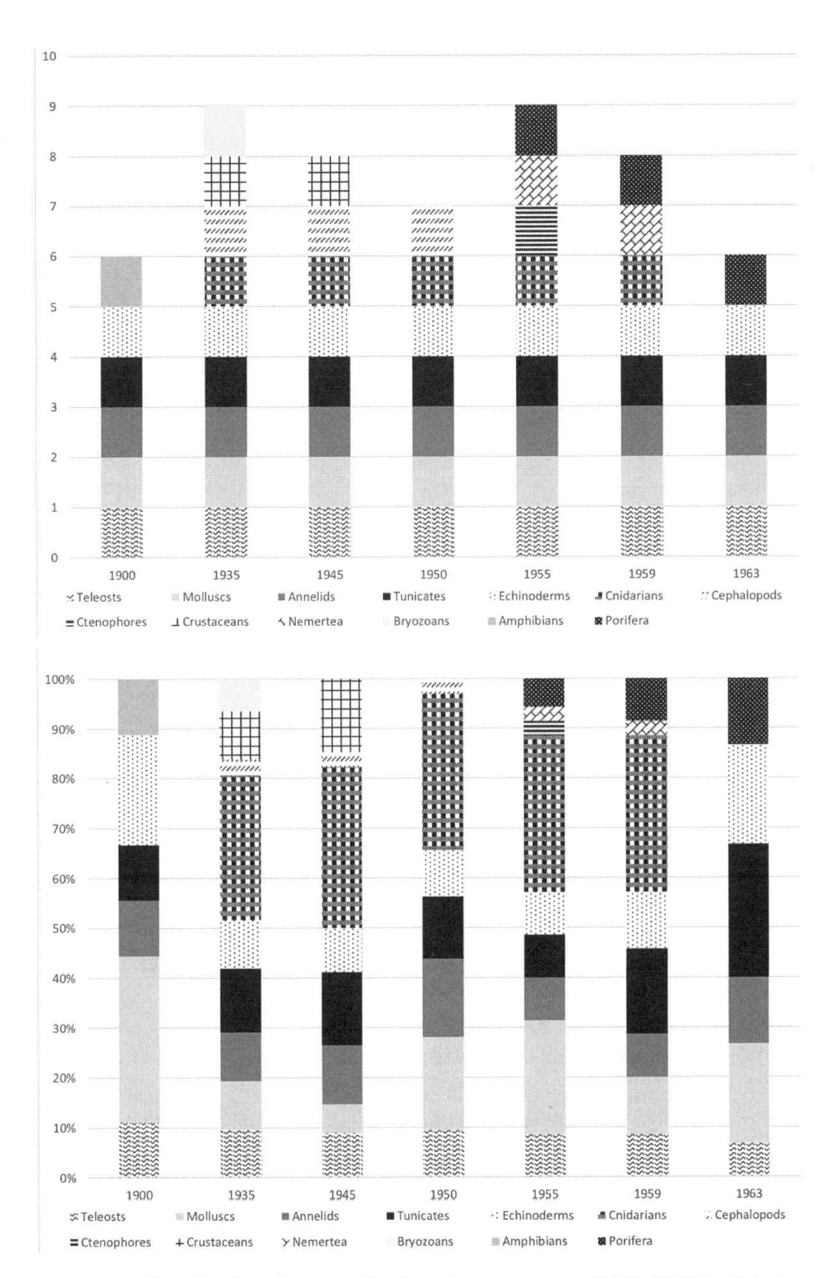

FIGURE 7.3. *Top*: Number of taxa used in the embryology course (1900–1963), both lectures and laboratory work, under various course directors: 1900 (Lillie), 1935 (Goodrich), 1945 (Hamburger), 1950 (Costello), 1955 (Rose), 1959 (Edds), 1963 (Ebert). Information about taxa was obtained from both MBL annual reports and embryology course books. *Bottom*: Relative and changing proportions of taxa used in the embryology course (1900–1963), in both lectures and laboratory work.

to continue to grow and enhance its educational efforts and research despite periods of financial instability and internal division and as other marine stations chose to bolster one or the other? The answer to this question, as indicated throughout this essay, is that Whitman and those whom he trained, like Lillie, created an environment at the MBL in which both research and education were crucial to the scientific integrity of the institution. Like many other American marine stations, the MBL could probably have survived by emphasizing solely one or the other, but its richness as a convening place where diverse perspectives converged on the big questions of biology would undoubtedly have suffered.

Notes

1. For extensive descriptions of European stations and their activities, see Dean (1893) and Kofoid (1910).
2. For more in-depth information on the fates of other marine laboratories founded in the late nineteenth century and the early twentieth, see Benson (1988).
3. For a list of the individuals and institutions who attended the inaugural year of the MBL, see MBL (1888b, 18–19). The circular provided to colleges is also available in MBL (1888b).
4. From the outset, the MBL tended to run at a budget deficit. In the early years, this deficit was covered by donations of various kinds and by members of the board of trustees. For further details, see Maienschein (1985a).
5. The corporation of the MBL was a body made up almost exclusively of scientists that administered and largely governed the MBL from 1897 until its formal affiliation with the University of Chicago on July 1, 2013.
6. Instructors often gave multiple lectures, were involved in daily operations, and helped run labs for the course, while lecturers were called on to give one or two lectures.
7. In an opinion piece in *Science*, Whitman notes that, at the time, the MBL spent $13,000 annually and often did not charge investigators for research space, whereas the Stazione Zoologica in Naples spent between $40,000 and $50,000 annually and charged investigators $500 a year for a table. For this and further information about finances, see Whitman (1902b).
8. Eight directors of the MBL (Whitman, Lillie, Packard, Armstrong, Ebert, Gross, Ruderman, and Patel) have been involved with the embryology course throughout its history.
9. For example: "In some cases cooperative monographical work is undertaken. A suitable form is selected, and the various problems of structure, development, and life-history are distributed among a number of

researchers. . . . Such a work on a marine annelid (*Arenicola*) is now in progress" (MBL 1901, 152).

10. In 1895, the third year the embryology course was offered, students continued to study primarily fish eggs, but the eggs of amphibians, other vertebrates, and some invertebrates were studied as well.
11. The only information preserved about the course under Lillie comes from MBL (1900). The availability of over twelve different taxa and the use of biodiversity to allow students to make comparisons of developmental processes between taxa were great feats.
12. The tradition of investigating the early cleavage and cell lineage of different organisms in the Woods Hole area continued throughout the early twentieth century, including by instructors in the embryology course like Samuel Jackson Holmes (Holmes 1900).
13. Female and male students were identified for each year by categorizing their first names as either female or male. Student information was not available for 1893 or 1903.
14. The reason for the four-year hiatus is not mentioned in any institutional documents and, as such, remains a mystery worth investigating. During these gap years, the MBL was under great financial strain, but that did not affect other course offerings. For more information about the economic problems at the MBL during this time period, see Maienschein (1985a).
15. In 1908, Whitman stepped down as director of the MBL, and Lillie assumed the position. Gilman Drew, professor of biology at the University of Maine, took Lillie's former post as assistant director.
16. For more details about the history of the embryology course and the course directors, see Maienschein (2007).
17. Take, e.g., Joan Ruderman, who was an embryology course student in 1974, became the MBL's first female course director in 1978, and the MBL's first female director in 2012. For more information about some of the content and people in the embryology course, see Jiang, MacCord, and Maienschein (2015).
18. For further readings on the shift from embryology to developmental biology in the mid-twentieth century, see Oppenheimer (1966), Horder, Witkowski, and Wylie (1986), Hopwood (2009), Horder (2010), and Crowe et al. (2015).
19. The next listed female instructors in the embryology course did not come until Barbara Hamalko and Joan Ruderman joined the teaching team in 1976.
20. Data for the gender ratios of students was accumulated only through 1979, so nothing can be said about the participation of women in the course after 1979.
21. There have been several iterations of *The Collecting Net*. The first run was published during the summer months from 1926 through 1953, with a gap in publication from 1944 through 1945.

22. Data for course staff were aggregated by comparing instructors and lecturers for each year of the embryology course to the set of MBL summer investigators listed in the MBL's annual reports.
23. This shift toward utilizing researchers outside the MBL community roughly coincided with the advent of course funding provided by the National Institutes of Health (NIH). However, the alignment of funding and movement toward external investigators is imperfect—the course was first awarded funding from NIH in 1971, but, for the majority of the decade, external investigators made up 25 percent or less of course staff.

References

Allard, Dean C. 1990. "The Fish Commission Laboratory and Its Influence on the Founding of the Marine Biological Laboratory." *Journal of the History of Biology* 23, no. 2:251–70.

Benson, Keith R. 1988. "Why American Marine Stations? The Teaching Argument." *American Zoologist* 28, no. 1:7–14.

Conklin, Edwin Grant. 1897. *The Embryology of Crepidula: A Contribution to the Cell Lineage and Early Development of Some Marine Gasteropods*. Boston: Ginn.

Crowe, Nathan, Michael R. Dietrich, Beverly S. Alomepe, Amelia F. Antrim, Bay Lauris ByrneSim, and Yi He. 2015. "The Diversification of Developmental Biology." *Studies in History and Philosophy of Biological and Biomedical Sciences* 53:1–15.

Dean, Bashford. 1893. "The Marine Biological Stations of Europe." *Biological Lectures* 2:211–34.

Elliott, Steve. 2009. "Jacques Loeb (1859–1924)." *Embryo Project Encyclopedia*. http://embryo.asu.edu/handle/10776/1678.

Galtsoff, Paul S. 1962. *The Story of the Bureau of Commercial Fisheries, Biological Laboratory, Woods Hole, Massachusetts*. Circular no. 145. Washington, DC: US Department of the Interior.

Hamburger, Viktor. 1942. "The Embryology Course at Woods Hole in 1942." *Collecting Net* 17, no. 148:6.

Holmes, Samuel J. 1900. "Preliminary Account of the Cell Lineage of *Planorbis*." *Biological Bulletin* 1, no. 2:95–101.

Hopwood, Nick. 2009. "Embryology." In *Cambridge History of Science*, vol. 6, *Modern Life and Earth Sciences*, ed. Peter J. Bowler and John V. Pickstone, 285–315. Cambridge: Cambridge University Press.

Horder, Timothy. 2010. "History of Developmental Biology." In *Encyclopedia of Life Sciences*. https://doi.org/10.1002/9780470015902.a0003080.pub2.

Horder, Timothy, Jan Witkowski, and Christopher Wylie, eds. 1986. *A History of Embryology*. Cambridge: Cambridge University Press.

Jiang, Lijing, Kate MacCord, and Jane Maienschein. 2015. "The MBL Embryology Course 1939." MBL History Project digital exhibit. https://history.archives.mbl.edu/exploring/exhibits/mbl-embryology-course-1939.

Kofoid, Charles Atwood. 1910. "The Biological Stations of Europe." *US Bureau of Education Bulletin*, no. 4, whole no. 440. Washington, DC: US Government Printing Office.

Lillie, Frank Rattray. 1893. "Preliminary Account of the Embryology of *Unio complanta*." *Journal of Morphology* 8, no. 3:569–78.

———. 1895 *The Embryology of the Unionidae*. Boston: Ginn.

———. 1911. "Charles Otis Whitman." *Journal of Morphology* 22:xv–lxxvii.

———. 1944. *The Woods Hole Marine Biological Laboratory*. Chicago: University of Chicago Press.

Loeb, Jacques. 1899. "On the Nature of the Process of Fertilization and the Artificial Production of Normal Larvae (Plutei) from the Unfertilized Eggs of the Sea Urchin." *American Journal of Physiology—Legacy Content* 3, no. 3: 135–38.

MacCord, Kate, and Jane Maienschein. 2018. "The Historiography of Embryology and Developmental Biology." In *The Historiography of Biology*, ed. Michael R. Dietrich, Mark Borrello, and Oren Harman, 1–23. Dordrecht: Springer.

Maienschein, Jane. 1981. "Shifting Assumptions in American Biology: Embryology, 1890–1910." *Journal of the History of Biology* 14, no. 1:89–113.

———. 1985a. "Early Struggles at the Marine Biological Laboratory over Mission and Money." *Biological Bulletin* 168:S192–S196.

———. 1985b. "First Impressions: American Biologists at Naples." *Biological Bulletin* 168:S187–S191.

———. 1991. *Transforming Traditions in American Biology, 1880–1915*. Baltimore: Johns Hopkins University Press.

———. 2007. "The Marine Biological Laboratory Embryology Course." *Embryo Project Encyclopedia*, n.d. http://embryo.asu.edu/handle/10776/1794.

Manning, Kenneth R. 1983. *Black Apollo of Science: The Life of Ernest Everett Just*. Oxford: Oxford University Press.

Marine Biological Laboratory (MLB). 1888a. Agreement of Association. https://hpsrepository.asu.edu/handle/10776/12938.

———. 1888b. *Annual Report*. https://hpsrepository.asu.edu/handle/10776/1451.

———. 1893. *Annual Report*. https://hpsrepository.asu.edu/handle/10776/1456.

———. 1894. *Annual Report*. https://hpsrepository.asu.edu/handle/10776/1457.

———. 1895. *Annual Report*. https://hpsrepository.asu.edu/handle/10776/1458.

———. 1900. *Annual Report*. https://hpsrepository.asu.edu/handle/10776/1459.

———. 1901. *Annual Report*. https://hpsrepository.asu.edu/handle/10776/1459.

———. 1937. *Annual Report*. https://hpsrepository.asu.edu/handle/10776/1490.

Morgan, Thomas Hunt. 1891. *A Contribution to the Embryology and Phylogeny of the Pycnogonids*. Baltimore: Isaac Friedenwald.

Oppenheimer, Jane M. 1966. "The Growth and Development of Developmental Biology." In *Major Problems of Developmental Biology*, ed. Michael Locke, 1–27. New York and London: Academic Press.

Pauly, Philip J. 1984. "The Appearance of Academic Biology in Late Nineteenth-Century America." *Journal of the History of Biology* 17, no. 3:369–97.

———. 1987. *Controlling Life: Jacques Loeb and the Engineering Ideal in Biology*. New York: Oxford University Press.

Roux, Wilhelm. 1894. "The Problems, Methods and Scope of Developmental Mechanics." Translated by William Morton Wheeler. *Biological Lectures* 3:149–90.

Whitman, Charles Otis. 1878. *The Embryology of Clepsine*. London: J. E. Adlard.

———. 1898. "Some of the Functions and Features of a Biological Station." *Science* 12, no. 159:37–44.

———. 1902a. Letter to Edmund Beecher Wilson, November 13. Box 3, folder 1c, Charles Otis Whitman Collection, Marine Biological Laboratory Archives, Woods Hole, Massachusetts. https://hpsrepository.asu.edu/handle/10776/11838.

———. 1902b. "The Impending Crisis in the History of the Marine Biological Laboratory." *Science* 16, no. 405:529–33.

———. 1902c. "Some Matters of Fact Overlooked by Professor Whitman." *Science* 16, no. 408:665–67.

Wilson, Edmund Beecher. 1892. *The Cell-Lineage of Nereis*. Boston: Ginn.

———. 1902. "The Marine Biological Laboratory and the Carnegie Institution. Some Matters of Fact." *Science* 16, no. 406:591–92.

EIGHT

Francis O. Schmitt: At the Intersection of Neuroscience and Squid

KATHRYN MAXSON JONES

The disciplinary maturation of the neurosciences ranks as one of the most significant developments in twentieth-century science (e.g., Vidal and Ortega 2011; Choudhury and Slaby 2012; Rose and Abi-Rached 2013; Stahnisch 2017, 2019). Nervous systems allow animals to sense and respond to their environments, carry out integrated physiological processes such as temperature regulation and sleep, and, in some species, experience the phenomena of learning and memory and states associated with consciousness, such as self-awareness. Research with marine organisms has helped make the historical development of the neurosciences possible—especially investigations using *squid*, a term referring to various orders, families, genera, and species of cephalopod mollusks generally appearing in the order Teuthida (Gilbert, Adelman, and Arnold 1990; Llinás 1999).[1] In this essay, I build on existing historical interpretations of marine biology and the neurosciences (e.g., Rasmussen 1997a, 1997b; Stadler 2009; de Sio 2011; Muka 2014, 2016; Stahnisch 2019). In particular, I investigate the closely intertwined histories of squid biology and the physiological study of neuronal functions from the 1930s to the 1960s.

The protagonist is Francis O. Schmitt (1903–95), whose career culminated in founding one of the world's first academic neuroscience institutions, the Neurosciences Research Program (NRP) at the Massachusetts Institute of Technology (MIT) (Young 1975; Adelman and Smith 1998; Adelman 2010). Created in 1963, the NRP resembled an "invisible college" and consisted of a network of scientist "Associates" knitted together by meetings on directed research topics as well as a journal, the *NRP Bulletin* (Abi-Rached and Rose 2010; Adelman 2010). Ultimately, this institution helped train ten of the first twelve presidents of the Society for Neuroscience (SfN), founded in 1969. The NRP and the SfN formed part of a tidal wave of "neuro" institutions also inaugurated in the 1960s and 1970s, including the Department of Neurobiology that Stephen W. Kuffler founded at Harvard Medical School in 1966 (McMahan 1990; Stahnisch 2017, 367). The disciples of both the NRP at MIT and the Department of Neurobiology at Harvard lay claim to the birth of neuroscience. While such claims contradict one another, they underscore Schmitt's prominence in this history while strengthening the point that the rise of *neuroscience* should be told as various, more local histories of the *neurosciences* (de Sio 2011). This essay examines how the Schmitt laboratory's biophysical and biochemical studies of structure and function in nerve cells, or neurons, relied on marine institutions and organisms.

I make two related arguments. First, from the 1930s through the 1960s, marine stations including the Marine Biological Laboratory (MBL) in Woods Hole, Massachusetts, offered brick-and-mortar places where Schmitt and his colleagues could adopt new experimental systems (as discussed in Rheinberger 1997, 110–13). Working from these stations, Schmitt and his MIT laboratory could obtain their chosen experimental organisms, squid, which in turn supplied the enormous axons, or long nerve cell extensions, necessary for their work. An important result was great midcentury growth in neurophysiology and, in turn, the neurosciences. Second, Schmitt and his laboratory's research with squid axons both required and facilitated comparative studies, expanding neurophysiologists' understandings of the varieties of neurons that exist in the natural world. Schmitt's ensuing discipline-building efforts onshore—not only in the neurosciences but also in biophysics, his preferred avenue for investigating the characteristics of nerve cells—have eclipsed the importance of marine studies in scientific and historical analyses of his career (Rasmussen 1997a, 1997b; Adelman and Smith 1998). Yet this earlier phase of work under Schmitt's direction provides new insights into how experiments with neurons, particularly those dissected from marine organisms, both preceded and connected to discipline building in this pivotal era for the neurosciences.

Neurophysiology before Schmitt

From the early years of the twentieth century through the 1930s, as Schmitt's experimental career was taking off, neurobiologists found themselves in the thick of debates over whether nervous systems are cellular or reticular (continuous) in nature and also whether all neurons (if, indeed, they existed) were essentially similar in structure and function (Jones 1999; Guillery 2005, 2007). In 1902, the German physiologist Julius Bernstein postulated that membranes surrounding neurons could block the flux of charged particles, or ions, across them, helping give rise to the electrical action potentials through which they communicated when this blockage broke down during excitation (Lenoir 1986; Stadler 2009). But, like many of his late nineteenth- and early twentieth-century colleagues (Holmes 1993; Logan 2002), Bernstein was working primarily with frogs, whose axons, or nerve fibers, were too small to detect the absolute values of charge differentials across their membranes directly. By around 1928, the elite group of physiologists calling themselves *axonologists* had ascertained how to use various recording and amplifying technologies, such as cathode ray tube oscilloscopes (Gasser and Erlanger 1922), to track voltage changes during excitation, at least in some isolated axons.[2] Yet recordings gleaned from electrodes (tiny needles that conduct electricity) placed inside neurons remained elusive, and dissecting out cells from frogs and earthworms, another laboratory staple, proved very difficult (Young 1975).

The situation changed rapidly in the mid-1930s. In 1929, the English microscopist and anatomist John Z. Young noticed a curious organ while dissecting octopuses at the Stazione Zoologica di Napoli (Young 1929; de Leo 2008).[3] In a retrospective account honoring Schmitt's career and the NRP in 1975, Young recalled:

> At the hind end of the stellate ganglion of an octopus there is a yellow spot. . . . There were no hypotheses about it at all. Moved by simple curiosity I cut sections to see what it was and found a vesicle in which ended processes arising from what seemed to be nerve cells. . . . I thought it was a sort of adrenal gland. . . . Looking in squids for this epistellar body, as we had called it, I found instead the giant nerve fibres, though for years I was foolish enough to think that they were veins and neglected to examine them closely. (Young 1975, 19)

This finding in squid specifically came first in *Loligo forbesi*, a species inhabiting the North Atlantic waters off the English coast. In 1936, Young published two comparative analyses of large nerve fibers in crustaceans, cephalopods, and several other invertebrates, detailing their microanatomy, and postulating physiological functions (Young 1936a, 1936b). His discovery enabled the first direct, intracellular measurements of voltage changes during action potentials, a development with enormous implications for neurobiology.[4]

Young explained that the stellate ganglion, where a squid's largest "giant" fibers arise, contains thousands of nerve cell bodies, many of which give rise to the stellar nerves connecting the ganglion to the mantle muscles (Young 1936a, 368–70; see fig. 8.1). In the stellate ganglion, axons from these cell bodies fuse, in syncytia, to form postganglionic giant axons, which then function like single cells. Each of the stellar nerves contains a single one of these giant axons, whereas other giant fibers, such as the preganglionics, pass "backwards" out of the ganglion inside the pallial nerve. As Young wrote: "It is not easy to say exactly how many cells go to the making up of each giant fibre. . . . The number of post-ganglionic giant fibres varies in different individuals of *Loligo forbesi* between nine and fifteen, and estimates of the number of cells in the giant fibre lobe, obtained by counting the cells in a small area, vary between 5,000 and 15,000." His conclusion was, thus, that "from 300 and 1,500 cells fuse to form each giant fibre" (Young 1936a, 370–71). A giant axon in *Loligo* can also reach nearly 1 millimeter (or 1,000 microns) in diameter, as compared to just 200 microns in a cuttlefish or just 1–20 microns in a typical vertebrate neuron (Young 1936b, 323; Purves et al. 2018). The giant axons in the squid proved relatively easy to dissect and manipulate during experiments, in contrast to the smaller neurons (and entire nerves) that had long been studied in invertebrates (such as earthworms) or in vertebrates (such as frogs) (Eccles, Granit, and Young 1932; Lenoir 1986).

By the mid-1940s, most neurobiologists had accepted that the squid's giant axons were indeed neuronal in function and that they could also be studied as single cells.[5] In the 1930s and 1940s, Young and a relatively circumscribed group of Anglo-American actors studied the basic biophysical features of these axons, precipitating a flourishing midcentury research tradition in axonology. For example, based at the MBL in the late 1930s, and making use of the local squid *Loligo pealeii* (now *Doryteuthis pealeii*), the Americans Howard Curtis and Kenneth Cole partially confirmed Bernstein's 1902 hypothesis that the action potential arose from increased membrane permeabilities to ions,

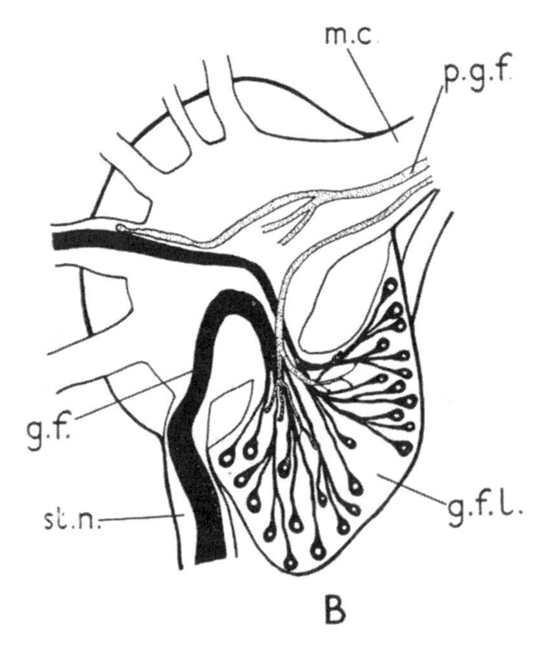

FIGURE 8.1. Diagram of the giant fiber system of the Atlantic squid, *Loligo forbesi*. Adapted from Young (1936a, 369, fig. 1*b*) with permission of the *Quarterly Journal of Mircoscopical Science* (now the *Journal of Cell Science*, The Company of Biologists Ltd.). The original figure contained three panels comparing the neuronal anatomy of the "stellate ganglia of Cephalopods": (*a*) *Sepia*, (*b*) *Loligo*, and (*c*) Octopods. From Young's lettering: *m.c.*, mantle connective (consisting of the pallial nerve); *p.g.f.*, "pre-ganglionic" giant fiber (running "backwards" in the pallial nerve, 368); *g.f.*, "post-ganglionic" giant fibers (forming the stellar nerves); *g.f.l.*, giant fiber lobe; *st.n.*, stellar nerve. In this diagram, the head of the squid would appear at the top of the figure, while the mantle would appear below. Notice two sets of giant fibers, one passing toward the mantle (*g.f.*) and the other away (*p.g.f.*).

correlating with a drastic decrease in membrane resistance (Cole and Curtis 1939). Working with R. J. Pumphrey at the Marine Biological Association (MBA) in Plymouth, England, Young found that, as in a metal wire, the larger the diameter of an (unmyelinated) squid giant axon, the lower its resistance and the faster an electrical signal can be propagated along it (Pumphrey and Young 1938).[6] Also at the MBA, Alan L. Hodgkin and Andrew F. Huxley measured voltage changes across the axonal membrane, employing glass and metal electrodes inserted directly into the giant axons of *Loligo forbesi* (Hodgkin and Huxley 1939, 1945). Soon thereafter, Curtis and Cole achieved this feat in *Loligo pealeii* (Curtis and Cole 1940). By 1952, Hodgkin and Huxley had mathematically modeled how a squid giant axon produces an action potential via timed, differential permeabilities to sodium and potassium ions (Hodgkin and Huxley 1952). The Hodgkin and Huxley papers, together with the squid giant axon, opened the neuron to novel kinds of biochemical and biophysical analyses (Craver 2008; Stadler 2009; Levy 2014).

Schmitt and Squid

Schmitt was a colleague of Hodgkin, Huxley, Cole, Curtis, and Young. In 1927, he completed his PhD under the axonologist Joseph Erlanger in the Department of Physiology of the medical school of Washington University in St. Louis (Adelman and Smith 1998, 343–45). As a graduate student and a postdoctoral researcher, Schmitt studied kidney and epithelial physiology, alongside the metabolism, electrical properties, and crystallography of frog nerves (e.g., Grave and Schmitt 1924; White and Schmitt 1925; Schmitt, Bear, and Clark 1935). He was also introduced to marine invertebrate biology early in his career, at the MBL's 1923 summer physiology course.[7] He assumed his first professorship in the Washington University Biology Department in 1930 and met Young there in 1935 (Young 1975, 17; Adelman and Smith 1998).

Schmitt and Young collaborated on squid giant axons in Woods Hole and the Biological Laboratory in Cold Spring Harbor, New York, in 1936. While Cole and Curtis were investigating the bioelectrical properties of *Loligo pealeii* neurons (Cole and Curtis 1939), Schmitt, Young, and Schmitt's student Richard Bear studied the axons' structural and chemical characteristics. This collaboration resulted in three papers, their appearance in the *Proceedings of the Royal Society of London* marking the strength of the professional ties among US and British marine and neurobiologists at this time (Bear, Schmitt, and Young 1937a, 1937b, 1937c). Most importantly, the 1937 papers strongly suggested that—given their microscopic and biochemical features and in comparison with known neurons in crustaceans and cows—the giant fibers Young had rediscovered were nervous in function. In conjunction with Young (1936a, 1936b), the Bear, Schmitt, and Young work also suggested that the fibers were single functional units, and could be studied as single cells. Schmitt's 1935 encounter with Young in St. Louis and his research at the MBL and in Cold Spring Harbor in 1936 also set his professional life on a new trajectory, one he pursued for the remainder of his experimental career.

By his own description, Schmitt was a general physiologist: he focused on fundamental life functions, as opposed to studying problems with obvious medical applications (Maienschein 1987; Pauly 1987a, 1987b; Rasmussen 1997a, 1997b). In the late 1930s, squid axons were the "right tools for the job" (Clarke and Fujimura 1992) for general physiologists because they could be excised from host animals and their axoplasm, or cytoplasm, coaxed out for staining, X-ray diffraction, biochemical analysis, and microscopic study. Schmitt followed these basic lines of research through the 1960s, adopting new tools such as electron microscopy, ultracentrifugation, and gel electrophoresis as they became available to well-funded laboratories such as his at MIT (Rasmussen 1996, 1997a, 1997b; Creager 2001, 2013; Chiang 2009). While investigators in this period such as Hodgkin and Huxley focused on the membranes of squid giant axons, however, Schmitt and his lab worked mainly on the axoplasm.

Throughout this process, the MBL and similar coastal laboratories helped make possible Schmitt and his colleagues' transitions to what Hans-Jörg Rheinberger called new *experimental systems* (Rheinberger 1997). These included the squid and, as we shall see, other large marine axons, coupled to technologies that included electrodes, amplifiers, oscilloscopes, radioisotopes, Geiger counters, ultracentrifuges, electrophoretic gels, and electron microscopes. Whether the work of these midcentury axonologists focused on the neuronal membrane or on the axoplasm, however, the epistemic object researchers hoped to study was usually the action potential. The nervous impulse was characterized by injecting current into the axon using an electrode and then plotting induced voltage changes against time or distance (Schmitt and Schmitt 1940). Other ways of studying the impulse included calculating the rates of mobility, diffusion, and active transport of sodium and potassium radioisotopes along and across both squid and cuttlefish giant axons (Hodgkin and Keynes 1953, 1955). Yet further experimental systems probed action potentials more structurally and statically, employing microscopy or biochemistry.

In Schmitt's MIT laboratory through the 1950s, for example, typical outputs from squid experiments tended to be pictorial or tabular in nature. The laboratory was best known for its electron micrographs, such as of fibrous entities that could be visualized in extruded axoplasm (Rasmussen 1997b). But these outputs also included tables of ionic concentrations, such as the "acid-base balance in squid nerve axoplasm" measured with methods including ion exchange titration (Koechlin 1954a, 61; Koechlin 1955). Such research, crucially, depended on neurons obtained at marine stations, which in turn attracted researchers, raised the prestige of neurophysiology, and helped keep the subfield of general physiology afloat. Squid giant axons enabled this process most directly, but, in laboratories throughout the world, giant nerve fibers from additional marine invertebrates also played important roles as experimenters searched for large neurons they could study in abundance. Neurophysiologists from the 1940s through the 1960s asked new, molecular questions of nerve cells. These extended to synapses and simple neural networks as well (Kandel and Tauc 1965), owing not only to evolving instrumentation but also to the sheer size of many marine neurons and the availability of organisms at marine laboratories.

The MBL and Montemar

Schmitt's laboratory interests in squid took him beyond just Woods Hole and Cold Spring Harbor. In 1941, right before the United States entered World War II, he had relocated from St. Louis to MIT, where he had been recruited as the inaugural head of the Department of Biology and Biological Engineering (Adelman and Smith 1998, 6). MIT built up world-class expertise in electron microscopy during the war, focusing on macromolecules in various cell types and also wound healing (Rasmussen 1997b). In the immediate postwar years, alongside its analyses of large protein fibers such as muscle and collagen, the Schmitt group revamped its investigations of nerves. While some of this research focused on vertebrates, such as chicks and frogs (Geren and Raskind 1953; Geren and Schmitt 1954), Schmitt also hoped to investigate the cytoplasm of invertebrate neurons, dovetailing with the vertebrate work conducted in his laboratory by Betty Ben Geren. Schmitt's intended physical and chemical analyses of the cytoplasm, however, required an abundant quantity of squid axons (Schmitt 1957c, 172), an anxiety reflected in his frenetic correspondence throughout the 1950s.

A letter to Clinton Atkinson, of the US Fisheries Laboratory in Beaufort, North Carolina (Schmitt 1951), was typical. "There are two types of work we do with squid nerve," Schmitt wrote. The first: "Structural studies with the electron microscope. This requires only relatively few squid." The second: "Protein chemistry of axoplasm. This requires many squid (40–50 at a time)." Axoplasm extruded from squid giant fibers was "immediately rushed to the biophysical laboratory" and "examined with the ultracentrifuge, electrophoresis, etc." "The latter type" of analyses could "be done only very near our Laboratory in Boston," but electron microscopic work "could be done elsewhere." Time was of the essence particularly with the biochemistry because nerve proteins degraded with time and distance, especially in the heat. *Loligo pealeii* were obtained in Woods Hole or Sakonnet, Rhode Island, from July through September (Smith and Hanson 1952; Rasmussen 1997b, 162–63). When MIT's specific facilities were required, the group carried specimens from the coast to MIT, making use of a refrigerated "squid truck" driven by Ben Geren with water and animals sloshing about and frequent mechanical breakdowns to boot (Rasmussen, 1997b, 162–63). Yet Schmitt soon grew frustrated with this system because the squid yields were often low and the animals were small.

An alternative quickly presented itself. In the early 1950s, the big game anglers Louis Marron and his wife, Eugenie, Schmitt's Chilean graduate student Mario Luxoro, and Luis Rene Rivas of the University of Miami Marine Laboratory drew Schmitt's attention to the Humboldt squid, *Dosidicus gigas*, which could be found off the coast of Chile (Schmitt 1990, 162–65; Bacigalupo and Kakuljan 1997; Scully 2008). As he wrote Rivas (Schmitt 1955), Schmitt estimated that these meter-long squid could yield between twenty-five to thirty times more axoplasm than New England squid and would be available during the Southern summer, corresponding to winter in the Northern Hemisphere, when it was nearly impossible to obtain animals from Woods Hole or Rhode Island. In 1955, with his promotion to MIT Institute Professor and thus relief from most of his teaching duties, Schmitt initiated a partnership with a Chilean laboratory to obtain a reliable supply of *Dosidicus gigas*. He secured funding from the US National Institutes of Health (NIH) as well as the Office of Naval Research and several private sources. His South American "large squid operations" launched as early as the summer of 1957 (Schmitt 1957c, 174) from the Marine Biological Station on the Montemar coast, near the Chilean city of Viña del Mar and just north of the city of Valparaiso (Bezanilla 2018, 1464). Others associated with the squid collection project were Luxoro, who completed his doctorate in 1956, and several administrators and professors at the University of Chile.

Luxoro furnished the MIT-funded laboratory and managed its daily operations. It soon became apparent, however, that this venture in Chile held many logistic challenges, including weather, squid migratory patterns, and shipping whole or pieces of squid nerves to MIT in the Northern winter. Luxoro needed a "freeze-drying apparatus, refrigerator, magnifying lens, fine dissecting instruments, compressing pump, [and] CO_2 tanks," Schmitt wrote to Carlos Ruiz at the Rector's Office of the University of Chile (Schmitt 1957b). Some of the samples arrived intact and functionally active, but others were delivered with the proteins denatured or the axoplasm jumbled with blood and other tissues (Schmitt 1957a; Scully 2008, 935). Before shipping, the samples of axoplasm were generally freeze-dried or frozen, "packaged into boxes with dry-ice for transport back to the United States by air" (Scully 2008, 935). There was also the problem of explaining to Chilean and US customs just what these packages contained. Finally (or, perhaps, first), the catching and dissecting of the squid had to be arranged, a task that required teams of people and introduced considerable unforeseen complexities. Hired fishers generally completed the first step and Luxoro's scientific team, which included technicians, the second; a pamphlet presumably from the period, complete with photographs, shows the instructions Luxoro provided to those doing *Dosidicus* dissections (Luxoro, n.d.). Schmitt wrote Luxoro, wondering "whether the squid were dead for a long time before the nerves were dissected": "Had the squid been caught some considerable time before this and brought back in one of the open boats, dead? Has any progress been made with the problem of trying to keep the squid alive until they are brought in?" (Schmitt 1957a). The timing and follow-through for both the catching and the dissection steps were crucial for subsequent analyses. In a handwritten list from the 1950s (fig. 8.2), Schmitt fretted over all these problems and more.

Then, of course, there were the challenges of the biology itself. Schmitt and the members of his laboratory had long been interested in a variety of structure-function relationships. These involved biophysical and biochemical studies mostly of axoplasm but also of myelin (a fatty, insulating substance covering some neurons) and various additional macromolecules associated with neurons and other cells, including muscle cells (Rasmussen 1997a, 1997b). "It is desirable," Schmitt wrote, "that the biophysical and biochemical investigation of the neuron not be limited to bioelectric processes, but [rather] be systematic and exhaustive so as to reveal 'what is there,' particularly at the molecular and macromolecular level" (Schmitt 1959b, 455). His predominant goal was to understand how macromolecules in solution (such as proteins, carbohydrates, and lipids) gave rise to action potentials as well as other complex neuronal properties, including axoplasmic ultrastructure. In squid specifically, a major achievement of the Schmitt team was the early characterization of the predominant fibrous proteins in axoplasm, which cell biologists later called *intermediate filaments* (Schmitt 1957c, 170; Rasmussen 1997b, 169).

Yet the Schmitt laboratory's efforts in marine research quickly

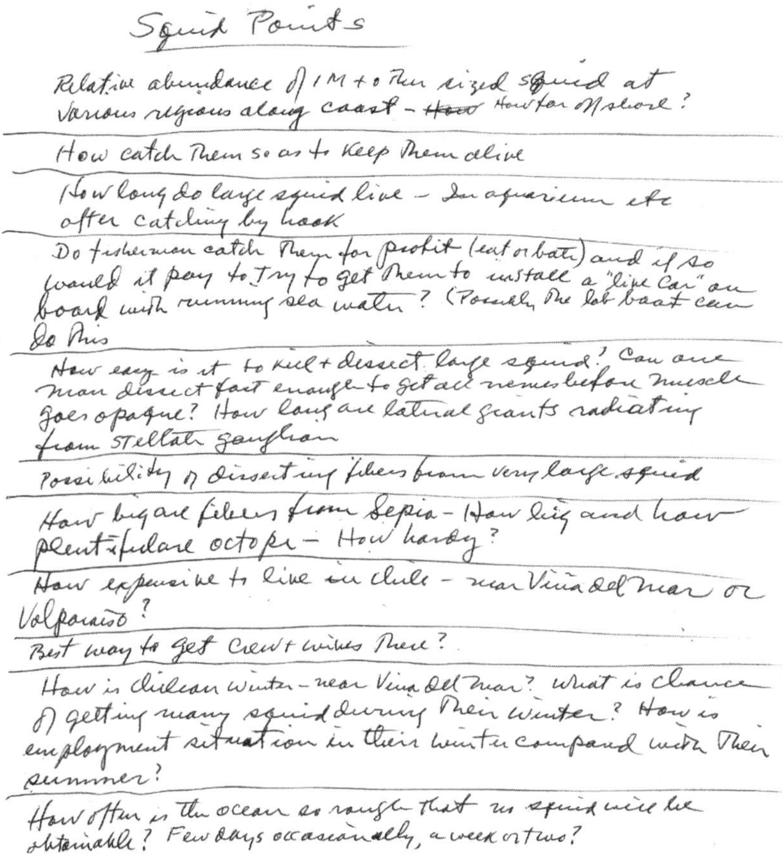

Squid Points

Relative abundance of 1M + other sized squid at various regions along coast – ~~How~~ How far offshore?

How catch them so as to keep them alive

How long do large squid live – In aquarium etc after catching by hook

Do fisherman catch them for profit (eat or bait) and if so would it pay to try to get them to install a "live car" on board with running sea water? (Possibly the lab boat can do this

How easy is it to kill + dissect large squid? Can one man dissect fast enough to get all nerves before muscle goes opaque? How long are lateral giants radiating from stellate ganglion

Possibility of dissecting fibers from very large squid

How big are fibers from Sepia – How big and how plentiful are octopi – How hardy?

How expensive to live in Chile – near Viña del Mar or Valparaiso?

Best way to get crew + wives there?

How is Chilean winter – near Viña del Mar? What is chance of getting many squid during their winter? How is employment situation in their winter compared with their summer?

How often is the ocean so rough that no squid will be obtainable? Few days occasionally, a week or two?

FIGURE 8.2. "Squid Points," a list compiled by Francis Schmitt in the 1950s. When launching his Chilean squid procurement program from MIT, Schmitt worried about such practical problems as, "How long do large squid live . . . after catching by hook," and "How is [the] Chilean winter—near Viña del Mar?" Folder 16, box 3, Francis Otto Schmitt Papers (MC0154), Institute Archives and Special Collections, Massachusetts Institute of Technology, Cambridge, MA. Reproduced with permission of the Institute Archives and Special Collections, Massachusetts Institute of Technology, Cambridge, MA.

grew comparative. One impetus for these comparative analyses was to test how findings with squid could be generalized. "In the light of our findings with squid axoplasm it became of interest to reinvestigate the anion [negatively charged] pattern of lobster and crab nerves and . . . determine by a more complete and accurate analysis the similarities as well as the differences . . . in the hope to establish some common denominator which might permit conclusions as to the significance of the anions for nerve function," one laboratory member, the biochemist Bernard Koechlin, wrote in a progress report on his New England squid work (Koechlin 1954b). Koechlin focused on how the relative concentrations of organic and inorganic ions contributed to the acid-base balance of axoplasm and how this compared across numerous marine species (Koechlin 1953, 1954a, 1955). Schmitt reported five years later, after initiating the Chilean squid procurement program, that *Dosidicus gigas* contained "five peptides whose composition is now being investigated and whose function is not known" (Schmitt 1959b, 457). The follow-up comparisons to the amino acids in *Loligo* axoplasm required "large amounts (10 to 20 g) of freeze-dried axoplasm" so that the peptides could be "isolated in quantities sufficient to permit determination of their composition and biological properties" (Schmitt 1959b, 457). For this work, focused on how far biochemical findings could be generalized, the Schmitt lab turned to the giant nerve fibers of various marine organisms. In so doing, these experimenters relied on the abundance of the sea.

Yet the Schmitt laboratory's agenda also called for comparative work when, owing to the Humboldt squid's migration patterns, weather, or other natural reasons, the animals simply were not available for study. In the handwritten list discussed above, Schmitt pondered whether *Sepia* (cuttlefish) or octopuses, which also boasted some larger fibers, were obtainable in Montemar (fig. 8.2). And, while sometimes this research with different organisms (in New England and Chile) left Schmitt and his colleagues grappling with the limitations of organismal specificity, at other times Schmitt seems to have felt more comfortable with general claims. When publicizing his group's research in the *Rhode Islander Providence Sunday Journal Magazine* in the 1950s, for instance, he emphasized how studies of squid shed light on all types of nerves, even human nerves, and might hold applicability to understanding human cancer (Smith and Hanson 1952). As we shall see with the NRP, Schmitt tended to follow this line of reasoning and emphasize the similarities among neurons he studied or at least common denominators: the more general conclusions he could draw from applying reductionist methods widely (Rasmussen 1997b, 194). This was true even as working with marine organisms comparatively forced him, especially in scientific papers, to address their differences directly (Schmitt 1957c).

By around 1960, both Luxoro and Schmitt had become frustrated with their Chilean arrangement. The reasons were complex and probably related to difficulties involved in paying workers and managing equipment from afar as well as the degradation and damage incurred by samples of axoplasm during their long journeys from Montemar to MIT. Financial disparities between MIT, a privileged US institution, and Viña del Mar almost certainly also played a role, as did Luxoro's waning patience with what he viewed as "only a factory for axoplasm" (Scully 2008, 935). The benefits for Chilean scientists and institutions in the arrangement were unclear, and Schmitt's appropriation of Chilean labor, space, and natural resources produced a colonial feel. Luxoro recalled: "They [the MIT people] were not interested in us [the Montemar people] doing pure research" (Scully 2008, 935). Ultimately, Luxoro petitioned the chancellor of the University of Chile for an electrophysiology facility to be coupled to the Montemar laboratory that by 1962 had become the Laboratory of Cellular Physiology under Luxoro (Scully 2008; Bezanilla 2018).[8] Nagged by logistic challenges, and undoubtedly pulled back to Cambridge by the nascent NRP, Schmitt's team, in contrast, withdrew its presence from Montemar in the mid-1960s.

The NRP

Yet Schmitt had been multitasking extensively throughout the 1950s, formalizing the discipline of biophysics (Rasmussen 1997b, 186–96) alongside managing the squid program in Chile. In 1954, Ernest Allen, the head of the Division of Research Grants at the NIH, had asked him to organize a study section that would delineate the scope of research problems in biophysics via the strategic awarding of extramural grants (Rasmussen 1997b, 188). In 1958, as a part of this initiative, Schmitt chaired an intensive study program in Boulder, Colorado, designed "to aid and encourage the further blending of concepts and methods of physical science with those of life science in the investigation of biological problems" (Oncley, Schmitt, Williams, Rosenberg, and Bolt 1959, 1). The four-week congress, which included dozens of international experts discussing questions ranging from hydrogen bonding and electrolytes to blood coagulation and hormone regulation, also produced a volume, *Biophysical Science—a Study Program* (Oncley, Schmitt, Williams, Rosenberg, and Bolt 1959), that proceeded to define biophysics for a generation (see also Abelson et al. 1959).

The scope of the new discipline under Schmitt's definition was breathtaking, and, consistent with the Schmitt's laboratory's research program, it granted neurobiology pride of place. "Particular emphasis centered upon muscle and nerve," with topics like "the molecular organization of the nerve fiber, the nature of the nerve impulse, nerve metabolism, sensory performance of organisms, receptor mechanisms, and integrated responses of higher neural centers" receiving considerable attention (Oncley, Schmitt, Williams, Rosenberg, and Bolt 1959, 1). The *Study Program* placed nearly all neurobiology then practiced—and much of the rest of experimental biology—under the heading *biophysics*, with Schmitt himself writing on "molecular biology and the physical basis of life processes" and "molecular organization of the nerve fiber" (Oncley, Schmitt, Williams, Rosenberg, and Bolt 1959, 5–10, 455–65; Schmitt 1959a, 1959b). Theodore Bullock, a comparative neurobiologist and marine zoologist in California (French, Lindsley, and Magoun 1984), wrote on receptors and the central nervous system, and Bernard Katz, a German-born physiologist working in England (Stahnisch 2017, 362–63), wrote on synapses (Oncley, Schmitt, Williams, Rosenberg, and Bolt 1959, 504–14, 524–31). The *Study Program* manifesto, outlining Schmitt's vision of biophysics, was not subtle, and it spilled into his agenda for neurobiology.

Soon after the Boulder conference, Schmitt organized two seminar series at MIT. The first, "Fast Fundamental Transfer Processes in Aqueous Biomolecular Systems," occurred in 1960 (Schmitt 1962), with "Molecular Specificity and Biological Memory" following in 1961 (Schmitt 1961). Both invited experts from outside MIT, engaged graduate students, and lasted for full academic terms. The second, in particular, built on the notion that, since many rapidly occurring complex organic processes, like neuronal action potentials, arise from assemblies of molecules in aqueous (often saltwater) solutions, "perhaps other 'higher' functions of the brain" such as learning and memory "may have a macromolecular as well as a neuronal and systems basis" (Schmitt 1961). Enabled by MIT's singular postwar intellectual climate, in these seminars (if not before) Schmitt began thinking about neurons in a more expansive, theoretical, and less experimental way. His Humboldt squid procurement program had been ongoing in Chile for at least three years, but he was transitioning rapidly from axons to brains.

Dovetailing with his various other grants for research into the biomolecular properties of nerves, Schmitt first applied to the NIH in 1962 to begin what became the Neurophysical Sciences Study Program (Schmitt 1962). (This was the same year that Luxoro took over the Montemar Laboratory of Cellular Physiology.) He proposed, with curious diction, "to investigate the 'wet and dry' biophysics of central nervous system function, i.e., to study the physical basis of long-term memory, learning, and other components of conscious, cognitive behavior, by

. . . utilization of the biophysical and biochemical sciences, from the physical chemistry of neuronal and glial constituents (wet biophysics) through bioelectric studies (moist biophysics) to studies of fast transfer of elementary charged particles, organized microfields, stochastic models, and . . . computer science (dry biophysics)" (Schmitt 1962). In the grant, he listed the 1960 and 1961 seminars as previous work. The idea was to create a wide-ranging, formal infrastructure for the investigation of nervous systems clustered under the neologism *neurophysical sciences* and building on the biophysical fundamentals so familiar to him and his laboratory. In the following decades, Schmitt's project (and other similar projects) would carry the moniker, *neuroscience*. The MIT program built on the Schmitt lab's general physiological research with neurons, but the goal now was to unravel the mysteries of vertebrate brains, from molecules to memory, learning, and consciousness (Schmitt 1990). The NRP began officially in 1963, dovetailing directly with the Neurophysical Sciences Study Program. The inaugural meeting of the NRP "associates," which included the core social and scientific network of the program, took place at the New York home of Louis and Eugenie Marron, the anglers who had helped alert Schmitt to the presence of the Humboldt squid (Schmitt 1962).

Conclusions

In this essay, I have made two related arguments. First, from the 1930s through the 1960s, marine stations—especially the MBL in Woods Hole but also the Marine Biological Station in Montemar—offered brick-and-mortar places where Schmitt and his colleagues could adopt new experimental systems. They built these systems around giant axons from squid and other marine invertebrates, expanding the numbers of investigators and organisms in neurophysiology and, in turn, helping keep the subfield of general physiology afloat.[9] This led to studies of both the axon membrane and the axoplasm, the latter being the Schmitt laboratory's specialty, and ultimately the NRP. Second, Schmitt and his lab's research with squid required and facilitated comparative studies, expanding neurophysiologists' understandings of the varieties of neurons that exist in the natural world. In the existing interpretations of Schmitt's career (Rasmussen 1997a, 1997b; Adelman 2010), his more administrative roles as a discipline builder in biophysics and then in the NRP have overshadowed the marine biological work he directed before. But this earlier phase provides new insights into how experiments with neurons, particularly those from marine organisms, connected to discipline building in this pivotal era for the neurosciences.

One such insight is that, when the NRP coalesced at MIT following the conclusion of Schmitt's Chilean squid procurement program, the details of its development obscured some of the marine infrastructure that helped enable it. There was not only a literal retreat from the coastline, from Montemar, Woods Hole, and Rhode Island to Cambridge. There was also a scaling move from neurons to brains that—especially because Schmitt remained consistent in the biophysical lens he applied—covered up the fact that much of his earlier research on neurons had taken place with marine invertebrates. A general physiologist, Schmitt always hoped to glean broad principles from specific inquiries. His papers with Richard Bear and J. Z. Young had been early cases in point (Bear, Schmitt, and Young, 1937a, 1937b, 1937c) as diction such as "*the* ultrastructure" and "*the* protein constituents" of nerve implied a universality in scope that rippled through almost the entirety of Schmitt's corpus and into the founding ethos of the NRP. As was common among axonologists in this era, Schmitt never much prioritized as a primary experimental objective the contextualized marine biology of squid.[10] When he did, the concerns came mostly in attempts to catch organisms he needed, such as when squid were not available. He preferred to extract creatures from the sea, axoplasm from axons, and general principles from local experiments. But this does not minimize how squid axonology nevertheless mandated comparison, or the marine biological moorings that buttressed his career. Indeed, that his marine studies proved necessarily comparative gave grist to the broad claims he wished to make, to the "common denominators" he framed as *similarities* among neurons that helped convince him he could build the NRP on the foundation of his biophysical work.

The comparative nature of neurobiological research with wild-caught marine organisms thus deserves historical attention. As this essay has shown, from the 1940s to the 1960s doing neurophysiology with marine species promoted comparison, whether by necessity, convenience, or design.[11] Many of the neurosciences in the 1970s through the early years of the twenty-first century, particularly cellular electrophysiology (Bezanilla 2008, 2018), axonal transport (Matlin, in this volume), and some developmental biology (de Chadarevian 1998), resembled and even directly reflected the biophysical, biochemical, and macromolecular focus that Schmitt helped pioneer in the 1960s

(Rose and Abi-Rached 2013). With the rise of genetically standardized model species such as *C. elegans* (Ankeny 1997) and *Aplysia californica*, this "neuromolecular gaze" was reinforced (Abi-Rached and Rose 2010; Rose and Abi-Rached 2013). Yet the relative reliance on wild-caught marine species decreased, one consequence being a parallel decline in comparative studies focused on *differences* at the levels of synaptic and simple neural network organization (Bullock 1984, 1995). Historians should continue to probe such trajectories through marine studies, answering Fabio de Sio's call for understanding how individuals, both animals and people, contributed to more local histories of the plural *neurosciences* (de Sio 2011).

Notes

1. The cephalopods also include the octopuses, cuttlefish, and nautiluses. These marine invertebrates are known for their large heads, arm-like and sucker-speckled tentacles, creative intelligences, and ink-squirting, color-changing, and shape-shifting abilities (Gilbert, Adelman, and Arnold 1990; Llinás 1999).
2. The axonologists were based primarily at Oxford and Cambridge Universities in England and the University of Chicago and Washington University in St. Louis in the United States (Marshall 1983, 1987). Their research, from the early years of the twentieth century through the 1940s, focused on the chemical and physical properties of axons.
3. By the early years of the twentieth century, the Naples Station had become a center for cephalopod research, including in neurobiology (Groeben and de Sio 2006; de Sio 2011; de Bont 2014; Muka 2014; Stahnisch 2019; Groeben, in this volume; Steiner, in this volume).
4. Young's breakthrough was actually a rediscovery (Young 1985, 155). Earlier, the Harvard Medical School anatomist Leonard W. Williams described the giant fibers as parts of the New England squid's nervous system (Williams 1909). There are various potential reasons why contemporary investigators ignored this point. The focus on frogs in neurophysiology may have played a role.
5. There were some exceptions. For instance, the prominent Spanish neurophysiologist Rafael Lorente de Nó "never accepted that the . . . giant fiber of squid is a confluence of processes from many neuron somata [cell bodies] of ordinary size" (Bullock 1995, 216). I thank Fabio de Sio for the advice to pay attention to this history.
6. This also provided an evolutionary logic for why a squid's giant axons, which enable crucial functions like startling and escaping responses, are unusually large.

7. "Physiology 1923," MBL Dataset, History of the Marine Biological Laboratory, MBLWHOI Library, Special Collections and MBL Archives, accessed 3 May 2019, https://history.archives.mbl.edu/course/903.
8. By the 1970s, Luxoro's center in Montemar was world renowned, training a generation of students who also became leaders in neurophysiology (Bacigalupo and Kakuljan 1997; Scully 2008; Bezanilla 2018). The laboratory fell into steep decline after 1973, however, a victim of Augusto Pinochet's dictatorship alongside unanticipated changes in the migratory patterns of the Humboldt squid.
9. Put another way, there is a strong argument to be made that, even though this biophysical and biochemical subfield fragmented among the proliferating biological sciences of the late twentieth century, one of the greatest legacies of early twentieth-century general physiology was in the neurosciences of the 1960s (Rasmussen 1997b). Other aspects of this subfield found their ways elsewhere, such as into molecular biology (Pauly 1987b).
10. In the mid-twentieth century, a powerful group of biologists studying marine organisms pushed to define yet another discipline, *marine biology*, which they envisioned as experimental biology, placed in ecological and evolutionary context, and aligned with biological oceanography (Ellis 2007). Schmitt was not one of them.
11. I have taken inspiration for this diction from Alan Hodgkin, whose autobiographical account, "Chance and Design in Electrophysiology," highlighted the roles of both factors in the development of neurophysiology (Hodgkin 1976).

References

Abelson, Philip H., et al. 1959. "Biophysical Science—A Study Program." *Reviews of Modern Physics* 31, nos. 1–2:1–256, 269–563.

Abi-Rached, Joelle M., and Nikolas Rose. 2010. "The Birth of the Neuromolecular Gaze." *History of the Human Sciences* 23, no. 1:11–36.

Adelman, George. 2010. "The Neurosciences Research Program at MIT and the Beginning of the Modern Field of Neuroscience." *Journal of the History of the Neurosciences* 19, no. 1:15–23.

Adelman, George, and Barry Smith. 1998. "Francis Otto Schmitt, November 23, 1903—October 3, 1995." In *Biographical Memoirs* (vol. 75), 342–54. Washington, DC: National Academies Press.

Ankeny, Rachel A. 1997. "The Conqueror Worm: An Historical and Philosophical Examination of the Use of the Nematode *C. elegans* as a Model Organism." PhD diss., University of Pittsburgh.

Bacigalupo, Juan, and Manuel Kakuljan. 1997. "Professor Mario Luxoro, Honorary Member of the Chilean Society of Physiological Sciences." *Biological Research* 30, no. 3:91–94.

Bear, Richard S., Francis O. Schmitt, and John Z. Young. 1937a. "Investigations on the Protein Constituents of Nerve Axoplasm." *Proceedings of the Royal Society of London Series B* 123 (833): 520–29.

———. 1937b. "The Sheath Components of the Giant Nerve Fibres of the Squid." *Proceedings of the Royal Society of London Series B* 123, no. 833: 496–504.

———. 1937c. "The Ultrastructure of Nerve Axoplasm." *Proceedings of the Royal Society of London Series B* 123, no. 833:505–19.

Bezanilla, Francisco. 2008. "Ion Channels: From Conductance to Structure." *Neuron* 60, no. 3: 456–68.

———. 2018. "Influences: The Cell Physiology Laboratory in Montemar, Chile." *Journal of General Physiology* 150, no. 11:1464–68.

Bullock, Theodore H. 1984. "Comparative Neuroscience Holds Promise for Quiet Revolutions." *Science* 225, no. 4661:473–48.

———. 1995. "Neural Integration at the Mesoscopic Level: The Advent of Some Ideas in the Last Half Century." *Journal of the History of the Neurosciences* 4, nos. 3–4:216–35.

Chiang, Howard H. 2009. "The Laboratory Technology of Discrete Molecular Separation: The Historical Development of Gel Electrophoresis and the Material Epistemology of Biomolecular Science, 1945–1970." *Journal of the History of Biology* 42, no. 3:495–527.

Choudhury, Suparna, and Jan Slaby, eds. 2012. *Critical Neuroscience: A Handbook of the Social and Cultural Contexts of Neuroscience*. New York: Wiley-Blackwell.

Clarke, Adele E., and Joan H. Fujimura, eds. 1992. *The Right Tools for the Job: At Work in Twentieth-Century Life Sciences*. Princeton, NJ: Princeton University Press.

Cole, Kenneth S., and Howard J. Curtis. 1939. "Electric Impedance of the Squid Giant Axon During Activity." *Journal of General Physiology* 22, no. 5:649–70.

Craver, Carl F. 2008. "Physical Law and Mechanistic Explanation in the Hodgkin and Huxley Model of the Action Potential." *Philosophy of Science* 7, no. 5:1022–33.

Creager, Angela N. H. 2001. *The Life of a Virus: Tobacco Mosaic Virus as an Experimental Model*. Chicago: University of Chicago Press.

———. 2013. *Life Atomic: A History of Radioisotopes in Science and Medicine*. Chicago: University of Chicago Press.

Curtis, Howard J., and Kenneth S. Cole. 1940. "Membrane Action Potentials from the Squid Giant Axon." *Journal of Cellular and Comparative Physiology* 15:147–57.

de Bont, Raf. 2014. *Stations in the Field: A History of Place-Based Animal Research, 1870–1930*. Chicago: University of Chicago Press.

de Chadarevian, Soraya. 1998. "Of Worms and Programmes: *Caenorhabditis elegans* and the Study of Development." *Studies in History and Philosophy of Biological and Biomedical Sciences* 29, no. 1:81–105.

de Leo, A. 2008. "Enrico Sereni: Research on the Nervous System of Cephalopods." *Journal of the History of the Neurosciences* 17, no. 1:56–71.

de Sio, Fabio. 2011. "Leviathan and the Soft Animal: Medical Humanism and the Invertebrate Models for Higher Nervous Function." *Medical History* 55, no. 3:369–74.

Eccles, John C., Ragnar A. Granit, and John Z. Young. 1932. "Impulses in the Giant Nerve Fibres of Earthworms." *Journal of Physiology* 77 (suppl.): 23P–25P.

Ellis, Erik. 2007. "What Is Marine Biology? Defining a Science in the United States in the Mid 20th Century." *History and Philosophy of the Life Sciences* 29, no. 4:469–93.

French, John D., John B. Lindsley, and Horace W. Magoun. 1984. *An American Contribution to Neuroscience: The Brain Research Institute, UCLA, 1959–1984.* Los Angeles: University of California, Los Angeles, Brain Research Institute.

Gasser, Herbert S., and Joseph Erlanger. 1922. "A Study of the Action Currents of Nerve with the Cathode Ray Oscillograph." *American Journal of Physiology* 62, no. 3:496–524.

Geren, Betty B., and Josephine Raskind. 1953. "Development of the Fine Structure of the Myelin Sheath in Sciatic Nerves of Chick Embryos." *Proceedings of the National Academy of Sciences of the U.S.A.* 39, no. 8:880–84.

Geren, Betty B., and Francis O. Schmitt. 1954. "The Structure of the Schwann Cell and Its Relation to the Axon in Certain Invertebrate Nerve Fibers." *Proceedings of the National Academy of Sciences of the U.S.A.* 40, no. 9:863–70.

Gilbert, Daniel L., William J. Adelman, and John M. Arnold, eds. 1990. *Squid as Experimental Animals.* New York: Plenum.

Grave, Caswell, and Francis O. Schmitt. 1924. "A Mechanism for the Coordination and Regulation of the Movement of Cilia of Epithelia." *Science* 60, no. 1550:246–48.

Groeben, Christiane, and Fabio de Sio. 2006. "Nobel Laureates at the Stazione Zoologica Anton Dohrn: Phenomenology and Paths to Discovery in Neuroscience." *Journal of the History of the Neurosciences* 15, no. 4:376–95.

Guillery, Rainer W. 2005. "Observations of Synaptic Structures: Origins of the Neuron Doctrine and Its Current Status." *Philosophical Transactions of the Royal Society B* 360, no. 1458:1281–1307.

———. 2007. "Relating the Neuron Doctrine to the Cell Theory: Should Contemporary Knowledge Change Our View of the Neuron Doctrine?" *Brain Research Reviews* 55, no. 2:411–21.

Hodgkin, Alan L. 1976. "Chance and Design in Electrophysiology: An Informal Account of Certain Experiments on Nerve Carried Out between 1934 and 1952." *Journal of Physiology* 263, no. 1:1–21.

Hodgkin, Alan L., and Andrew F. Huxley. 1939. "Action Potentials Recorded from Inside a Nerve Fibre." *Nature* 144, no. 3651:710–11.

———. 1945. "Resting and Action Potentials in Single Nerve Fibres." *Journal of Physiology* 104, no. 2:176–95.

———. 1952. "A Quantitative Description of Membrane Current and Its Application to Conduction and Excitation in Nerve." *Journal of Physiology* 117, no. 4:500–544.

Hodgkin, Alan L., and Richard D. Keynes. 1953. "The Mobility and Diffusion Coefficient of Potassium in Giant Axons from *Sepia*." *Journal of Physiology* 119, no. 4:513–28.

———. 1955. "Active Transport of Cations in Giant Axons from *Sepia* and *Loligo*." *Journal of Physiology* 128, no. 1:28–60.

Holmes, Frederic L. 1993. "The Old Martyr of Science: The Frog in Experimental Physiology." *Journal of the History of Biology* 26, no. 2:311–28.

Jones, Edward G. 1999. "Golgi, Cajal and the Neuron Doctrine." *Journal of the History of the Neurosciences* 8, no. 2:170–78.

Kandel, Eric R., and Ladislav Tauc. 1965. "Mechanism of Heterosynaptic Facilitation in the Giant Cell of the Abdominal Ganglion of *Aplysia depilans*." *Journal of Physiology* 181, no. 1:28–47.

Koechlin, Bernard A. 1953. "The Amino Acid Composition of a Protein Isolated from Lobster Nerve." *Journal of Biological Chemistry* 205:597–604.

———. 1954a. "The Isolation and Identification of the Major Anion Fraction of the Axoplasm of Squid Giant Nerve Fibers." *Proceedings of the National Academy of Sciences of the U.S.A.* 40, no. 2:60–62.

———. 1954b. "Progress Report." November 23. Folder 16, box 3, Francis Otto Schmitt Papers (MC0154), Institute Archives and Special Collections, Massachusetts Institute of Technology, Cambridge, MA.

———. 1955. "On the Chemical Composition of the Axoplasm of Squid Giant Nerve Fibers with Particular Reference to Its Ion Pattern." *Journal of Biophysical and Biochemical Cytology* 1, no. 6:511–29.

Lenoir, Timothy. 1986. "Models and Instruments in the Development of Electrophysiology, 1845–1912." *Historical Studies in the Physical and Biological Sciences* 17, no. 1:1–54.

Levy, Arnon. 2014. "What Was Hodgkin and Huxley's Achievement?" *British Journal for the Philosophy of Science* 65, no. 3:469–92.

Llinás, Rodolfo R. 1999. *The Squid Giant Synapse: A Model for Chemical Transmission*. New York: Oxford University Press.

Logan, Cheryl A. 2002. "Before There Were Standards: The Role of Test Animals in the Production of Empirical Generality in Physiology." *Journal of the History of Biology* 35, no. 2:329–63.

Luxoro, Mario. n.d. "Instrucciones para Disecar el Nervio Medial de la Jibia." Francis Otto Schmitt Papers (MC0154), folder 16, box 3, Institute Archives and Special Collections, Massachusetts Institute of Technology, Cambridge, MA.

Maienschein, Jane. 1987. "Physiology, Biology, and the Advent of Physiological Morphology." In *Physiology in the American Context, 1850–1940*, ed. Gerald Geison, 177–94. Baltimore: American Physiological Society.

Marshall, Louise H. 1983. "The Fecundity of Aggregates: The Axonologists at Washington University, 1922–1942." *Perspectives in Biology and Medicine* 26, no. 4:613–36.

———. 1987. "Instruments, Techniques, and Social Units in American Neurophysiology, 1870–1950." In *Physiology in the American Context, 1850–1940*, ed. Gerald Geison, 351–70. Baltimore: American Physiological Society.

McMahan, Uel J., ed. 1990. *Steve: Remembrances of Stephen W. Kuffler*. Sunderland: Sinauer.

Muka, Samantha K. 2014. "Working at Water's Edge: Life Sciences at American Marine Stations, 1880–1930." PhD diss., University of Pennsylvania.

———. 2016. "The Right Tool and the Right Place for the Job: The Importance of the Field in Experimental Neurophysiology, 1880–1945." *History and Philosophy of the Life Sciences* 38, no. 7:1–28.

Oncley, John L., Francis O. Schmitt, Robley C. Williams, M. D. Rosenberg, and Richard H. Bolt, eds. 1959. *Biophysical Science—a Study Program*. New York: Wiley.

Pauly, Philip J. 1987a. *Controlling Life: Jacques Loeb and the Engineering Ideal in Biology*. Baltimore: The Johns Hopkins University Press.

———. 1987b. "General Physiology and the Discipline of Physiology, 1890–1935." In *Physiology in the American Context, 1850–1940*, ed. Gerald Geison, 195–208. Baltimore: American Physiological Society.

Pumphrey, R. J., and John Z. Young. 1938. "The Rates of Conduction of Nerve Fibres of Various Diameters in Cephalopods." *Journal of Experimental Biology* 15, no. 4:453–66.

Purves, Dale, et al., eds. 2018. *Neuroscience*, 6th ed. New York: Oxford University Press.

Rasmussen, Nicolas. 1996. "Making a Machine Instrumental: RCA and the Wartime Origins of Biological Electron Microscopy in America, 1940–1945." *Studies in History and Philosophy of Science* 27, no. 3:311–49.

———. 1997a. "The Mid-Century Biophysics Bubble: Hiroshima and the Biological Revolution in America." *History of Science* 35, no. 109:245–93.

———. 1997b. *Picture Control: The Electron Microscope and the Transformation of Biology in America, 1940–1960*. Stanford, CA: Stanford University Press.

Rheinberger, Hans-Jörg. 1997. *Toward a History of Epistemic Things: Synthesizing Proteins in the Test Tube*. Stanford, CA: Stanford University Press.

Rose, Nikolas, and Joelle M. Abi-Rached, eds. 2013. *Neuro: The New Brain Sciences and the Management of the Mind*. Princeton, NJ: Princeton University Press.

Schmitt, Francis O. 1951. Letter to Clinton Atkinson. March 5. Folder 16, box 3, Francis Otto Schmitt Papers (MC0154), Institute Archives and Special Collections, Massachusetts Institute of Technology, Cambridge, MA.

———. 1955. Letter to Luis Rene Rivas. January 19. Folder 16, box 3, Francis Otto Schmitt Papers (MC0154), Institute Archives and Special Collections, Massachusetts Institute of Technology, Cambridge, MA.

———. 1957a. Letter to Mario Luxoro. March 5. Folder 16, box 3, Francis Otto Schmitt Papers (MC0154), Institute Archives and Special Collections, Massachusetts Institute of Technology, Cambridge, MA.

———. 1957b. Letter to Carlos Ruiz. March 11. Folder 16, box 3, Francis Otto Schmitt Papers (MC0154), Institute Archives and Special Collections, Massachusetts Institute of Technology, Cambridge, MA.

———. 1957c. "The Fibrous Protein of the Nerve Axon." *Journal of Cellular and Comparative Physiology* 49:165–74.

——— 1959a. "Molecular Biology and the Physical Basis of Life Processes." *Reviews of Modern Physics* 31, no. 1:5–10. Also in *Biophysical Science—a Study Program*, ed. John L. Oncley, Francis O. Schmitt, Robley C. Williams, M. D. Rosenberg, and Richard H. Bolt, 5–10 (New York: Wiley).

———. 1959b. "Molecular Organization of the Nerve Fiber." *Reviews of Modern Physics* 31, no. 2:455–65. Also in *Biophysical Science—a Study Program*, ed. John L. Oncley, Francis O. Schmitt, Robley C. Williams, M. D. Rosenberg, and Richard H. Bolt, 455–65. New York: Wiley.

———. 1961. "Molecular Specificity and Biological Memory, Molecular Biology Seminar 7.98." Seminar syllabus. Folder 14, box 1, Francis Otto Schmitt Papers (MC0154), Institute Archives and Special Collections, Massachusetts Institute of Technology, Cambridge, MA.

———. 1962. "Neurophysical Sciences Study Program." March 5. Grant application to the NIH. Folder 27, box 10, Records of the Neurosciences Research Program (AC107), Institute Archives and Special Collections, MIT, Cambridge, MA.

———. 1990. *The Never-Ceasing Search*. Philadelphia: American Philosophical Society.

Schmitt, Francis O., Richard S. Bear, and George L. Clark. 1935. "The Role of Lipoids in the X-Ray Diffraction Patterns of Nerve." *Science* 82, no. 2115: 44–45.

Schmitt, Francis O., and Otto H. Schmitt. 1940. "Partial Excitation and Variable Conduction in the Squid Giant Axon." *Journal of Physiology* 98, no. 1: 26–46.

Scully, Tony. 2008. "The Great Squid Hunt." *Nature* 454, no. 7207:934–36.

Smith, Henry H., and Edward C. Hanson. 1952. "It's a Case of Pure Nerves: Squid Nerves for Research." *Rhode Islander Providence Sunday Journal Magazine*, August 3, folder 16, box 3, Francis Otto Schmitt Papers (MC0154), Institute Archives and Special Collections, Massachusetts Institute of Technology, Cambridge, MA.

Stadler, Max. 2009. "Assembling Life: Models, the Cell, and the Reformations of Biological Science, 1920–1960." PhD diss., Imperial College, London.

Stahnisch, Frank W. 2017. "How the Nerves Reached the Muscle: Bernard Katz, Stephen W. Kuffler, and John C. Eccles—Certain Implications of Exile for the Development of Twentieth-Century Neurophysiology." *Journal of the History of the Neurosciences* 26, no. 4:351–84.

———. 2019. "Catalyzing Neurophysiology: Jacques Loeb, the Stazione Zoologica di Napoli, and a Growing Network of Brain Scientists, 1900–1930s." *Frontiers in Neuroanatomy* 13:1–14.

Vidal, Fernando, and Francisco Ortega, eds. 2011. *Neurocultures: Glimpses into an Expanding Universe.* New York: Wiley-Blackwell.

White, H. L., and Francis O. Schmitt. 1925. "Observations on Kidney Function in *Necturus Maculosus.*" *Science* 62, no. 1606:334.

Williams, Leonard W. 1909. *The Anatomy of the Common Squid Loligo Pealii, Lesueur.* Leiden: Brill.

Young, John Z. 1929. "Fenomeni Istologici Consecutivi alla Sezione dei Nervi nei Cefalopodi." *Bollettino della Società italiana di biologia sperimentale* 4: 741–44.

———. 1936a. "The Giant Nerve Fibres and Epistellar Body of Cephalopods." *Quarterly Journal of Mircoscopical Science* 311:367–86.

———. 1936b. "The Structure of Nerve Fibres in Cephalopods and Crustacea." *Proceedings of the Royal Society of London B* 121, no. 823:319–37.

———. 1975. "The F. O. Schmitt Lecture: Sources of Discovery in Neuroscience." In *The Neurosciences: Paths of Discovery,* ed. Frederic G. Worden, Judith P. Swazey, and George Adelman, 15–46. Cambridge, MA: MIT Press.

———. 1985. "Cephalopods and Neuroscience." *Biological Bulletin* 168, no. 3: S153–S158.

NINE

Microscopes and Moving Molecules: The Discovery of Kinesin at the Marine Biological Laboratory

KARL S. MATLIN

Every spring, the squid *Loligo pealei*[1] migrates from the edge of the continental shelf into warmer waters near the coast of Cape Cod and other regions on the eastern shore of the United States (Arnold et al. 1974, 28). Today, and likely in the past, the squid migration is reason for excitement because striped bass and bluefish, favorites of anglers and restaurantgoers alike, are attracted to the Cape waters by this abundant source of food. Squid are shellfish that lost their shells during evolution. For protection, they instead developed the capacity to swim freely in the ocean and accelerate rapidly away from predators, propelled by a jet of water that is often accompanied by a smoke screen of black ink.

Since the late nineteenth century, another migration accompanied that of the squid to New England. Beginning in the early summer, biologists from around the world travel to the Marine Biological Laboratory (MBL) in Woods Hole, Massachusetts, to set up laboratories beside the sea. Since the MBL was established in 1888, many of those biologists came to Woods Hole expressly to conduct research on squid. While prolific, squid have a particularly delicate skin. When they are captured and placed in holding tanks, their skin is easily damaged, and, if it is, they die within days. For this reason as well as seasonal availability, biologists dependent on live animals must go to the source for their experiments.

Historically, squid are among neurobiology's most important experimental animals. In 1936, while working at the MBL, John Z. Young described a series of large nerve fibers emanating from the stellate ganglia of *Loligo pealei* near the head and running to the muscles of the mantle (Young 1936).[2] Apparently, such large nerves are necessary for the quick transmission of signals to the squid muscles in the mantle responsible for generating the propulsive jet of water needed to escape predators (Young 1936). Nerves are composed of nerve cells or neurons, single cells consisting of a cell body with a nucleus and organelles similar to other eukaryotic cells and long processes projecting from the cell body called *axons* and *dendrites*. Significantly, the nerve fibers that Young observed consisted of not a bundle of thin axons enervating the mantle but instead a single giant axon several centimeters in length with a diameter of 300–700 microns (μm) (Arnold et al. 1974).[3] Within a few years of Young's report, the British scientists Alan Hodgkin, who had learned how to dissect the squid giant axon from Kenneth Cole at the MBL, and Andrew Huxley had used electrodes inserted into the cut end of the axon to establish how ions move across the axonal membrane to generate action potentials that transmit signals in the nervous systems of squid and higher organisms (Schwiening 2012).

This essay describes the identification and description of the molecular motor protein kinesin at the MBL, a discovery essential to understanding not just the biology of neurons and the squid giant axon but all eukaryotic cells. Axons in all higher organisms are generally not of the diameter found in the squid but are often at least as long, if not longer. The cytoplasm of the axon, or axoplasm, is populated with organelles, cytoskeletal elements, and myriad small vesicles similar to those found in the cytoplasm of other cells. Most originate in the cell body and, to perform their functions, must be moved enormous distances through the axoplasm to nerve terminals. Until the mid-1980s, nobody had a good idea of how this occurred rapidly enough to be biologically meaningful. Within a short time, this situation changed through the simultaneous efforts of several competing research groups, all working with the squid giant axon at the MBL.[4]

This story, however, is not only about kinesin but also about the unique working atmosphere at the MBL. Although the scientists described here all needed to be at the seaside to work with squid, they were also drawn to the MBL by the opportunities to collaborate with other scientists they normally might never otherwise encounter. Another important aspect of this story is how critical the microscope was to this work. Although the discovery was molecular in nature, what made it possible was not some advanced biochemical technique but rather a new version of the light microscope. As we shall see, the microscope enabled the molecular studies to be directly linked to the axon, ensuring that the molecular and chemical activities of kinesin never supplanted its biological function.

In the first section, I provide some background on the light microscope and its inherent limitations. Next, I introduce Robert D. Allen, who, with his collaborator and wife Nina Strömgren Allen, discovered while teaching at the MBL a novel type of video microscopy that circumvented some of the light microscope's limitations. Allen, with his collaborators Scott Brady and Ray Lasek, then began to use this microscope to look at the movement of vesicles and organelles on filaments in the squid axoplasm. Their work was interrupted by the arrival at the MBL of two scientists who were not regular visitors to the lab. Collaboration among these research groups turned into competition, and, in the end, a team composed of the recently arrived individuals was, with the help of other MBL scientists, able to prove that vesicles moved on microtubule filaments in the axon driven by the molecular motor kinesin.

But first to the light microscope, which is where the story begins.

Microscopes

The study of cells has depended on the light microscope since Robert Hooke viewed cork with a primitive instrument in the seventeenth century and named the holes in the cork *cells*. Early microscopes had problems resolving small objects owing, in part, to distortions caused by lens curvature and the glass used to construct the lenses (Bradbury 1967; James 1976; Bracegirdle 1978; Harris 1999; Maienschein 2018; Oldenbourg 2018; Schickore 2018). Solutions to these problems began to appear in the beginning of the nineteenth century with the design of achromatic lens combinations. By the 1830s, improved microscopes were widely available in Europe, and detailed observations of tissues and cells multiplied. At about this time, Schleiden and Schwann suggested that cells are the fundamental building blocks of both plants and animals, and observations of cells expanded to create the field of cytology (Harris 1999; Maienschein 2018; Schickore 2018).

In the second half of the nineteenth century, routinely available microscopes could reach the near theoretical limit of resolution governed by the wavelength of light (Bradbury 1967; James 1976; Bracegirdle 1978; Oldenbourg 2018). Resolving power is defined as the smallest distance at which two objects can be distinguished as separate. In visible light, this distance is approximately 0.2 μm (200 nanometers [nm]) (James 1976). Even with maximal resolution, many parts of cells are not discernible because cells lack contrast and appear transparent (James 1976; Schickore 2007, 2018; Oldenbourg 2018). To solve this problem, scientists applied chemicals both to preserve cells and tissues and to stain cytoplasmic and nuclear components differently to make them visible (Schickore 2018). While the use of stains was a great advance, it largely precluded microscopic observations of living cells because the staining procedures killed the cells, except in certain instances in which naturally occurring pigments provided contrast or so-called vital stains could be applied (Schickore 2018).

This situation began to change in the twentieth century when microscopists developed optical methods to improve contrast that were applicable to living cells. As light passes through transparent objects, little is absorbed, but there is a change in the propagation velocity or phase of the transmitted light waves. The degree of phase change varies according to the refractive index of the object.[5] In 1934, the Dutch physicist Fritz Zernicke invented phase contrast microscopy, which essentially makes phase differences appear as differences in contrast (Bradbury 1967; James 1976). Because individual parts of even living cells have different refractive indexes, they are visible by phase contrast microscopy. Other methods followed. Among these was differential interference contrast or DIC. DIC in a sense resembles phase contrast in that phase changes are converted into changes in contrast. In the case of DIC (and other forms of interference microscopy), the beam of light directed toward the object is split in two by a prism and then recombined after passing through the object in such a way that the two beams interfere with each other to generate contrast (James 1976). DIC was invented by the French scientist Georges Nomarski in 1952 and turned into a practical instrument in collaboration with the microscope manufacturer Carl Zeiss in the 1960s (Allen, David, and Nomarski 1969).

Robert Allen

Robert Day Allen was a biologist, not a microscope developer, although one could be forgiven for thinking otherwise. His research program was driven by his wish to understand how cells and their contents moved, not the desire to create new instruments, even though he became famous for the latter.[6] He began his career as an embryologist working on fertilization (Allen 1953). By the mid-1950s, he was studying protoplasmic streaming in amoebas as it related to the supposed mechanisms driving cell motility. He continued to work with a variety of organisms, such as the foraminifer *Allogromia* (a kind of marine protozoan), throughout his career (Allen, Travis, et al. 1982). This work depended not only on microscopy but also on the ability to film movies through the microscope to capture and analyze cell movement (Allen and Roslansky 1959). *Microcinematography,* as this technique was called, allowed investigators like Allen to record slowly moving cells over many hours and then essentially compress time by speeding up these time-lapse movies so that the details of cell movement became apparent (Landecker 2011).

From the beginning of his career, Allen was closely associated with the MBL. Some of his doctoral research was conducted there, and he returned nearly every summer, eventually buying a house in Woods Hole and joining the MBL governing board. Like other summer investigators, each year he transported instruments and reagents from his permanent university laboratory at Dartmouth College to a rented space at the MBL. There he and his students continued their work, focusing on easily available marine organisms. Often Allen and other MBL regulars spent the entire summer at the lab, forming close professional and personal relationships. In addition to his scientific skills, Allen was an outstanding cellist, and he frequently performed for the Woods Hole community during the summer, continuing a musical tradition as old as the MBL itself (Rebhun 1986). The scientific atmosphere at the MBL was one of friendly competition; researchers respected the boundaries of individual projects, but they shared instruments and ideas and frequently formed spontaneous collaborations. At the end of each summer, research groups presented their accomplishments in a series of short talks. To encourage openness, the *Biological Bulletin,* the MBL house journal, published abstracts of the talks, which protected the priority of discovery and no doubt also helped scientists justify the summer's expenditures to funding agencies and home institutions.

By the 1970s, Allen's work on protoplasmic streaming led him to focus on the cellular cytoskeleton (Dietrich 2015). Protoplasmic streaming, or the "circulation of the cell sap," had been observed since the eighteenth century, but its basis was not understood (Dietrich 2015). By the mid-twentieth century, scientists proposed mechanisms of streaming that involved a process analogous to muscle contraction. Muscle contraction is dependent on the interaction of actin filaments with the motor protein myosin. Motor proteins use the energetic molecule ATP to drive motility. Because both actin and myosin are found not only in muscle but also in nonmuscle cells, including the amoeba and other organisms that Allen studied, their proposed involvement in cytoplasmic streaming was plausible (Cheney, Riley, and Mooseker 1993; Dietrich 2015).

In addition to actin and myosin, Allen also gradually became interested in motility linked to microtubules, another cytoskeletal element in cells, one that is composed of the protein tubulin (McGee-Russell and Allen 1971; Berlinrood, McGee-Russell, and Allen 1972). In the 1970s, it was known that the beating of cilia and flagella was caused by the bending and sliding of microtubules powered by the protein dynein, the only known microtubule motor. Microtubules were also thought to be involved in the separation of chromosomes during mitosis, although, in this case, the required motor protein was unknown (Cytoskeleton 2008).

Although both actin filaments and microtubules were visible in the electron microscope, an instrument capable of very high magnification, their diameters (8 nm and 25 nm, respectively) were too small to be resolved by the conventional light microscope. Because the electron microscope could not be used on living cells, investigators carried out a variety of experiments using chemical inhibitors in conjunction with light microscopy to indirectly implicate both actin filaments and microtubules in not only muscle contraction and the beating of cilia and flagella but also the movements of particles through the cytoplasm of cells, a process presumably also driven by motor proteins.

In 1972, Allen used microcinematography to follow the movement of particles in bundles of amphibian embryo nerve fibers in organ culture, speculating that microtubules were involved. In parallel electron micrographs, he identified individual microtubules (which he called *neurotubules*), but he was unable to resolve either these or the individual neurites (thin processes resembling axons and dendrites) that made up the nerve fibers in the light microscope (Berlinrood, McGee-Russell, and Allen 1972). Despite this venture into the study of nerve cells, he continued over the next few years to concentrate his efforts on amoebas and his other favorite nonneuronal experimental motility models.

In addition to spending summers at the MBL, Robert and Nina Allen began teaching an advanced microscopy course there in the winter. In the December 1978 course, they included a demonstration of video microscopy. Video cameras had been used on microscopes as early as the 1950s following the invention and development of television (Allen, Allen, and Travis 1981; Inoué 1986). However, by the late 1970s, miniaturization and improved sensitivity as well as the production of images compatible with computer manipulation and the availability of relatively affordable video recorders made video microscopy more feasible. Video microscopy was attractive because images were recorded and could be displayed on monitors immediately without any additional processing. This was in contrast to the photographic movie film used in microcinematography, which not only was expensive but also needed to be developed in the darkroom, a tedious and time-consuming process.

Although Allen was aware of prior attempts to use video cameras mounted on microscopes, he was never sufficiently impressed with the resolution to try it himself. This changed after his father-in-law, a well-known astronomer, informed the Allens that astronomers were successfully using the new improved video cameras on telescopes and urged them to try them in microscopy.

In the 1978 course, the Allens connected a video camera loaned by the Japanese manufacturer Hamamatsu to a Zeiss DIC microscope.[7] While one motivation for using a video camera may have been to make it easier for students in the course to see microscopic images without looking through the eyepieces, Nina Allen also discovered another, more significant advantage. While viewing a preparation of live cells with the students, she found serendipitously that, if she altered the microscope illumination in such a way that the image field was flooded with light to the point where almost nothing was visible and then manually adjusted the black level on the video camera, she was able to see details of cell structure on the video monitor that had never been seen before.

Over the next two years, the Allens figured out what had happened in class and improved the technique, which they named AVEC-DIC for Allen video-enhanced contrast–DIC microscopy. As they explained in one of the two papers announcing their discovery (Allen, Allen, and Travis 1981; Allen, Travis, Allen, and Yilmaz 1981), visually perceived contrast (*C*) is related to the brightness or intensity of the background

(I_B) and that of the specimen (I_S) such that $C = (I_B - I_S)/I_B$. With the addition of a video camera in which the background brightness can be adjusted, an additional factor, I_V, is included in the equation such that $C = (I_B - I_S)/(I_B - I_V)$. The result is that the video-perceived image now has higher contrast than can be achieved with the microscope alone (Allen, Allen, and Travis 1981; see also Salmon 1995).[8]

The most important, even revolutionary advantage of the AVEC-DIC method was that investigators could now detect objects below the resolution of normal light microscopy (Allen, Allen, and Travis 1981). In particular, they could see microtubules, either in cells or assembled from the purified microtubule protein tubulin, even though their diameter is 25 nm, well below the 200-nm resolving power of the light microscope. Significantly, AVEC did not improve the resolution of the light microscope but only made small things visible; microtubules and tiny cellular particles could be distinguished in the video image, but they all appeared to be about 200 nm in diameter, corresponding to the minimal resolvable distance. The limit of resolution was still in force, but AVEC had made the diffracted light footprint of very small objects detectable.

Prior to publishing the AVEC method in 1981, Allen described it in a series of presentations. The first of these was at the meeting of the New York Microscopical Society on November 6, 1980.[9] Apparently, the video that Allen showed of submicroscopic particles moving around in cells led to an explosion of excitement in the popular press, possibly ignited by a brief report on the meeting in the *New York Times* (Sullivan 1980).[10] Within a few weeks, Allen and other members of his laboratory appeared on the CBS Morning News, ABC News, and the Today show.[11]

The attention also generated a negative reaction. After Allen presented AVEC at the annual meeting of the American Society for Cell Biology (ASCB) in Cincinnati, also in November 1980, his longtime MBL colleague Shinya Inoué reportedly stated in the session that Allen had been "ruining [his] images and then rescuing them electronically."[12] Inoué was a contemporary of Allen's, and the two were somewhat mirror images: while Allen considered himself a biologist who "dabbled" in microscopy (and could not handle the mathematical formalisms of optics), Inoué was a rigorous microscope developer who used his polarizing microscopes to make major discoveries about the mitotic spindle and other biological processes (Oldenbourg 2018). At the MBL they were friendly competitors. Inoué had been in the room when the Allens made their accidental discovery that led to AVEC and had independently pursued a different high-contrast video microscope system that he described in the same session of the ASCB meeting (Inoué 1981).[13] Allen bragged that AVEC did not need the expensive attachments that Inoué favored and did not even need to produce a good-quality visual image through the microscope. His claim may have offended Inoué's rigorous sensibilities at the same time that the attention paid to Allen by the press made him a bit jealous. Apparently, however, a letter from Allen to Inoué after the Cincinnati meeting calmed the waters, and the two remained cordial.[14]

Axoplasm

In the early summer of 1981, Allen attended a symposium at the Cold Spring Harbor Laboratory in New York on the organization of the cytoplasm. His presentation focused on the movements of "ultrastructural elements" in different types of cells that were now visible with his new AVEC technique (Allen, Travis, et al. 1982). In addition to observations made in *Allogromia* and some other cell types, he mentioned preliminary experiments examining the movement of particles in chick neurons. Later that summer at the MBL, Allen was approached by Jan Metuzals, an electron microscopist working with the neurophysiologist Ichiji Tasaki, who urged him to apply his microscopic technique to the squid giant axon. Clearly, the axon's size would give Allen much more to look at than the tiny chick neurons. Allen agreed, and in short order Tasaki was dissecting out the squid axon for Allen's experiments. The operation resembled the Keystone Cops: Tasaki's lab was on the third floor of the Whitman (now Rowe) building at the MBL, while Allen's microscope was on the third floor of the Loeb building across a small quadrangle. After dissecting the axon, Tasaki handed it to Metuzals, who ran it immediately to Allen's lab for observation. They were not disappointed; with AVEC, particles could be seen moving quickly up and down the axon. The only problem was that motion continued for only about a minute.

While at Cold Spring Harbor earlier that summer, Allen had also spoken with Ray Lasek and his postdoctoral fellow Scott Brady, who presented at the meeting as well (Lasek and Brady 1982). Allen and Lasek already knew each other since, like Allen, Lasek had been coming to the MBL for years. Lasek, from Case Western Reserve University, was one of the leading authorities on axon transport; he had published an influential paper as a graduate student and made a number of important contributions subsequently (Lasek 1967, 1970).

The idea that considerable cellular material moves through the axon arose during early studies of nerve regeneration motivated by nerve injuries during World War II (Grafstein and Forman 1980; Ochs 2004, 217–20). In the course of this work, scientists observed that axons swell to three times their normal diameter when constricted, suggesting that fluid and other components of the axon cytoplasm or *axoplasm* moved through the interior of the axon and accumulated at the point of the obstruction. This finding was consistent with earlier observations by Young that axoplasmic fluid spilled out from the cut end of the squid giant axon (Ochs 2004, 222). When radioactive isotopes became available after the war, scientists, including Lasek, began to study axoplasmic transport by injecting radioactive precursors of proteins and nucleic acids into ganglia, which are mainly collections of nerve cell bodies. After injection, the movement of radioactive molecules was followed down the axon by cutting out axon segments at different times and analyzing labeled molecules while also localizing radioactive molecules in the microscope by radioautography (Lasek 1967; Hoffman and Lasek 1975; Grafstein and Forman 1980; Ochs 2004, 223–29).[15]

By the early 1980s, experiments of this type demonstrated that there are generally two rates of axonal transport, called *fast* and *slow*. Membranous organelles move by fast transport at several hundred to a thousand millimeters per day. This rapid rate makes sense because some axons can be a meter long, but it is, nevertheless, dramatic when one considers that most nonneuronal cells are only about 20 μm (0.02 millimeters [mm]) across. On the other hand, slow transport, which carries soluble proteins and other molecules, proceeds at only one or fewer millimeters per day (Grafstein and Forman 1980; Lasek, Garner, and Brady 1984). Scientists also recognized that transport occurs in both directions in the axon; movement toward the distal (nerve terminal or synaptic) end is called *anterograde*, while movement toward the cell body is called *retrograde*. The mechanism of fast transport was unknown at the time of the 1981 Cold Spring Harbor meeting, but investigators suspected that cytoskeletal elements, particularly actin filaments and microtubules, might be involved.

Ray Lasek and Scott Brady were also at the MBL the summer that Allen turned his AVEC microscope on the squid giant axon. Lasek had been studying axoplasmic transport in squid at the MBL since the late 1960s when he was a postdoctoral fellow at the nearby McLean Hospital (Lasek 1970). In the summer of 1981, he and Brady were continuing the work. They soon learned of Allen's observations, and Brady helped Allen design a new microscope chamber that allowed transport to continue for an extended period by keeping the isolated axon under more physiological conditions. Allen, Brady, and Lasek saw the advantages of working together, and a formal collaboration was initiated. On August 27, 1981, Allen wrote to Lasek, who had returned home to Case Western in Cleveland, laying out a plan for dividing up the work, addressing, among other things, authorship of two anticipated papers to be submitted to *Science* based in part on work already accomplished that summer.[16] Allen also volunteered to share his technical knowledge to help Lasek write a grant so that his laboratory could obtain its own microscope and begin working independently. He also suggested that Brady return to the MBL in December to take Allen's microscopy course, which he subsequently did.

In December 1982, Allen, Brady, and Lasek published two important papers in the same issue of *Science* (Allen, Metuzals, Tasaki, Brady, and Gilbert 1982; Brady, Lasek, and Allen 1982). The first, in which Allen was joined by Metuzals, Tasaki, Brady, and Susan Gilbert, Allen's technician at the time, reported observations of moving particles in intact squid giant axons dissected from the organism and viewed in the AVEC-DIC microscope. They observed particles moving in both the anterograde and the retrograde directions at rates averaging 2.5 μm per second. Small particles moved linearly along the axon axis, possibly along cytoskeletal elements, including microtubules. None of these things had been seen before and could not have been seen without Allen's AVEC method.

Brady's paper went one step further to establish a cell-free system for studying axonal transport (Brady, Lasek, and Allen 1982; see also Brady, Lasek, and Allen 1985). Richard Bear, Francis O. Schmitt, and John Z. Young had earlier reported that the axoplasm of the squid giant axon could be isolated through a very straightforward procedure (Bear, Schmitt, and Young 1937). When the axon is dissected, its cut ends are tied with thread to facilitate its transfer to experimental setups (fig. 9.1*B*). After such a transfer, Bear and his colleagues simply cut the axon and then used fine forceps to squeeze the axoplasm out, like toothpaste from a tube (fig. 9.1*C*). In their case, they transferred the axoplasm to a solution so that it could be analyzed biochemically. Brady, using similar procedures, recognized that the cylindrical form of the axon was retained by the extruded axoplasm. When examined under Allen's microscope, transport of particles along the cytoskeleton in the axoplasm still occurred and persisted for several hours. Brady's observations immediately demonstrated that axoplasmic transport did not require the plasma membrane, which had been discarded, or action

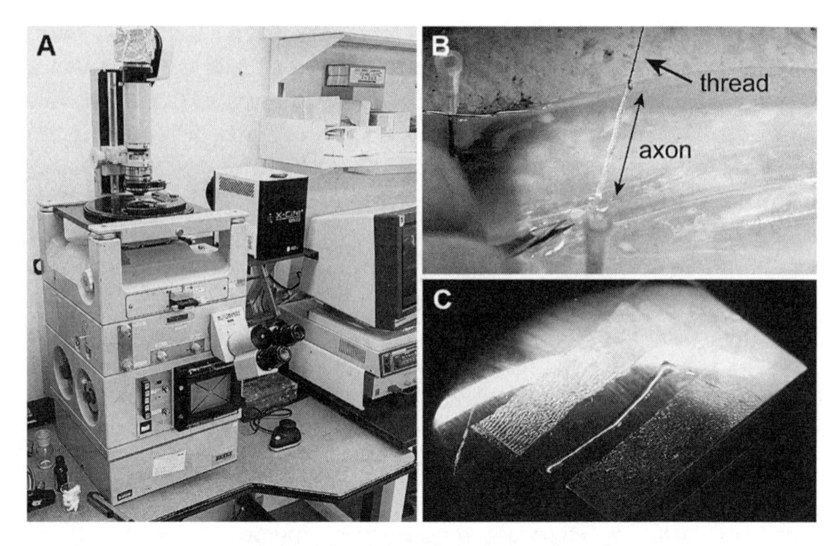

FIGURE 9.1. Preparing and examining the squid giant axon. *A*, A video microscope setup very similar to Allen's original AVEC-DIC using a Zeiss Axiomat microscope. Scott Brady first observed the effects of AMP-PNP on axonal transport using this microscope (further details are given in the text). *B*, The squid giant axon partially freed from squid tissue, with the upper portion tied off with thread. *C*, Extruded axoplasm from the same giant axon seen in panel *B*, ready for observation in the video microscope. Photographs from 2017 by the author and courtesy of Scott Brady.

potentials since the latter are dependent on the membrane. More importantly, the accessibility of this cell- and membrane-free system to both pharmacological and in situ biochemical manipulation suggested strategies to determine the molecular mechanisms of transport.

Irregulars

In the summer of 1983, two new and unexpected visitors, Ron Vale and Mike Sheetz, arrived at the MBL looking for squid. Vale was an MD/PhD student from the Neurobiology Department at Stanford University. Sheetz, on the other hand, had just completed a sabbatical in James Spudich's laboratory in the Structural Biology Department at Stanford, where he had met Vale. Sheetz, who was on the faculty at the University of Connecticut Health Center in Farmington, Connecticut, was an expert on the red blood cell membrane with a particular interest in how the cytoskeleton underlying the membrane was involved in shape changes undergone by the red cell as it traversed the circulation.

The Spudich lab was focused on the role of myosin in the motility of cells other than muscle. While there, Sheetz took up a project to see whether myosin was involved in the movement of vesicles in cells. To do this, he used *Nitella axillaris*, a green alga with unusually large cells that could be cut open to expose a cytoplasmic surface covered with oriented actin cables. When he added tiny plastic beads coated with purified myosin to the exposed surface, the beads moved along the actin cables in an ATP-dependent manner, suggesting that actin and myosin were responsible for moving vesicles around the cytoplasm (Sheetz and Spudich 1983). This result, along with further developments that showed bead movement on purified actin (Spudich, Kron, and Sheetz 1985), made Sheetz well-known in the motility field.

Even though Vale was based in the Neurobiology Department, his interest in cytoskeletal attachment of vesicles containing the nerve growth factor (NGF) receptor had already led him to Spudich's lab for advice. Signals from the NGF receptor need to be transmitted long distances along nerve cell processes to the cell body. Because Sheetz's work with *Nitella* suggested a way of moving NGF receptor–containing vesicles, Vale and Sheetz soon began discussing the possibility that actin/myosin-mediated transport might be important in fast axonal transport in neurons. Aware of the long history of work in neurobiology with the squid giant axon, Vale suggested that he and Sheetz obtain squid to see if myosin-coated beads might move in the axon in a manner similar to that observed in *Nitella*.

The natural place for Vale and Sheetz to obtain squid was the Hopkins Marine Station, an affiliate of Stanford located ninety miles away on Monterey Bay. Unfortunately, in the spring and summer of 1983, squid did not migrate into the Monterey area because of unusually warm ocean waters. Because Sheetz was, in any case, soon due to return to Connecticut, he and Vale made a spur-of-the-moment decision to rent a laboratory at the MBL, where squid were readily available. Within weeks they were there.

Vale found the MBL to be very different from Stanford. At the time, Stanford and its neighbor, the University of California, San Francisco, were at the forefront of work on recombinant DNA, a set of technologies that was rapidly accelerating the study of individual molecules. The focus on recombinant work had pushed aside more traditional biological approaches. At Stanford, light microscopy was largely considered to be outmoded. Even the setup used by Sheetz and Spudich in the *Nitella* experiments was relatively primitive because the large myosin-coated beads were visible in a low-power dissecting microscope without any sophisticated optics. In contrast, at the MBL, research dependent on the light microscope had never been pushed aside, even by the postwar rise of electron microscopy within the cell biology community. Allen and Inoué's use of advanced optical methods and cutting-edge video, which foreshadowed a renaissance in light microscopy at the end of the twentieth century and the beginning of the twenty-first, were emblematic of the significance of light microscopy at the MBL.

Neither Vale nor Sheetz had previously worked with squid, and the two began to visit other laboratories at the MBL, including those of Lasek and Brady, Allen, and Tom Reese, to figure out how to dissect the giant axon. Reese was a structural neurobiologist very well known for his development and use of innovative techniques of electron microscopy (Heuser, Reese, and Landis 1976; Heuser et al. 1979). Despite his ready interactions with Lasek and Allen in summers at the MBL, Reese tended to keep to himself to pursue his own interests in, particularly, the structure and dynamics of the synapse. Nevertheless, the advantages of working at the MBL were significant, and, by 1983, he was in the process of setting up a year-round laboratory in Woods Hole.

One technique that Reese used was called *freeze etching*. It involved rapidly freezing tissues to very low temperatures and then breaking them open (fracturing them) to expose internal surfaces. Specimens were then "etched" in a vacuum to sublimate superficial ice, making minute structural details visible. In all cases, specimens were then "shadowed" with a fine spray of electrically volatized platinum and other metals to create an exact copy of the biological surface, which was called a *replica*. The replica was then examined in the electron microscope, revealing the dramatic topology of the cell interior in all its complexity (Heuser et al. 1979; Schnapp and Reese 1982).

The arrival of Vale and Sheetz and their energetic and somewhat aggressive efforts to get started with work on the squid giant axon were considered by some regular visitors to the MBL to be almost rude. After their 1982 *Science* papers, the Lasek, Brady, and Allen collaborations had continued, and the appearance of outsiders interested in roughly the same problems and wishing to use the same experimental system may have caused some alarm. Nevertheless, Allen agreed to allow them to use his AVEC microscope when he did not need it with the understanding that they would be testing whether their myosin-coated beads move in the squid axoplasm. At the time, Brady was systematically investigating extruded axoplasm as a way of determining the biochemical parameters of transport, while Allen was beginning to extend his work on microtubules in nonneuronal cells to vesicle transport in the squid axon.

Problems quickly developed. Once they had learned how to dissect and manipulate the squid giant axon, it was straightforward for Vale and Sheetz to test their myosin hypothesis by applying their myosin-coated beads to extruded axoplasm. The result was negative; the beads did not move. Faced with this outcome, Vale and Sheetz realized that they had to consider other mechanisms of axonal transport and approach their work on the squid axon more systematically. The myosin bead experiment made it seem unlikely that actin alone was responsible for vesicle transport. This left microtubules or some combination of microtubules and actin as prime candidates. However, Allen saw no reason to add two new collaborators who had just begun working with squid to his research team, not to mention that microtubules were his main focus. He angrily revoked permission for Vale and Sheetz to use his microscope.[17]

At about this time, Sheetz approached Bruce Schnapp, an associate of Tom Reese's, and showed him some video images of axoplasm taken in Allen's lab. That summer Allen had been observing the movement of vesicles on filaments dissociated from the periphery of extruded squid axoplasm by dilution in a "buffer X,"[18] as he reported at the end of summer presentations at the MBL in August 1983 (Allen, Brown, Gilbert, and Fujiwake 1983). Schnapp thought that the images taken on Allen's microscope that Sheetz showed him were of poor quality, but the observations hinted at something promising. Without knowing about the schism with Allen, Schnapp agreed to work with Sheetz and Vale on axoplasmic transport (fig. 9.2). With Reese's agreement, the collaboration began. Without a microscope of their own, Sheetz and Vale were now in a very fortuitous position.

The involvement of Schnapp turned out to be of crucial importance to the project because of his breadth of experience. While in graduate school at the University of Connecticut several years earlier, Schnapp became an expert in electron microscopy and the freeze etching of neurons. After graduate school, he did postdoctoral research at Harvard Medical School in an electrophysiology laboratory, learning along the way the electronics necessary for sensitive measurements of ion channels. One summer, when he accompanied his lab to the MBL, he met Reese and learned about his experiments with John Heuser using a novel technique similar to freeze etching to visualize the fusion of synaptic vesicles with the membrane by rapid freezing (Heuser et al.

FIGURE 9.2. Students and faculty from the 1984 neurobiology course at the MBL. Circled (*from top to bottom*): Tom Reese, Ron Vale, and Bruce Schnapp. Photograph courtesy of the MBL History Project and Historical Archives, https://history.archives.mbl.edu.

1979). Although Schnapp had already arranged a second postdoc in England, prior to his departure he managed to spend the summer of 1981 with Reese at the MBL. Both had become intrigued with the problem of axonal transport and had decided to use freeze etching to look at the axoplasm in turtle optic nerves. They believed that examination of the detailed structure of axoplasm in the electron microscope would shed light on the transport mechanism. While they were able to see the abundance and organization of microtubules and neurofilaments, the latter a type of cytoskeletal element not believed to be involved in transport, their observations told them nothing about how transport worked (Schnapp and Reese 1982). In November 1981, Schnapp left for England. The following summer, after Reese saw Allen's images of moving vesicles in the squid axoplasm for the first time, he excitedly called Schnapp in England and invited him back to pursue axoplasmic transport again using both video and electron microscopy. Reese offered Schnapp a position at the National Institutes of Health, Reese's home institution, and Schnapp left England immediately.

By the time that Vale and Sheetz initiated their collaboration with Schnapp and Reese at the MBL in late summer 1983, Schnapp had already been playing with video microscopy, trying to duplicate and improve on Allen's AVEC system, and had learned how to isolate the squid giant axon from Tasaki. With both Inoué and Allen working at the MBL, the place was a mecca for light microscopy. Not only was cutting-edge research being conducted with their respective video microscopes, but there were also both summer and winter courses focused on advanced microscopy.[19] Companies that manufactured microscopes, video equipment, and associated instruments recognized the marketing possibilities and descended on the MBL in the summer with their latest products.[20] Schnapp recognized an opportunity and assembled a video microscope capable of visualizing microtubules in axoplasm entirely from parts loaned by manufacturers.

Microtubules

As described previously, microtubules were for many years suspected of involvement in the transport of organelles and vesicles in cells (Berlinrood, McGee-Russell, and Allen 1972; Hyams and Stebbings 1979; Hayden, Allen, and Goldman 1983). Early on, scientists observed that particles visible in the light microscope stopped and restarted as they moved through the cell, a type of motion that can be described as *saltatory* (Rebhun 1972). However, the pathways of motion were linear or curvilinear and not erratic, suggesting that filaments might be important (Hayden, Allen, and Goldman 1983). Although Allen had long been interested in microtubules and particle movement, he made little progress until the invention of AVEC-DIC microscopy (Berlinrood, McGee-Russell, and Allen 1972; Allen, Allen, and Travis 1981; Travis and Allen 1981). In his original paper describing the method, he observed what he believed were individual microtubules in the reticulopodial network visible in a flattened region of the *Allogromia* cytoplasm (Allen, Allen, and Travis 1981). However, because single, double, or bundles of microtubules all produced a 200-nm-wide diffracted image by AVEC, he could not be sure.

In papers published in 1983 and 1984, Allen attempted to prove definitively that vesicles seen by AVEC were moving along microtubules. In keratocytes from the corneal stroma of the frog *Rana pipiens*, Allen and his postdoc John Hayden used an antibody against tubulin to label microtubules fluorescently, reporting that the pattern of "linear elements" observed with AVEC resembled that of the labeled microtubules viewed by fluorescence microscopy (Hayden, Allen, and Goldman 1983). Next, they tried to correlate AVEC video images of particles moving on filaments with electron microscopy of the same sample. To do this, they cultured keratocytes on gold indicator "grids," tiny sample holders suitable for insertion into an electron microscope.[21] After noting the exact location on the grid of moving particles and filaments seen by AVEC, they viewed the same location in the electron microscope after the sample was fixed and stained (Hayden and Allen 1984). With electron microscopy, they were able to discriminate between single and double microtubules, but their images were exceedingly complex because the entire cell as a whole had been extracted and viewed, revealing many more filaments of different types than were detectable by video microscopy. In addition, they were unable to clearly identify the particles they had seen on or adjacent to the microtubules. Although in the end they concluded that particles were moving on single microtubules, the evidence still fell somewhat short of substantiating the claim.

After initiating his collaboration with Lasek, Allen had extended his study of particle movement and microtubules to the squid giant axon (Allen, Metuzals, Tasaki, Brady, and Gilbert 1982). Nevertheless, the obstacles to determining microtubule involvement presented by keratocytes and *Allogromia* were no different in the intact axon. In his paper with Allen in 1982, Brady noted that filaments could be dissociated from the edges of extruded axoplasm using a fine needle and that transport of particles on these separated filaments continued (Brady, Lasek, and Allen 1982). In August 1983, Allen gave a presentation at the annual MBL general scientific meeting that was summarized in an abstract published later that year in the *Biological Bulletin* (Allen, Brown, Gilbert, and Fujiwake 1983). In it, he described a refinement of Brady's procedure: axoplasm was sheared after dilution in buffer X, producing short filaments with moving vesicles. Some of the filaments themselves seemed to move on the coverslip. Allen noted at the end of his abstract that his results "suggest that reconstitution experiments involving biochemically defined, interactive filaments may shed light on the mechanisms of fast axonal transport" (Allen, Brown, Gilbert, and Fujiwake 1983). By this time, Vale and Sheetz had joined Schnapp and Reese to pursue a similar approach.

Schnapp, Vale, Sheetz, and Reese realized that the question about the exact identity and nature of the filaments supporting vesicle movement was still unresolved. They had not seen Allen's paper with Hayden trying to correlate video images with electron microscopy; it would not appear until the end of 1984 (Hayden and Allen 1984). They faced the same challenges as Allen did: filaments capable of vesicle transport needed to be separated from the bulk of the axoplasm so that they could be clearly observed. Then, the filaments seen by video microscopy had to be located and examined in the electron microscope at sufficient magnification and resolution to tell clearly whether they were microtubules and, if so, solitary or bundled.

Schnapp's new video microscope in Reese's lab was comparable to, if not better than, Allen's AVEC system for visualizing vesicle movement on filaments (Vale, Schnapp, Reese, and Sheetz 1985a; Schnapp 1986). Schnapp also had an idea about how he could use the freeze etching techniques he and Reese had employed previously to resolve both the filaments and the vesicles. He placed an electron microscope finder grid with an asymmetrical pattern of squares on a coverslip and shadowed it with a fine spray of gold and palladium (Schnapp, Vale, Sheetz, and Reese 1985). When the grid was removed, the pattern of the metal spray remained. He and Vale then suspended extruded squid axoplasm in half-strength buffer X, which dissociated and separated filaments and attached vesicles, and then added the suspension to the coverslip over the grid pattern (Vale, Schnapp, Reese, and Sheetz 1985a). When they added ATP, they saw vesicles moving up and down the filaments by video microscopy. Schnapp noted the location on the grid pattern, and he and Reese prepared the sample for the freeze etching technique to visualize microtubules and vesicles in the electron microscope. They quickly froze the sample by slamming it against a copper block cooled with liquid helium. Then, after sublimating ice to reveal fine details, they shadowed the sample with a platinum spray. They then removed the resulting metal replica of the filaments and vesicles from the glass coverslip and observed it in the electron microscope. Fortuitously, the platinum replica stuck to the gold/palladium grid pattern, and they could easily find the location that they had observed by video microscopy (Schnapp, Vale, Sheetz, and Reese 1985; see fig. 9.3).

By this time in the early autumn of 1983, Vale and Schnapp were

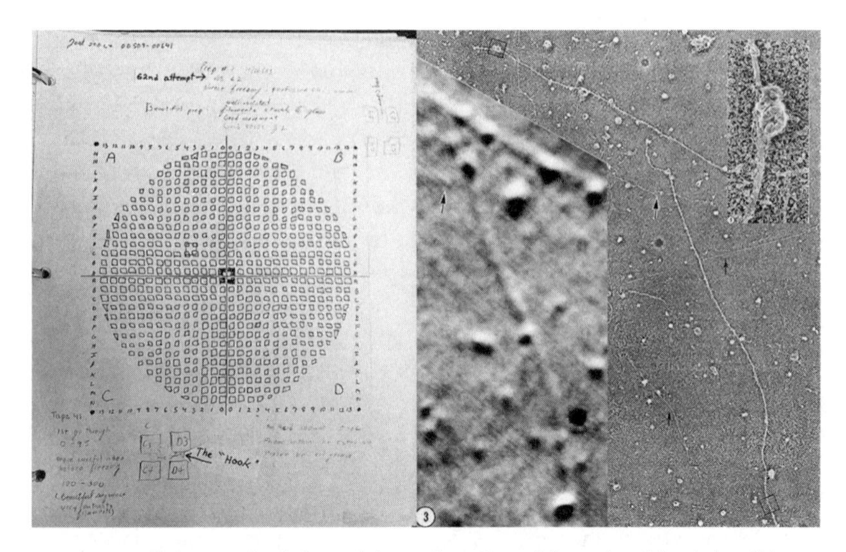

FIGURE 9.3. Sixty-two tries. *Left panel*: A page from Bruce Schnapp's notebook describing and illustrating the sixty-second attempt on November 16, 1983, to visualize the exact image of moving vesicles and filaments seen by video microscopy in the electron microscope (courtesy of Bruce Schnapp). The round cross-checked pattern corresponds to the pattern on the coverslip derived from the finder grid (see the text). *Right panel*: Images from the published sixty-second try experiment. Figures 3 and 5 (*inset*) from Schnapp's paper showing the video record of moving vesicles and the electron micrograph of the same area. The video record and the electron micrograph are at the same magnification (×13,500), but the microtubules appear 200 nm wide in the video micrograph owing to the resolution limit of the video microscope. The electron micrograph is of the metal replica (see the text). The inset shows the area in the small box on the left of the electron micrograph magnified 220,000 times. The linear element is a single microtubule, and the two oblong objects are vesicles. The right panel is reprinted and slightly modified from Schnapp, Vale, Sheetz, and Reese (1985, figs. 3 and 5) with permission of Elsevier.

the ones mainly conducting the experiments, although Reese helped with freeze etching. Reese's lab now operated throughout the year at the MBL instead of just summers, but to a large extent the MBL was deserted once the seasonal investigators and students left. Allen, Lasek, and Brady, in particular, had returned to their home institutions. Vale and Schnapp began working eighteen- to twenty-hour days in Schnapp's tiny, windowless lab in the basement of the Loeb laboratory building. Finding the filaments and vesicles in the electron microscope that they had observed by video microscopy was exceedingly difficult, though they were often teased by near success. With electron microscopy, they could frequently correlate distributions of filaments and vesicles with locations on the grid pattern identified by video microscopy, but, each time, the areas of interest were obscured by dirt contamination picked up during processing. Eventually, Vale was forced for a time to go back to Stanford, where the prospect of beginning his training in clinical medicine was hanging over his head.

Schnapp continued trying, sometimes with assistance from Sheetz, who visited from Connecticut, and Reese, who helped with production of the replicas. On the sixty-second try, in November 1983, he succeeded. Looking in the electron microscope, he found the exact pattern of filaments that he had seen in the video and that he described in his notes as the "hook" (Schnapp, Vale, Sheetz, and Reese 1985; see fig. 9.3). When he increased magnification from 13,500 to 220,000, he observed a mass corresponding to a vesicle that was closely associated with a filament the dimensions of a single microtubule (fig. 9.3). At a magnification of 500,000, he was even able to resolve the individual subunits of the filament corresponding to the helical pattern of the microtubule protein tubulin (Schnapp, Vale, Sheetz, and Reese 1985). Now there was no doubt that vesicles were moving on single microtubules. But how?

After further delaying his return to medical school at Stanford, Vale returned to the MBL in the summer of 1984 intent on trying to reconstitute movement using purified microtubules and vesicles from squid axoplasm. He had a background in biochemistry from college, and his education at Stanford, which had a strong biochemical tradition, reinforced this. Traditional biochemistry requires the purification of proteins, and the ability to purify proteins is dependent on having a good assay—a method to track the protein during purification. While many such assays are based on chemical changes induced by the protein (often an enzyme) that are followed spectrophotometrically, in this case Vale's assay was the movement of vesicles on purified microtubules *observed in the microscope.* Thus, in a way, *the biological process itself* was the end point used to search for the molecular mechanism of movement.

The immediate problem was that Vale had to learn how to purify microtubules and do the cell fractionation necessary to isolate axoplasmic vesicles. Microtubules, which are polymers of tubulin, are commonly purified by disrupting tissues rich in microtubules and then adjusting conditions so that the microtubules cycle between the polymerized and the unpolymerized state with the help of the drug taxol, which stabilizes the polymers. After each polymerization, the microtubules are isolated by centrifugation, depolymerized, and then polymerized again. Through multiple cycles, this process yields very pure microtubules because polymerization is dependent only on tubulin and contaminating proteins are gradually diluted out. The tissue most people used as a microtubule source was cow brain because of its availability from slaughterhouses and abundant, microtubule-rich neurons. Without cow brains available at the MBL, Vale used the optic lobes of squid, a normally discarded side product of giant axon dissections that is also loaded with microtubules.

To purify axoplasmic vesicles, Vale employed a variation on classic cell fractionation techniques that rely on centrifugation at high speed to separate cellular components of different size and density.[22] To start, he pooled axoplasm from several giant axons, dispersed it by homogenization, and then centrifuged it to remove larger fragments of undissociated and aggregated axoplasmic components. The remaining supernatant was then treated with taxol to polymerize any tubulin and stabilize existing microtubules. These microtubules were then removed by centrifugation, yielding a final suspension consisting of both small vesicles similar to those seen moving in the microscope and soluble axoplasmic components. The suspension was added to the top of a tube in which three sucrose-containing solutions of different density had been layered. When this gradient was then centrifuged at very high speed, the components in the suspension were distributed in the gradient. The very top of the tube contained only soluble material, while vesicles were found at the interface between two of the more dense sucrose layers (Vale, Reese, and Sheetz 1985; Vale, Schnapp, Reese, and Sheetz 1985b).

By August 1984, Vale and Sheetz were now ready to do the first experiments to try and identify the molecular motor responsible for moving vesicles on microtubules in the axon.[23] When the vesicle and soluble fractions from the gradient were combined on a coverslip with purified microtubules from the squid optic lobe and ATP, the vesicles moved along the microtubules. To make sure that they were really looking at vesicles and not some aggregated proteins, Vale and Sheetz did a control experiment in which only the soluble fraction and microtubules were combined, expecting nothing to happen. Instead, the microtubules began crawling across the coverslip (Vale, Schnapp, Reese, and Sheetz 1985b)! Suddenly, a series of nagging questions were resolved.

When Vale was traveling to the MBL for the first time in 1983, he read a paper from Richard Adams and Dennis Bray in London, who reported that plastic beads and other foreign particles injected into giant crab axons moved (Adams and Bray 1983). At the time, this finding made no sense to him because he believed that any motor driving vesicle transport would be attached to the vesicles. If so, why would bare plastic beads move? After all, in Sheetz's *Nitella* experiment, he needed to coat beads with the motor protein myosin to get them to move on actin filaments. Now, after seeing microtubules crawling on the coverslip without vesicles, Vale and Sheetz realized that any axoplasmic motor activity was in the soluble fraction, not bound to vesicles. The motor had likely stuck to the glass, and, when added microtubules bound to the immobilized motor, the microtubules were driven across the glass in the presence of ATP. To confirm this, they drew a circle on a glass slide and added a drop of the soluble fraction to the marked area. After washing away any unbound material, they added microtubules and ATP to both marked and unmarked areas. The microtubules moved, but only where the soluble fraction had been applied. Beads treated with the soluble fraction also moved on microtubules. Apparently, the motor could stick to either glass or beads and then cause movement when it bound to microtubules in the presence of ATP (Vale, Schnapp, Reese, and Sheetz 1985b).[24]

Motors

Purification of the motor from the soluble fraction was not likely to be an easy task. By this time, in the autumn of 1984, squid were becoming less available in Woods Hole owing to their seasonal migration. Sheetz and Vale discovered that they could freeze dissected squid axons and squeeze out the axoplasm later without loss of activity. However, the amount of axoplasm obtained in this manner was still unlikely to be enough for purification of the motor. The squid optic lobes were another possible source, not only for microtubules, but also, perhaps, for the motor. As Sheetz and Vale dissected axons for storage, they also cut out optic lobes and froze as many of them as possible.

Because Vale had a good assay for the motor, purification was conceptually straightforward. Proteins in the soluble fraction could be separated by column chromatography, a traditional but well-developed strategy using sequential separations based on different protein characteristics such as size and net charge. However, this approach is often a notoriously inefficient way to purify any protein, and Vale worried that the amount of stockpiled tissue might not be sufficient.

The key to solving this dilemma came from Scott Brady. Although Brady was still in Lasek's lab, Lasek had turned over the project on the biochemistry of axonal transport in extruded axoplasm to Brady, who was looking to establish his independent career. At the MBL in the summer of 1984, Brady pursued a systematic approach to the problem, applying a series of inhibitors to the extruded axoplasm to establish some overall chemical characteristics of transport. He tried inhibitors that blocked various aspects of energy metabolism because he knew that transport required ATP and presumed that the motor driving transport was an ATPase, an enzyme that catalyzes ATP hydrolysis. Some of the inhibitors blocked myosin, while others blocked dynein, still the only known microtubule-based motor protein. Hence, from the pattern of inhibition he could tell whether either of these was responsible for vesicle transport in axoplasm (Brady, Lasek, and Allen 1985). In August 1984, Brady presented some of the results from the summer to the general scientific meetings at the MBL (Lasek and Brady 1984).[25] In his talk, he reported that the nonhydrolyzable ATP analogue AMP-PNP[26] inhibited vesicle transport and produced a "stable intermediate" visible in the microscope. In the presence of AMP-PNP, vesicles stopped moving but remained attached to filaments. In a subsequent paper, Brady described the vesicles as "frozen" on the filaments in the presence of AMP-PNP and stated that, as new vesicles bound to the filaments, they also stuck until the filaments looked like a "string of pearls" (Lasek and Brady 1985; see fig. 9.4). This result, along with the finding that myosin and dynein inhibitors were not as effective as AMP-PNP, suggested to Brady that the motor driving axonal vesicle transport was something new (Lasek and Brady 1984; Brady, Lasek, and Allen 1985; Lasek and Brady 1985). If the motor protein was either myosin or dynein, he expected that AMP-PNP would dissociate vesicles from filaments, but just the opposite was observed.

Neither Vale nor his colleagues working at the MBL presented any work at the general scientific meetings in August that appeared in the published abstracts. However, it is likely that at least some of them were there when Brady presented his initial findings on AMP-PNP. Both Brady and Vale quickly realized that the ability of AMP-PNP to freeze the vesicles on filaments suggested a way to purify the unknown motor easily. Vale was certain from Schnapp's results as well as his own experiments that axonal vesicle transport occurred mainly on microtubules.[27] When put together, Brady's finding and Vale's experiments indicated that AMP-PNP might cause the motor to bind tightly to microtubules even in the absence of vesicles, and microtubules were easy

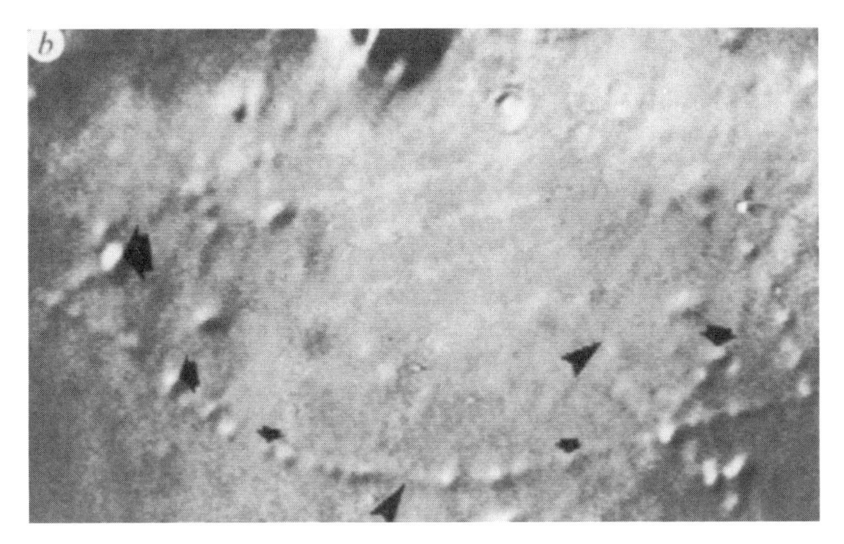

FIGURE 9.4. Scott Brady's "string of pearls," a video record of vesicles "frozen" on a filament by AMP-PNP. Reprinted from Lasek and Brady (1985, fig. 1*b*) with permission of Springer Nature.

to recover by centrifugation (Lasek and Brady 1984; Lasek and Brady 1985; Vale, Schnapp, Reese, and Sheetz 1985b).

With this information, Vale quickly proceeded to purify the motor. From his original reconstitution experiments, which had led to the crawling microtubule observations, he had plenty of purified microtubules. In addition, since he had negotiated another one-year delay on his return to medical school at Stanford, he now had the entire autumn and winter in Reese's lab at the MBL to look for the motor. He started with the soluble axoplasmic fraction from the top of the sucrose gradient, using microtubule crawling as an assay. Microtubules were incubated with the soluble fraction in the presence of AMP-PNP, and the microtubules recovered by centrifugation. When the microtubules were then treated with ATP to overcome the effects of AMP-PNP and the material released from the microtubules examined, Vale detected a protein with a molecular weight of about 110 kiloDaltons (kD) by SDS-gel electrophoresis (Vale, Reese, and Sheetz 1985). The same material also caused microtubule crawling when added back to fresh microtubules in the presence of ATP, suggesting that the 110 kD protein might be at least a component of the motor. Vale found the same protein in squid optic lobe supernatants and was able to purify and concentrate it further by cycles of microtubule binding and release. From this it became clear that two other smaller proteins of 65–70 kD accompanied the one of 110 kD. He confirmed this by further purifying the proteins using column chromatography, which also suggested that the 110 kD "heavy chain" and the 65–70 kD "light chains" formed a large multimeric complex with a total molecular weight of about 600 kD.[28] Vale named the motor protein *kinesin* from the Greek word *kinein*, "to move" (Vale, Reese, and Sheetz 1985).

In the meantime, Brady had returned to Cleveland and was also trying to find the motor. Compared to Vale, he had a competitive disadvantage. Not only did he no longer have access to squid, but he also had few research funds to support his effort. Furthermore, he was about to move to Dallas, Texas, to take up a faculty position. By November 1984, he was deep into the project, relying on extracts of radioactively labeled chick brain and microtubules purified from baboon and rat brains.[29] Using a purely biochemical approach that relied, like Vale's, on the motor binding to microtubules in the presence of AMP-PNP, he identified a protein of about 130 kD from chick brain.[30] While the preparation containing the protein was not pure, its enrichment correlated with an increase in ATPase activity that he believed made it a good candidate for the motor. In the spring, before he could complete the purification or test the preparation for its ability to induce either vesicle or microtubule motility, he received from a colleague Vale's unpublished manuscript describing the identification and purification of kinesin. He quickly wrote up his findings and submitted them to the journal *Nature*, requesting an expedited review (Brady 1985).

Vale's paper was accepted for publication in *Cell* in May 1985 and Brady's in *Nature* at the end of July 1985 (Brady 1985; Vale, Reese, and Sheetz 1985). Both research groups were at the MBL that summer, but communication between them was poor, likely a legacy of the schism with Allen two summers earlier. Because *Nature* realized that the discovery of a new molecular motor was very important and had just published Brady's paper on AMP-PNP, the key to the purification (Lasek and Brady 1985), it decided to publish a "News and Views" feature highlighting the discoveries in the same issue as Brady's paper. Brady and Vale and his colleagues were all delighted, but Sheetz, Reese, and Vale, who had obtained a copy of Brady's manuscript, wrote to *Nature* insisting that Brady's observation of an ATPase activity in soluble preparations, which they had not detected, be highlighted in the "News and Views" as an important discrepancy.[31] In the end, it was mentioned, but the emphasis was on kinesin and not on particular differences. Indeed, the "News and Views" presciently pointed out the possibility that kinesin might be involved in all sorts of motility events beyond axonal transport as well as the fact that, purification aside, the function of kinesin in the actual axon had not been directly demonstrated (Hollenbeck 1985).[32]

One last twist remained. Microtubules are polarized filaments because of the way in which the tubulin subunits polymerize to create so-called plus and minus ends. In axons, microtubules are oriented with the plus ends directed toward the nerve terminal. Experiments performed by Vale seemed to indicate that purified kinesin moved vesicles in only one direction on microtubules, but he was not certain which way (Vale, Reese, and Sheetz 1985; Vale, Schnapp, Reese, and Sheetz 1985a). However, vesicles moving on microtubules in actual axoplasm move in both directions, sometimes passing each other along the way (Schnapp, Vale, Sheetz, and Reese 1985). Around the spring of 1985, Tim Mitchison, a graduate student at the University of California, San Francisco, contacted Schnapp and Vale. He had heard about their results and thought he knew how to tell whether kinesin moved vesicles toward the plus or minus end of microtubules. He had developed a method to purify centrosomes from cells. Centrosomes are organelles related to so-called microtubule organizing centers that, when isolated from cells, can nucleate the growth of microtubules from the minus end, which remains attached to the centrosome, to the plus end. Mitchison reasoned that, if such astral arrays of polarized microtubules were used in a motility assay, Vale and his associates could determine which way kinesin moved. In initial experiments with Schnapp using purified kinesin, motility was indeed solely in the plus direction. Later, when Vale tried the same experiment with Mitchison, he decided to use not purified kinesin, which was precious, but instead the axoplasmic supernatant from the top of the sucrose gradient that he knew both contained kinesin and caused microtubule crawling and vesicle movement. When he added the supernatant and vesicles to the oriented astral arrays, vesicles moved in both directions! After going over his steps, he realized that he had pushed the pipet used to obtain the supernatant too deep into the gradient, penetrating the sucrose layer below. A second motor, later identified as a form of dynein, was in that layer. Dynein moved vesicles toward the minus ends of microtubules, whereas kinesin moved them toward the plus ends (Vale, Schnapp, Mitchison, et al. 1985; Vale 2003; see fig. 9.5).

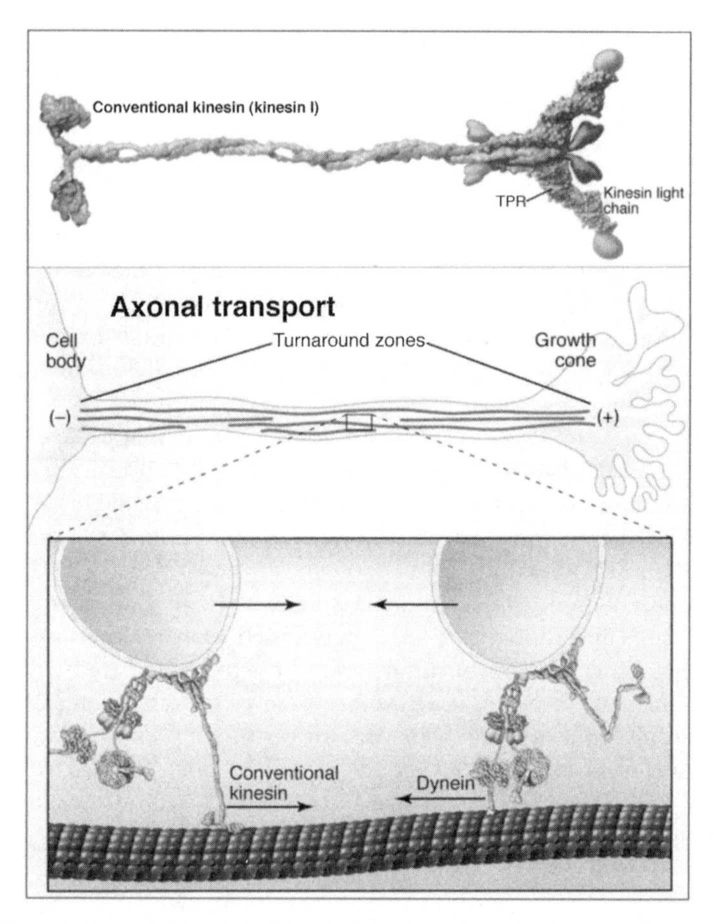

FIGURE 9.5. A model of kinesin based on its three-dimensional structure as of 2003, along with a cartoon showing kinesin and dynein carrying vesicles moving in opposite directions on a microtubule. Note the two heavy and the two light chains. Kinesin is known to "walk" step by step on the microtubule without losing contact, a type of movement called *processive* (Vale 2003). An entire family of different kinds of kinesins was discovered after 1985. Reprinted and slightly modified from Vale (2003, figs. 1 and 2) with permission of Elsevier.

Epilogue: Finding Molecules with the Microscope

The discovery of kinesin is a scientific story, but it is also a story about the distinctive culture of collaboration and competition at the MBL. While the discovery of kinesin was, in all likelihood, just a matter of time, there is no doubt that the conjunction of squid and scientists in summers at the MBL incubated and accelerated the discovery. In the years afterward, almost everyone involved in the discoveries moved on to successful careers at different institutions, and several continued to work every summer at the MBL. Robert Allen, who, with Nina Allen, serendipitously and crucially found a way to detect very small moving objects in the video microscope, was diagnosed with cancer at the end of 1984 and died in 1986, embittered by his belief that Vale and Sheetz had exploited his observations of microtubule motility while they worked with his microscope in the summer of 1983.[33]

Indeed, the microscope at the beginning of this story exhibits characteristics of an "epistemic thing" as video microscopy is investigated and video-enhanced contrast is discovered. Later, the microscope becomes the "technical object" or the key "experimental condition," while the moving axoplasmic vesicles and filaments visible in the microscope are the "scientific object" or "epistemic thing" (Rheinberger 1997, 28–29). According to Hans-Jörg Rheinberger, epistemic things are "material entities or processes—physical structures, chemical reactions, biological functions—that constitute the objects of inquiry. . . . [They] embody what one does not yet know." The technical object, on the other hand, encompasses the scientific object or epistemic thing to "embed" and "restrict and contain" it: "The technical conditions determine the realm of the possible representations of the epistemic thing" (Rheinberger 1997, 28–29).

The microscope first identified the biological process and then, by linking all further work to the initial observations, ensured that this work, no matter how molecular, still applied to the biological phenomena under investigation (Matlin 2016). Despite working in the end with a purified protein, purified microtubules, and synthetic beads, Vale and his collaborators were able to speculate with some confidence that kinesin was a protein responsible for vesicle movement in the axon, while others immediately postulated its involvement in other biological phenomena (Hollenbeck 1985; Scholey, Porter, Grissom, and McIntosh 1985; Vale, Reese, and Sheetz 1985). Why? Because the microscope linked axonal motility to axoplasm motility and, finally, to bead motility on microtubules. No step along the way to the discovery of kinesin depended on any chemical or structural assay divorced from the biological context made visible with the microscope. Seeing was believing.

Notes

I am most grateful for the cooperation of the following individuals who shared their recollections of the events described here: N. Allen, S. Brady, S. Gilbert,

P. Hollenbeck, L. Kerr, S. Kron, J. McIlvain, E. McNally, T. Reese, B. Schnapp, T. Schroer, R. Sloboda, R. Vale, and S. Zottoli. Holly Harte, Jane Maienschein, Rachel Ankeny, Peter Hollenbeck, Bill Green, Michael Dietrich, Hannah Worliczek, and Steve Zottoli provided critical comments on the manuscript, for which I am most appreciative. I am also grateful to Jennifer Walton, head librarian, and the entire staff of the MBL library for assistance with the Robert D. Allen Papers, and to Scott Brady for demonstrating the isolation of the squid giant axon, extrusion of axoplasm, and the AVEC-DIC microscope. This work was supported in part by grants from the National Science Foundation to the MBL History Project and by funds from the Edwin S. Webster Foundation for history of biology programs at the MBL.

1. Now known as *Doryteuthis pealeii* of the family Loliginidae. I have retained the original name in the text because it was used during the period covered by this essay.
2. In fact, the first recorded observation of the nerves was by Leonard Williams in 1909, as Young notes (Williams 1909; Young 1936).
3. In comparison, the diameter of a mammalian axon is typically just a few microns.
4. In addition to published articles, sources for this essay include documents from the Robert D. Allen Papers, MBL Library, Woods Hole, Massachusetts, and my conversations in 2017 and 2018 with individuals who had direct knowledge of the described events. A list of these individuals is provided in the acknowledgment note above.
5. *Refractive index* refers to the degree to which light waves are bent (refracted) when passing from one material to another, such as from the aqueous medium surrounding a cell to the cell itself or from one part of the cell to another.
6. In a 1982 letter to Elizabeth Hay, chair of the Department of Cellular Biology and Anatomy at Harvard Medical School, Allen stated: "As I am sure you know, my long-term interest is in cell motility. . . . You also presumably know that I dabble rather deeply in microscope development" (Allen Papers, box 4).
7. Letter from Nina Allen to Bill Hansen, February 12, 1981 (Allen Papers, box 4).
8. In his classic book on video microscopy, Shinya Inoué states things similarly: "Numerically, we adopted the definition of contrast of $\Delta I/I$, or the difference in intensity (ΔI) over the intensity of the background (I). Contrast is thus increased either by raising ΔI or by lowering I. In other words, the gain of the [video] camera need not necessarily be changed, but the contrast of the same scene can be increased by raising the threshold of the camera, or the level of light that the camera accepts as the black level. Reducing the background (I) without changing ΔI effectively raises the contrast" (Inoué 1986, 232).

9. Letter from Robert Allen to Ivan Manson, Boehringer-Ingelheim, November 25, 1980 (Allen Papers, box 4).
10. In a 1995 historical review, the biologist Ted Salmon stated: "It is hard in this article to convey the excitement generated when Allen, Brady, Lasek and their colleagues first showed their VE-DIC movies" (Salmon 1995, 155).
11. Transcripts of television interviews and correspondence between Robert Allen and television network representatives (Allen Papers, box 4).
12. Letter from Robert Allen to Shinya Inoué, November 21, 1980 (Allen Papers, box 16).
13. On the basis of a handwritten draft of the schedule, Allen appears to have chaired the session, entitled in the draft "New Vistas in Optical Microscopy," placing Inoué first and himself last (Allen Papers, box 4). The session reportedly attracted great attention at the meeting.
14. Letter from Robert Allen to Shinya Inoué, November 21, 1980 (Allen Papers, box 16).
15. Radioautography is a technique in which the radioactive specimen is prepared for microscopy and then coated with a photographic emulsion. After a period of time, radioactive components in the specimen show up when the emulsion is developed as black silver grains located in the part of the cell containing radioactive molecules.
16. Letter from Allen to Lasek, August 27, 1981 (Allen Papers, box 4).
17. This characterization of events is supported by many of the individuals who shared their recollections with me and denied by none.
18. Buffer X was Scott Brady's simplification of "buffer P," the latter formulated to match most components identified by complete chemical analysis of axoplasm (Brady, Lasek, and Allen 1985).
19. In fact, Schnapp had taken the Allens' advanced microscopy course in 1978.
20. As noted previously, the video camera that the Allens were using when they discovered the AVEC-DIC method was on loan to them from the Japanese company Hamamatsu (letters between Robert Allen and Hamamatsu, Allen Papers, box 4).
21. To view specimens in an electron microscope, one places them on tiny round screens with very fine holes called *grids*. Grids have a diameter of about 3 mm. With normal grids, the screen is a uniform mesh of very fine wires. Indicator or finder grids, such as those used by Allen (and later by Bruce Schnapp), have an asymmetrical pattern so that each location in the mesh can be uniquely identified. Grids with mounted specimens are inserted into the vacuum chamber of the electron microscope to expose them to the electron beam.
22. Cell fractionation was developed in the 1940s and 1950s to purify components from disrupted cells by separating them using very high speed centrifugation. One technique, differential centrifugation, separates

material by size, with the largest and more dense cellular components going to the bottom of the centrifuge tube to form a pellet, leaving behind smaller and less dense material in the supernatant, the liquid above the pellet. Another technical approach is to fill the centrifuge tube with layers of solutions of increasing density (usually created with high concentrations of sucrose, a sugar) and place the mixture of cell components on top. When this sucrose gradient is centrifuged at high speed for a long time, cell components migrate to the layer of sucrose corresponding to their own density. Vale used both approaches to purify vesicles and other components.

23. Schnapp became ill late that summer and by autumn was stuck in Boston, where he was being treated.
24. Vale and Sheetz were lucky. They later learned that the motor was inactive in solution, but when it bound to even synthetic surfaces like glass and plastic it became active.
25. The abstract from Brady's talk was published in October 1984 (Lasek and Brady 1984) in the *Biological Bulletin*. A more complete report was published the following year in *Nature* (Lasek and Brady 1985). Lasek is the first author on both papers because Brady was the senior (i.e., last) author.
26. AMP-PNP: adenyl imidodiphosphate. ATP releases energy when it is hydrolyzed to ADP by enzymes called, generically, *ATPases*. All molecular motors must hydrolyze ATP to work.
27. At the time that Brady presented his talk in August 1984, Schnapp's paper on single microtubules had not been submitted, even though the definitive experiment was conducted in November 1983. The results were presented at the MBL in the summer of 1984. The delay in publishing Schnapp's paper was due in part to the complications in producing a publishable image from the video monitor—typically, the monitor screen was just photographed because there were few ways to extract the video image electronically. Publication may also have been delayed because it was paired with Vale's paper reporting the crawling microtubules. Schnapp's paper was finally submitted on September 10, 1984, and published in February 1985 (Schnapp, Vale, Sheetz, and Reese 1985). At least some of the results from both papers were presented at meetings of the American Society for Cell Biology (Sheetz presented) and the New York Academy of Sciences (Schnapp presented) in late 1984.
28. SDS gel electrophoresis uses a detergent (SDS) to break apart protein complexes and give each protein subunit a net negative charge. When these are then applied to a polyacrylamide gel in an electric field, the protein subunits separate according to their individual molecular weights. One of the column chromatography steps used by Vale also separates proteins by molecular weight but does not break apart complexes. Thus, he was able to tell from the latter technique, called *gel filtration*, that the

size of the entire complex was about 600 kD (Vale, Reese, and Sheetz 1985).

29. Baboons were being used by other investigators at Case Western Reserve. Not only were the brains free, but they also provided a great deal of material for microtubule purification.
30. Because of species differences, that the chick motor might have a molecular weight slightly different from squid would not be surprising or indicate that the two proteins were dramatically different.
31. Letter from Mike Sheetz, Tom Reese, and Ron Vale to Maxine Clark at *Nature*, July 31, 1985, a copy of which was provided to me.
32. Indeed, that same year, a kinesin-like protein was also found in sea urchins and localized to the mitotic spindle (Scholey, Porter, Grissom, and McIntosh 1985).
33. In one of his final papers Allen described his own observations of gliding microtubules and bidirectional particle movement on microtubules (Allen et al. 1985). In a note added in proof, he acknowledged the just-published papers of Schnapp, Vale, and the others "confirming some of our results" and also, extraordinarily, quoted from a letter he solicited from Mike Sheetz afterward that acknowledged Allen's original contributions (Allen et al. 1985).

References

Adams, R. J., and D. Bray. 1983. "Rapid Transport of Foreign Particles Microinjected into Crab Axons." *Nature* 303, no. 5919:718–20. https://www.ncbi.nlm.nih.gov/pubmed/6190095.

Allen, R. D. 1953. "Fertilization and Artificial Activation in the Egg of the Surf-Clam, Spisula Solidissima." *Biological Bulletin* 105, no. 2:213–39.

Allen, R. D., N. S. Allen, and J. L. Travis. 1981. "Video-Enhanced Contrast, Differential Interference Contrast (AVEC-DIC) Microscopy: A New Method Capable of Analyzing Microtubule-Related Motility in the Reticulopodial Network of Allogromia laticolaris." *Cell Motility* 1, no. 3:291–302. https://www.ncbi.nlm.nih.gov/pubmed/6190095.

Allen, R. D., D. T. Brown, S. P. Gilbert, and H. Fujiwake. 1983. "Transport of Vesicles along Filaments Dissociated from Squid Axoplasm." *Biological Bulletin* 165:523.

Allen, R. D., G. B. David, and G. Nomarski. 1969. "The Zeiss-Nomarski Differential Interference Equipment for Transmitted-Light Microscopy." *Zeitschrift fur wissenschaftliche Mikroskopie und mikroskopische Technik* 69, no. 4:193–221. https://www.ncbi.nlm.nih.gov/pubmed/5361069.

Allen, R. D., J. Metuzals, I. Tasaki, S. T. Brady, and S. P. Gilbert. 1982. "Fast Axonal Transport in Squid Giant Axon." *Science* 218, no. 4577:1127–29. https://www.ncbi.nlm.nih.gov/pubmed/6183744.

Allen, R. D., and J. D. Roslansky. 1959. "The Consistency of Ameba Cytoplasm and Its Bearing on the Mechanism of Ameboid Movement, I: An Analysis of Endoplasmic Velocity Profiles of Chaos (L.)." *Journal of Biophysical and Biochemical Cytology* 6:437–46. https://www.ncbi.nlm.nih.gov/pubmed/13848774.

Allen, R. D., J. L. Travis, N. S. Allen, and H. Yilmaz. 1981. "Video-Enhanced Contrast Polarization (AVEC-POL) Microscopy: A New Method Applied to the Detection of Birefringence in the Motile Reticulopodial Network of Allogromia laticollaris." *Cell Motility* 1, no. 3:275–89. https://www.ncbi.nlm.nih.gov/pubmed/7348604.

Allen, R. D., J. L. Travis, J. H. Hayden, N. S. Allen, A. C. Breuer, and L. J. Lewis. 1982. "Cytoplasmic Transport: Moving Ultrastructural Elements Common to Many Cell Types Revealed by Video-Enhanced Microscopy." *Cold Spring Harbor Symposium on Quantitative Biology* 46:85–87. http://symposium.cshlp.org/cgi/doi/10.1101/SQB.1982.046.01.012.

Allen, R. D., D. G. Weiss, J. H. Hayden, D. T. Brown, H. Fujiwake, and M. Simpson. 1985. "Gliding Movement of and Bidirectional Transport along Single Native Microtubules from Squid Axoplasm: Evidence for an Active Role of Microtubules in Cytoplasmic Transport." *Journal of Cell Biology* 100, no. 5:1736–52. https://www.ncbi.nlm.nih.gov/pubmed/2580845.

Arnold, J. M., W. C. Summers, D. L. Gilbert, R. S. Manalis, N. W. Daw, and R. J. Lasek. 1974. *A Guide to Laboratory Use of the Squid Loligo Pealei*. Woods Hole, MA: Marine Biological Laboratory.

Bear, R. S., F. O. Schmitt, and J. Z. Young. 1937. "Investigations on the Protein Constituents of Nerve Axoplasm." *Proceedings of the Royal Society Series B—Biological Sciences* 123, no. 833:520–29. http://rspb.royalsocietypublishing.org/cgi/doi/10.1098/rspb.1937.0067.

Berlinrood, M., S. M. McGee-Russell, and R. D. Allen. 1972. "Patterns of Particle Movement in Nerve Fibres in Vitro. An Analysis by Photokymography and Microscopy." *Journal of Cell Science* 11, no. 3:875–86. http://jcs.biologists.org/content/11/3/875.abstract.

Bracegirdle, B. 1978. *A History of Microtechnique*. Ithaca, NY: Cornell University Press.

Bradbury, S. 1967. *The Evolution of the Microscope*. Oxford: Pergamon.

Brady, S. T. 1985. "A Novel Brain ATPase with Properties Expected for the Fast Axonal Transport Motor." *Nature* 317, no. 6032:73–75. https://www.ncbi.nlm.nih.gov/pubmed/2412134.

Brady, S. T., R. J. Lasek, and R. D. Allen. 1982. "Fast Axonal Transport in Extruded Axoplasm from Squid Giant Axon." *Science* 218, no. 4577:1129–31. https://www.ncbi.nlm.nih.gov/pubmed/6183745.

———. 1985. "Video Microscopy of Fast Axonal-Transport in Extruded Axoplasm—a New Model for Study of Molecular Mechanisms." *Cell Motility and the Cytoskeleton* 5, no. 2:81–101. https://www.ncbi.nlm.nih.gov/pubmed/2580632.

Cheney, R. E., M. A. Riley, and M. S. Mooseker. 1993. "Phylogenetic Analysis of the Myosin Superfamily." *Cell Motility and the Cytoskeleton* 24, no. 4:215–23. http://doi.wiley.com/10.1002/cm.970240402.

Cytoskeleton. 2008. *Nature Milestones*. http://www.nature.com/milestones/milecyto/timeline.html.

Dietrich, M. R. 2015. "Explaining the 'Pulse of Protoplasm': The Search for Molecular Mechanisms of Protoplasmic Streaming." *Journal of Integrative Plant Biology* 57, no. 1:14–22. http://doi.wiley.com/10.1111/jipb.12317.

Grafstein, B., and D. S. Forman. 1980. "Intracellular Transport in Neurons." *Physiological Review* 60, no. 4:1167–1283. http://physrev.physiology.org/content/60/4/1167.abstract.

Harris, H. 1999. *The Birth of the Cell*. New Haven, CT: Yale University Press.

Hayden, J. H., and R. D. Allen. 1984. "Detection of Single Microtubules in Living Cells—Particle-Transport Can Occur in Both Directions along the Same Microtubule." *Journal of Cell Biology* 99, no. 5:1785–93. https://www.ncbi.nlm.nih.gov/pubmed/6333427.

Hayden, J. H., R. D. Allen, and R. D. Goldman. 1983. "Cytoplasmic Transport in Keratocytes: Direct Visualization of Particle Translocation along Microtubules." *Cell Motility* 3, no. 1:1–19. https://www.ncbi.nlm.nih.gov/pubmed/6601992.

Heuser, J. E., T. S. Reese, M. J. Dennis, Y. Jan, L. Jan, and L. Evans. 1979. "Synaptic Vesicle Exocytosis Captured by Quick Freezing and Correlated with Quantal Transmitter Release." *Journal of Cell Biology* 81, no. 2:275–300. https://www.ncbi.nlm.nih.gov/pubmed/38256.

Heuser, J. E., T. S. Reese, and D. M. Landis. 1976. "Preservation of Synaptic Structure by Rapid Freezing." *Cold Spring Harbor Symposium on Quantitative Biology* 40:17–24. https://www.ncbi.nlm.nih.gov/pubmed/1065523.

Hoffman, P. N., and R. J. Lasek. 1975. "The Slow Component of Axonal Transport: Identification of Major Structural Polypeptides of the Axon and Their Generality among Mammalian Neurons." *Journal of Cell Biology* 66, no. 2:351–66. https://www.ncbi.nlm.nih.gov/pubmed/49355.

Hollenbeck, P. J. 1985. "Organelle Transport: A Third Front for Cell Motility." *Nature* 317, no. 6032:17–18.

Hyams, J. S., and H. Stebbings. 1979. "Microtubule Associated Cytoplasmic Transport." In *Microtubules*, ed. K. Roberts and J. S. Hyams, 487–530. London: Academic.

Inoué, S. 1981. "Video Image Processing Greatly Enhances Contrast, Quality, and Speed in Polarization-Based Microscopy." *Journal of Cell Biology* 89, no. 2:346–56.

———. 1986. *Video Microscopy*. New York: Plenum.

James, J. 1976. *Light Microscopic Techniques in Biology and Medicine*. The Hague: Martinus Nijhoff.

Landecker, Hannah. 2011. "Creeping, Drinking, Dying: The Cinematic Portal and the Microscopic World of the Twentieth-Century Cell." *Science in*

Context 24, no. 3:381–416. http://www.journals.cambridge.org/abstract_S0269889711000160.

Lasek, R. J. 1967. "Bidirectional Transport of Radioactively Labelled Axoplasmic Components." *Nature* 216, no. 5121:1212–14. https://www.ncbi.nlm.nih.gov/pubmed/6076067.

———. 1970. "The Distribution of Nucleic Acids in the Giant Axon of the Squid (*Loligo pealii*)." *Journal of Neurochemistry* 17, no. 1:103–9. https://www.ncbi.nlm.nih.gov/pubmed/5494035.

Lasek, R. J., and S. T. Brady. 1982. "The Axon: A Prototype for Studying Expressional Cytoplasm." *Cold Spring Harbor Symposium on Quantitative Biology* 46, pt. 1:113–24. https://www.ncbi.nlm.nih.gov/pubmed/6179689.

———. 1984. "Adenyl Imidodiphosphate (AMP-PNP), a Non-Hydrolyzable Analogue of ATP, Produces a Stable Intermediate in the Motility Cycle of Fast Axonal Transport." *Biological Bulletin* 167:503.

———. 1985. "Attachment of Transported Vesicles to Microtubules in Axoplasm Is Facilitated by AMP-PNP." *Nature* 316, no. 6029:645–47. https://www.ncbi.nlm.nih.gov/pubmed/4033761.

Lasek, R. J., J. A. Garner, and S. T. Brady. 1984. "Axonal Transport of the Cytoplasmic Matrix." *Journal of Cell Biology* 99, no. 1:212S–221S.

Maienschein, J. 2018. "Changing Ideas about Cells as Complex Systems." In *Visions of Cell Biology*, ed. K. S. Matlin, J. Maienschein and M. D. Laubichler, 15–45. Chicago: University of Chicago Press.

Matlin, K. S. 2016. "The Heuristic of Form: Mitochondrial Morphology and the Explanation of Oxidative Phosphorylation." *Journal of the History of Biology* 49, no. 1:37–94. http://link.springer.com/10.1007/s10739-015-9418-3.

McGee-Russell, S. M., and R. D. Allen. 1971. "Reversible Stabilization of Labile Microtubules in the Reticulopodial Network of Allogromia." In *Advances in Cell and Molecular Biology*, ed. E. J. DuPraw, 153–84. New York: Academic.

Ochs, S. 2004. *A History of Nerve Functions*. Cambridge: Cambridge University Press.

Oldenbourg, R. 2018. "Observing the Living Cell: Shinya Inoué and the Reemergence of the Light Microscopy." In *Visions of Cell Biology*, ed. K. S. Matlin, J. Maienschein, and M. D. Laubichler, 280–300. Chicago: University of Chicago Press.

Rebhun, L. I. 1972. "Polarized Intracellular Particle Transport: Saltatory Movements and Cytoplasmic Streaming." *International Review of Cytology* 32: 93–137.

———. 1986. "Robert Day Allen (1927–1986): An Appreciation." *Cell Motility and the Cytoskeleton* 6, no. 3:249–55. http://doi.wiley.com/10.1002/cm.970060302.

Rheinberger, H.-J. 1997. *Toward a History of Epistemic Things*. Stanford, CA: Stanford University Press.

Salmon, E. D. 1995. "VE-DIC Light Microscopy and the Discovery of Kinesin." *Trends in Cell Biology* 5, no. 4:154–58. https://www.ncbi.nlm.nih.gov/pubmed/14732150.

Schickore, J. 2007. *The Microscope and the Eye*. Chicago: University of Chicago Press.

———. 2018. "Methodological Reflections in General Cytology in Historical Perspective." In *Visions of Cell Biology*, ed. K. S. Matlin, J. Maienschein, and M. D. Laubichler, 73–99. Chicago: University of Chicago Press.

Schnapp, B. J. 1986. "Viewing Single Microtubules by Video Light-Microscopy." *Methods in Enzymology* 134:561–73. https://www.ncbi.nlm.nih.gov/pubmed/2881191.

Schnapp, B. J., and T. S. Reese. 1982. "Cytoplasmic Structure in Rapid-Frozen Axons." *Journal of Cell Biology* 94, no. 3:667–69. https://www.ncbi.nlm.nih.gov/pubmed/6182148.

Schnapp, B. J., R. D. Vale, M. P. Sheetz, and T. S. Reese. 1985. "Single Microtubules from Squid Axoplasm Support Bidirectional Movement of Organelles." *Cell* 40, no. 2:455–62. https://www.ncbi.nlm.nih.gov/pubmed/2578325.

Scholey, J. M., M. E. Porter, P. M. Grissom, and J. R. McIntosh. 1985. "Identification of Kinesin in Sea Urchin Eggs, and Evidence for Its Localization in the Mitotic Spindle." *Nature* 318, no. 6045:483–86. https://www.ncbi.nlm.nih.gov/pubmed/2933590.

Schwiening, C. J. 2012. "A Brief Historical Perspective: Hodgkin and Huxley." *Journal of Physiology* 590, no. 11:2571–75. http://doi.wiley.com/10.1113/jphysiol.2012.230458.

Sheetz, M. P., and J. A. Spudich. 1983. "Movement of Myosin-Coated Structures on Actin Cables." *Cell Motility* 3, nos. 5–6:485–89. https://www.ncbi.nlm.nih.gov/pubmed/6661767.

Spudich, J. A., S. J. Kron, and M. P. Sheetz. 1985. "Movement of Myosin-Coated Beads on Oriented Filaments Reconstituted from Purified Actin." *Nature* 315, no. 6020:584–86. https://www.ncbi.nlm.nih.gov/pubmed/3925346.

Sullivan, W. 1980. "New Microscope Gives Scientists Much Better View of Living Cells." *New York Times*, November 7.

Travis, J. L., and R. D. Allen. 1981. "Studies on the Motility of the Foraminifera, I: Ultrastructure of the Reticulopodial Network of Allogromia laticollaris (Arnold)." *Journal of Cell Biology* 90, no. 1:211–21. https://www.ncbi.nlm.nih.gov/pubmed/6894760.

Vale, R. D. 2003. "The Molecular Motor Toolbox for Intracellular Transport." *Cell* 112, no. 4:467–80. https://www.ncbi.nlm.nih.gov/pubmed/12600311.

Vale, R. D., T. S. Reese, and M. P. Sheetz. 1985. "Identification of a Novel Force-Generating Protein, Kinesin, Involved in Microtubule-Based Motility." *Cell* 42, no. 1:39–50. https://www.ncbi.nlm.nih.gov/pubmed/3926325.

Vale, R. D., B. J. Schnapp, T. Mitchison, E. Steuer, T. S. Reese, and M. P. Sheetz. 1985. "Different Axoplasmic Proteins Generate Movement in Opposite

Directions along Microtubules in Vitro." *Cell* 43, no. 3, pt. 2:623–632. https://www.ncbi.nlm.nih.gov/pubmed/2416467.

Vale, R. D., B. J. Schnapp, T. S. Reese, and M. P. Sheetz. 1985a. "Movement of Organelles along Filaments Dissociated from the Axoplasm of the Squid Giant Axon." *Cell* 40, no. 2:449–54. https://www.ncbi.nlm.nih.gov/pubmed/2578324.

———. 1985b. "Organelle, Bead, and Microtubule Translocations Promoted by Soluble Factors from the Squid Giant Axon." *Cell* 40, no. 3:559–69. https://www.ncbi.nlm.nih.gov/pubmed/2578887.

Williams, L. W. 1909. *The Anatomy of the Common Squid Loligo pealii, Lesueur.* Leiden: Brill.

Young, J. Z. 1936. "Structure of Nerve Fibers and Synapses in Some Invertebrates." *Cold Spring Harbor Symposium on Quantitative Biology* 4:1–6.

TEN

Using Repertoires to Explore Changing Practices in Recent Coral Research

RACHEL A. ANKENY AND SABINA LEONELLI

In the last three decades of the twentieth century, scientists working in coral reef biology documented unprecedented and extensive changes in and degradation of reefs worldwide (Bryant and Burke 1998), including global declines in live cover, species richness, and the condition of reef-building corals. Researchers conducted numerous assessments in the late 1990s showing that these changes were likely due to a combination of several factors, including global warming, ozone depletion, hypertrophication (an excess of nutrients in the water), and anthropogenic influences, including overfishing, habitat destruction, pollution runoff, and poor land use practices (for a review, see Richardson [1998]). Beginning in the 1970s, researchers tended to describe many of these conditions solely on the basis of their external characteristics, all of which involved changes in coloration and other patterns in the coral with distinct banding patterns and were observed to destroy corals at a rate of several millimeters per day (e.g., Antonius 1981). However, during the 1970s–1990s, most coral research was descriptive; efforts to identify any potential underlying pathogens were limited, and there was even disagreement among researchers as to whether it was correct to think of these conditions as diseases.

The increasing amounts and extent of damage to coral reefs documented worldwide in the closing years of the twentieth century created clear imperatives for researchers to find ways to control and mitigate it. However, the global community of coral reef biology researchers faced considerable challenges understanding the causes of the extensive changes that they were observing and documenting and devising actionable solutions. One key challenge arose from the confluence of several potential and competing explanations for these phenomena, ranging from climate change and to related environmental disturbances, human interventions and their impacts on marine ecology, and the nature of the microbiota themselves. One explanation that emerged in the 1990s and took increasing hold in the early years of the twenty-first century among many coral researchers was that the destructive changes being observed were best understood as symptoms of underlying coral disease rather than as resulting from external or environmental forces. This explanation was in many ways a simplification of a much more complex ecological phenomenon and unavoidably involved downplaying the potential usefulness of differently focused ecological approaches. Nevertheless, as we document below, the disease-related view proved invaluable to many leading researchers in the field when it came to rallying resources, deepening understanding of the changes being observed, and devising interventions.

This essay investigates the evolution of coral reef biology research during this critical period at the cusp of the twenty-first century, focusing on the emergence and use in the field of a "repertoire" (Ankeny and Leonelli 2016; Leonelli and Ankeny 2015) that, as we document, was borrowed from biomedical research and that we refer to as the *infection repertoire*. As we discuss in more detail below, a repertoire serves as a framework for research practices. The infection repertoire as we define it involves the ensemble of practices, strategies, tools, venues, and concepts used by biomedical researchers when they face the problem of local disease outbreaks potentially becoming global pandemics. Coral reef biology researchers borrowed and used this repertoire, which included recognizing and leveraging critical institutional factors such as strategies to align their research with various national and global funding priorities as well as managerial decisions concerning the setup, infrastructures, and technologies to be prioritized for the production and circulation of data. As we argue, these institutional and managerial characteristics were as crucial to emerging approaches in the field of coral reef biology as were the conceptual and methodological factors relating to the identification and investigation of the causes of the changes being observed.

In this essay, we argue that the fruitfulness of the disease-related explanation of reef damage was not a serendipitous outcome of the application of a particular theoretical framework but rather a well-engineered and deliberate choice made by a coalition of marine researchers who actively decided to reproduce a certain way of organizing and conducting research that had previously been utilized in biomedical research. We contend that the adoption of the infection repertoire in coral research was due as much to the practical know-how derived from imitating the structures of research, management, and intervention characterizing current reactions to global epidemics and the potential benefits from these structures as it was to the explanatory power of understanding many forms of reef damage as the result of infectious disease. The introduction of new concepts, such as dysbiosis, and an increased focus more generally on the role of microbes in health did provide conceptual rationales that allowed the application of this repertoire to both realms. Nevertheless, we contend that these factors do not suffice to explain the emergence of a novel approach to research in coral biology. To do so, we must analyze how comparability across insights, techniques, and data from these two communities was strategically construed and nurtured through a range of organizational, conceptual, institutional, and methodological innovations in ways that led to a sustained and fruitful research program in coral reef biology.

The field of coral reef research presents an intriguing domain to study in order to reflect on practices in marine biology, given its rapid evolution in recent years and because it has involved researchers from multiple disciplines working together, importing and adapting resources (including repertoires, as we discuss below) from other fields in ways that significantly affected ongoing research. We contend that the use of the infection repertoire in coral reef research had a critical impact not only on the conceptualization of coral health but also on the tailoring of investigative methods toward specific forms of public health–style interventions. Our analysis builds on early historical work on related topics by Jan Sapp (1999), who has explored this field up to the late 1990s. Sapp emphasizes the importance of the institutional growth of coral reef environmental science and management, the growing political emphasis in this period on global environmental issues, and increasing awareness among coral researchers of the need for baseline data in order to distinguish human-induced changes from long-term natural processes, all of which are critical to our story. However, our account takes the analysis of research developments in this field through the early years of the twenty-first century, a period during which coral reef biology underwent significant changes, and thus draws out different elements of the history of this field as well as utilizing a distinct analytic framework in order to reflect on some of the scientific practices that characterize contemporary marine biology.

Studying Corals in the Mid- to Late Twentieth Century

Roughly through the 1970s and the early 1980s, researchers generally saw corals as highly stable, and, thus, the field tended to focus on documenting and explaining this stability by measuring distribution patterns and exploring "biologically accommodating" ecological processes such as competition and predation (e.g., Endean 1977). Some went so far as to claim that the most unpredictable events in the lives of large corals might include only millennium-scale changes in sea level, which would be exceedingly rare (Potts 1984). Infrequent storms were thought to be responsible for what were considered to be intense and localized mortality among large reef-building corals (Hughes and Jackson 1985). In the 1980s, evidence started to mount that reefs were not stable, including key episodes of major change such as the mass pathogen-induced die-off of the long-spined urchin *Diadema antillarum* (the dominant herbivore in the Caribbean), resulting in a rapid increase in algal biomass and a "phase shift" from a coral-dominated to an algal-dominated system (Hughes 1994), which was described retrospectively by those in the field as a "complete surprise" (Mumby and Steneck 2008). During this period, coral researchers also began to suggest that negative changes had been anthropogenically induced or at least exacerbated rather than being a result of natural cyclic fluctuations (e.g., Richmond 1993; and Wilkinson 1999).

During the 1980s and 1990s, there were a number of larger-scale studies by conservation organizations and various national governments that began to reveal the significant threats facing coral reefs (for a more detailed summary of this history in a relatively contemporaneous source, see Sapp [1999]). For example, a survey conducted for the International Union for the Conservation of Nature in 1984–89 indicated that significant damage or destruction of reefs had occurred in ninety-three countries. Estimates suggested that approximately 10 percent of coral reefs globally were degraded beyond recovery, with an additional 30 percent likely to degrade over the next twenty years. However, questions still remained as to whether this damage should be classified as a form of natural stress, along with storm damage, earthquakes, and wave action (ICRI 1996, 2), and also about the relative influence of human activities on reefs.

Reports of increased coral bleaching, including major bleaching events in the late 1990s and the early years of the twenty-first century and related increases in disease incidence and prevalence (Hoegh-Guldberg 1999), also accelerated attention to these types of issues. Bleaching occurs when corals expel the algae (zooxanthellae) living in their tissues, which causes them to turn completely white. Corals can survive a bleaching event and recover, as bleaching itself does not cause death, but bleached coral is under more stress and, thus, is subject to higher rates of disease and mortality. The causes of bleaching events are related to stresses caused by changes in conditions related to light, temperature, and/or nutrients. One of the key issues under debate in the late 1990s, particularly following the initial report of the UN Intergovernmental Panel on Climate Change in 1990, related to the role of global warming and whether bleaching events were caused or merely accelerated by opportunistic infections. No one disputed that bleaching was, in fact, occurring and causing serious damage to and destruction of coral reefs; however, there were debates about the scope and, most importantly, the significance of these phenomena, particularly in relation to global warming, as well as about whether bleaching was an artifact of reporting and increased attention to coral reefs particularly as a result of the implementation of larger-scale monitoring programs (e.g., Glynn 1996).

Those in the field retrospectively note what they describe as a *major paradigm shift* (Woodley et al. 2003, 11)[1] during this period among scientists and environmental managers regarding the major causes of coral damage and decline, including disease as related in part to global climate change in addition to the factors that had traditionally been recognized (see also Knowlton and Rohwer 2003). In addition, there was growing recognition of the value of and need for monitoring research, which had been previously thought to be "mindless" and not adequately hypothesis driven (Sapp 1999, 334).

Another supposed major shift in this period was associated with abandoning long-held assumptions about corals being spatially uniform and temporally stable on the scale of millennia owing to mounting evidence of a number of emerging diseases (Mumby and Steneck

2008). Although by 2000 over thirty-five different coral diseases and syndromes had been reported globally, only a small number had been explored or addressed in the peer-reviewed scientific literature (Richardson 1998; Green and Bruckner 2000). Thus, despite the rapid emergence of new diseases, disease etiologies remained uncertain and underexplored, as did the causative agents underlying them (Woodley et al. 2003, 5). In addition, researchers in this period identified a number of gaps in key concepts relating to these new threats: for instance, previous models of bleaching had been primarily high level and phenomenological (e.g., induction of bleaching via light, heat, and other stimuli was the main focus, mostly occurring in laboratory settings) with limited attention paid to underlying cellular or molecular mechanisms; researchers also did not have the necessary tools to determine what causes bleaching in individual, natural reefs (Woodley et al. 2003, 11ff.). Furthermore, owing to a lack of detailed investigations of coral physiology, our limited understanding and limited baseline measurements prevented disease being distinguished from natural variation (Woodley et al. 2003, 17).

A useful analysis of trends in coral reef management and research notes that, while most work (58 percent) in the 1970s focused on patterns of diversity and habitat use, the majority of citations in the 1980s and into the 1990s focused on explanatory processes such as reproduction, recruitment, herbivory, and predation (Mumby and Steneck 2008). In the 1990s, disturbances and degradation of reefs (primarily due to disease and bleaching but also to anthropogenic impacts) became key topics, resulting in 35 percent of the citations in that decade. Most notably, from 2000 to 2008, 85 percent of citations focused on these disturbances, with publications on disease and management steadily increasing. As one team of researchers put it: "Clearly the scientific focus has shifted from small-scale, curiosity-driven basic research of a presumed stable system to larger-scale (even global) threats to coral reef ecosystems and how best to manage them" (Mumby and Steneck 2008, 557).

Borrowing the Infection Repertoire

At the turn of the century, driven partly by the increasing frequency with which major bleaching events and other forms of damage were occurring with increasingly devastating effects (as detailed in the last section), coral biologists turned to other fields for inspiration about how to tackle such threats. Biomedicine, and particularly attempts to tackle the spread of human disease around the globe, was an attractive source of inspiration on several counts. Given its potential impact on public health and the perceived urgency of the threat, the battle against infectious diseases was highly visible to funding bodies and public policy organizations and, thus, commanded immense human, financial, and technological resources. There is also a long biomedical tradition of attempting to monitor, study, and intervene with regard to pathogen spread in order to limit the potential damage of highly contagious, fast-spreading agents. Building on the recognition that contagion is impossible to contain without international coordination, the history of framing and intervening to stop disease has increasingly involved an emphasis on global networking and the establishment of a rich infrastructure for the circulation and commodification of disease-related information (Hinchcliffe, Bingham, Allen, and Carter 2016). Coral biologists noticed the parallels and started to collaborate with biomedical researchers to establish commonalities and identify lessons learned from their approaches (e.g., see the overview of these efforts in Work, Richardson, Reynolds, and Willis [2008]).

In 2002, in response to the US Coral Reef Task Force's National Action Plan to Conserve Coral Reefs, a network of field and laboratory scientists, coral reef managers, and federal agency representatives came together to found the collaborative Coral Disease and Health Consortium in order to consult a range of experts, including those working in ecology, biology, and coral disease as well as environmental microbiology and human and veterinary medicine. From this consultation, a series of guidelines and suggestions for structuring future research on coral biology emerged that were captured in a report on coral reef management published in 2003 by this group (Woodley et al. 2003, 1).[2] The report identifies four short-term objectives for coral biology: (1) the establishment of standard terminology, methodology, and protocols to enable effective communication among researchers, (2) the expansion of knowledge about pathogens and infection processes, (3) the establishment of model coral species on which basic research could be conducted to improve the understanding of disease mechanisms and infection processes, and (4) the development of centralized data/knowledge systems, Web sites, repositories, and core diagnostic facilities to facilitate comparisons of sites and systematic monitoring of reef damage around the globe. These objectives are presented in the report as building on interdisciplinary expertise and insights to produce an integrated, effective, and systematic approach to coral reef management and, in turn, a better way to monitor and prevent continued damage and mortality. Underpinning the implementation of these objectives was a series of assumptions and commitments to methods, conceptualizations, and ways of organizing and managing research many of which were newly imported into coral biology as fundamental ingredients of what we call here the *infection repertoire*.

In order to understand our use of the repertoires framework to analyze these developments in the field, we must briefly outline its key components (for more details, see Ankeny and Leonelli [2016]). A *repertoire* is a framework used to describe, analyze, and explain the conditions under which groups of researchers organize themselves and form relatively stable epistemic communities and systems of practice (see Chang 2012), particularly within large-scale, multidisciplinary projects such as the coral reef research analyzed in this essay. A critical component of any repertoire is the assemblage of the skills, behaviors, and material, social, and epistemic components that groups use to practice and manage certain kinds of science and train newcomers and whose enactment affects the methods and results of research. For our framework, we exploit the complementary character of two typical definitions of *repertoire*.[3] On the one hand, scientific repertoires include material and conceptual elements, such as specific technologies, methods, and theories. Indeed, the adoption and use of instruments and concepts is a crucial step within the establishment of a repertoire, which is why many twentieth-century philosophers have identified these elements as core components of research programs (e.g., Bachelard [1934] 1985; Lakatos 1970; and Laudan 1977). On the other hand, a repertoire emerges only when scientists establish what they perceive to be reliable and effective ways to work with these ideas and materials within and across groups, which typically means developing appropriate social structures and know-how (ways of distributing labor, norms, skills, and behaviors). Most importantly, the development of a repertoire involves the elaboration of strategies for coordinating and managing these conceptual, material, and social components so that, when they are combined, they produce the intended performance.

Repertoires are clearly distinct from Thomas Kuhn's notion of *paradigms*. Both notions aim to identify activities that are simultaneously conceptual, social, and material and that are constitutive of research communities. Kuhn (1962) points to "revolutionary" paradigmatic shifts as ways to identify and circumscribe such activities into coherent and stable assemblages.[4] This intertwining of conceptual, social, and material factors in research is a core idea that serves as a starting point for our own approach. However, Kuhn conceptualizes paradigms as static and inflexible entities in which change occurs only in dramatic fashion and on rare occasions, whereas much of science is, in fact, everyday and mundane, or what Kuhn terms *normal*. As many commentators have observed, this makes paradigms into unhelpful framing concepts for the analysis of fast-moving, dynamic, and interdisciplinary science, like much of recent and contemporary research in the life and environmental sciences. Furthermore, Kuhn's account and his choice of case studies gives undue primacy to theoretical knowledge as the primary output of science; because of this framing assumption, and despite his deep awareness of the significance of the material and social aspects of research, the idea of a paradigm does not provide guidance for those who wish to investigate and analyze the critical roles played by shifts in technologies, social and institutional resources and infrastructures, and procedures and norms specifically aimed at stimulating institutional and financial support for science, as we wish to do in our analysis.[5] This rather narrow focus encourages an excessively internalistic view of scientific practice, one in which strategies and activities aimed at attracting and retaining material, human, economic, and political resources tend to be viewed as external to the processes of scientific research and are typically acknowledged as significant only when they directly shape the content of the propositional knowledge derived from these processes.[6]

To return to practices in coral biology, during this period DNA, RNA, and protein data became essential as the evidence base for inferences about the mechanisms and potential spread of disease (for discussion, see Woodley et al. [2003]). Thus, expensive technologies for the rapid and automated production, analysis, and comparison of molecular data, such as high-throughput sequencing tools and related data repositories, as well as collaborators skilled in use of these technologies became crucial to research advancement and funding applications in the field. A sequencing program of the genome of a representative coral species and its symbiotic dinoflagellate algae was also proposed in order to permit researchers to interpret gene function in healthy and diseased colonies (Woodley et al. 2003).[7] Such information was incorporated, systematized, and aligned with environmental and climate data through international and national online databases.[8] These databases were geared as much to raising alerts about new or emerging coral reef crises as to providing evidence for ongoing and future studies; hence, they simultaneously had management and research implications, as would be expected on the basis of the repertoires framework.

The same was the case for techniques and tools that would facilitate rapid identification of bacteria, monitoring of the symptomatology of affected species, and faster diagnosis and an advance warning system for emerging infections. Again, we see the intertwining of managerial, organizational, and more traditional scientific aspects of the research in the choices of techniques and tools to adopt.

Furthermore, specific histological techniques and standard approaches to sampling (e.g., of cells and other diagnostic indicators) as used in clinical research were to be associated with easily applicable criteria to determine what counts as normal variation as opposed to disease or deviation in reef structure and ecology (Woodley et al. 2003, 18). Again, mirroring standardized clinical terminology for health and disease, the report on coral reef management also advocated improvements in existing methods of taking case histories and reporting field and laboratory observations so that the resulting data could be comparable across affected locations around the world (Woodley et al. 2003, 26). Such heavy-handed standardization efforts were accompanied by extensive work on implementing and enforcing related guidelines and the establishment of international venues and conferences to promote and strengthen international collaboration and coordination among those using related infrastructures and experimental systems. Importing these elements had obvious implications in terms of training new generations of coral researchers. Existing skills had to be complemented with specialized training in pathology, histology, and disease etiologies as well as the data science skills necessary for scientists to work with large data collections.

To obtain such resources and implement training programs on a large scale, the very rhetoric underpinning funding applications, public engagement, and policy programs associated with coral conservation shifted toward a discourse that mimicked the fight against disease typical of global health initiatives geared toward identifying and preventing the spread of potential pandemics. Within this repertoire, disease is typically conceptualized in relation to environmental stressors acting as triggers on particular biological mechanisms (Hinchcliffe, Bingham, Allen, and Carter 2016; Reiss and Ankeny 2016). The conceptual foundations and key theoretical commitments underpinning research on corals were strongly affected by this repertoire. Coral biology moved to a similar conception of disease grounded in knowledge of how coral responds to different pathogens under varying environmental conditions. This conception of disease prompted researchers to place increasing emphasis on the study of particular pathogen-host interactions so as to understand the biological causes and processes underpinning host susceptibility to particular pathogens, that is, the ways in which the presence or absence of given microorganisms affects corals. Furthermore, the conceptualization of what constitutes the relevant environment for the study of pathogen-host interactions became associated with the idea of microbiomes (the collective genomes of the microorganisms residing in an environmental niche or the microorganisms themselves) that could be identified and analyzed through emerging "omic" technologies. Experimental research on the biological characteristics of specific model systems and the ever-changing microbiome to which they are exposed thus became central to coral biology.[9]

This approach was in contrast with the more traditional observational studies of biodiversity and stable ecological relations within the reefs that had previously dominated the field. As outlined in the previous section, while much of twentieth-century biological thinking about coral environments had focused on their stability and resilience against external stressors, the emphasis now shifted to the extent to which corals were responsive to and interdependent on the marine environment and specifically its microbiome (Garren, Raymundo, Guest, Harvell, and Azam 2009; Bourne, Morrow, and Webster 2016). Several review papers published in the early years of the twenty-first century focus on possible explanations for seemingly disrupted responses in coral to what was previously assumed to be a relatively stable environment. Explanatory options provided by researchers include (1) the presence of undiscovered new pathogens, (2) internal coral malfunction, and (3) sudden shifts in the relation between the host and the environment (e.g., Knowlton and Rohwer 2003; Work, Richardson, Reynolds, and Willis 2008; Bourne et al. 2009).

The first two of these options became strongly associated with the use of omic technologies to investigate molecular mechanisms that can trigger disease within corals or as a result of the interaction between corals and specific components of the microbiome. Indeed, it is in this period that coral biologists adopted the notion of *holobionts*, a term used particularly in reference to hard corals as a way to indicate the extent to which they are dynamic, multidomain assemblages (Knowlton and Rohwer 2003). Corals themselves come to be increasingly referred to as *metaorganisms*, that is, as multicellular organisms consisting of a macroscopic host and multiple microorganisms that interact synergistically to shape the ecology and evolution of the entire association (Bosch and McFall-Ngai 2011). Increasing amounts of research also began to be focused on the susceptibility of coral hosts to bacteria.

The third explanatory option—sudden shifts in the relation between the host and the environment—was explored through the adoption of the notion of dysbiosis, which is often meant to indicate a disturbance or shift in the microbiome of an organism resulting in chronic disease (or, more generally, as discussed in Hooks and O'Malley [2017], the interdependence between microbiota patterns and disease states). The term emerged in the late twentieth century in relation to the biomedical study of intestinal diseases in humans (e.g., Tamboli, Neut, Desreumaux, and Colombel 2004) as a counterpoint to the notion of *symbiosis* and as an innovative way to conceptualize disease and therapeutic interventions in biomedicine. The concept of dysbiosis has recently become a pillar for a broad biological understanding of disease caused by environmental imbalances rather than single etiological agents, not just in corals, but more broadly in marine ecosystems (see Egan and Gardiner 2016, table 1).

This terminology of *dysbiosis* and associated concepts acquired increasing prominence given its coherence with the material, technological, and institutional elements of the repertoire that we outlined earlier in this section. Later research continued to build on this approach and became ever more focused on the idea that using biomedical concepts and approaches coming both from human and from veterinary medicine was crucial to tackling bleaching and other threats to coral reefs and their health. Last but not least in our inventory of elements central to the infection repertoire is the central role played by reliable and efficient diagnostic procedures as investigative tools for researchers working within the infection repertoire. As noted also by Cheryl Woodley et al.: "Diagnostics are biological tests used to define boundaries of disease, and to define the state of health. It is a process by which potential causes are eliminated. The steps involved include: *a*) collecting comprehensive case histories; *b*) recording gross observations; *c*) developing a flowchart of steps used to identify patterns; *d*) implementing diagnostic tools; and *e*) deriving a final diagnosis from total body information" (Woodley et al. 2003, 23). This process occurs at two levels, that of communities and that of individuals, mirroring human diagnostic efforts in public health and infectious disease. Community-level approaches provide historical information typically after an outbreak or deaths have occurred, whereas analysis of individual organisms focuses on the identification of causative agents and characterization of mechanisms of disease and transmission patterns to identify probable causes and suggest preventative strategies or treatments.

Table 10.1 summarizes the core elements of the infection repertoire

Table 10.1 Elements of the infection repertoire

Type of element	Elements in the repertoire that moved from biomedicine to marine biology
Methodological	Microbiome tool kit and ways of producing, analyzing, and comparing data Identification of bacteria Focus on symptomatology and fast diagnosis/warning Role of histology and sampling Criteria for what counts as normal variation vs. disease or deviation Establishment of standardized clinical terminology for health and disease, improved methods for taking case histories and reporting field and laboratory information to produce comparable data across worldwide sites Setup of data collections and repositories as systems of monitoring, geared to raising alerts and providing evidence for ongoing and future studies
Managerial	Implementation and enforcement guidelines/practices for standards (terminology, composition of case histories, and data collection) Development of platforms and venues for international collaboration Coordination of data collections and infrastructures Coordination of adoption and handling of model organism of reference
Financial and institutional	Focus on interdisciplinary and specialized training in pathology, histology, and disease etiologies Resourcing, which involves the acquisition of instruments for microbiome sequencing and genetic (re)classification as well as infrastructural elements as above Framing of social relevance of research as fight against disease
Conceptual	Overthrow of fundamental assumptions around stability of environment Adoption of the term *dysbiosis* Awareness of host susceptibility to bacteria Criteria for defining what counts as a disease View of animals as metaorganisms Adoption of term *holobiont* to indicate that corals are dynamic, multidomain assemblages

borrowed from biomedicine that we have identified in coral biology at the turn of the twenty-first century.[10] The extent to which these components consistently align with each other as a specific approach to research was noted by several commentators writing in the early years of the twenty-first century. A review paper by David G. Bourne and his colleagues, for instance, starts with a statement in which the authors pinpoint precisely which elements of strategy, vision, and method are being imported from human/veterinary research into coral biology:

We review our current understanding of the role of microorganisms in coral health and disease, and highlight the pressing interdisciplinary research priorities required to elucidate the mechanisms of disease. We advocate an approach that applies knowledge gained from experiences in human and veterinary medicine, integrated into multidisciplinary studies that investigate the interactions between host, agent and environment of a given coral disease. These approaches include robust and precise disease diagnosis, standardized ecological methods and application of rapidly developing DNA, RNA and protein technologies, alongside established histological, microbial ecology and ecological expertise. Such approaches will allow a better understanding of the causes of coral mortality and coral reef declines and help assess potential management options to mitigate their effects in the longer term. (Bourne, Morrow, and Webster 2009, 554)

Conclusions: Scientific Change, Repertoires, and Coral Research

Recent changes in scientific practices associated with coral research have been claimed by those involved to have required dramatic shifts in the way in which various key concepts are understood: "The paradigm of widespread healthy, stable coral reef ecosystems has evolved to one that views them as patchy, unstable and fragile" (Mumby and Steneck 2008, 562).[11] Many also contend that the field has been deeply affected by revised understandings of the roles of climate change and other anthropogenic impacts on coral, as discussed above. Our account allows a more nuanced view of this critical period in coral research by using the repertoires framework as a way to understand scientific change; that framework also permits us to make better sense of a range of factors that have contributed to this change without requiring defense of any radical notions of discontinuity, paradigm shifts, etc.

Over the last decade, the infection repertoire has acquired increasing prominence within coral research, most substantially by intersecting with research on the crucial role of microbial colonies in reef health and resistance. Technological innovations in microscopy, such as the ability to use high-speed confocal microscopy on live coral, are providing new evidence for the role of microbiome regulation (through actions such as shedding bacteria) in coral health and disease (Garren and Azam 2012). The conceptual and methodological implications of this move are a reframing of the ideas around ecology and environment in marine research and a resulting shift away from studies of wider environmental diversity and observational approaches together with adoption of a more mechanistic focus on specific model systems and related microbiomes.

It is crucial to emphasize the extent to which this shift has opened up opportunities for new kinds of experimental interventions on coral and for the formulation and testing of preventative, diagnostic, and therapeutic measures based on the manipulation of the microbiome (Glasl, Webster, and Bourne 2017). As in the case of pandemics, and particularly given the recent catastrophic, rapid, widespread, and seemingly accelerating nature of reef destruction, the timeliness of such measures is of the essence. Researchers working on coral reefs are not necessarily rejecting or even explicitly critiquing observation-based modes of research focused on macroecological factors and the effects of human intervention on reef ecology. In fact, these elements are noted in many narratives within reef research and remain equally plausible ways to investigate and explain the phenomena currently.

In other words, what has occurred in coral research as documented here was not a conceptual and methodological change that could be explained simply as a paradigm shift in a Kuhnian fashion, requiring a radical change in the theory central to the field in question. Not only were there no obvious crises in explanatory resources within coral biology, but many conceptual elements present for decades also actually persisted throughout the 1980s, 1990s, and even into the twenty-first century. Our analysis presents a case that it is more plausible to understand emerging developments in coral biology as a response to a biological crisis, requiring a complex apparatus and strategy that coral biology research did not possess, including the intersection of specific conceptual apparatuses with techniques and tools (including databases, sequencing of microbial species, identification of species in microbiota, and microbiome analysis). This requirement is the main reason why the infection repertoire was adopted: it provided a set of easily applicable tools and strategies that helped researchers confront the current emergency in the short term and in ways that could be internationally coordinated and funded.

Indeed, as in biomedicine, the infection repertoire has proved useful for facilitating translational research, in the sense that it helps researchers couple their evolving understandings of coral biology with sophisticated forms of intervention in marine ecosystems (for a characterization of translational research that fits this account, see Leonelli [2013] and Rajan and Leonelli [2013]). Figure 10.1, which was adapted and published by reef biologists, indicates precisely how traditional

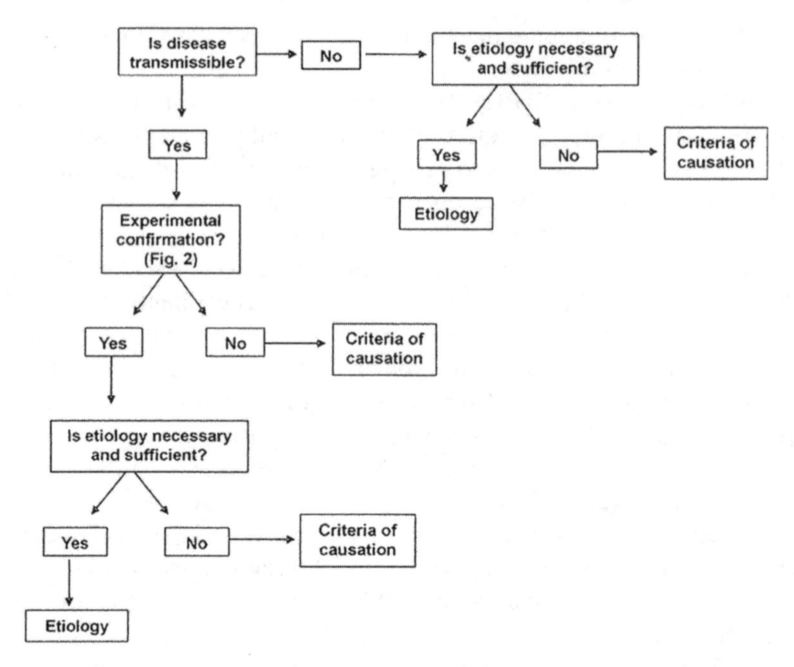

FIGURE 10.1. What is termed a *conceptual road map*, from a review paper intended to demonstrate to marine researchers what can be learned from biomedical notions of disease and related interventions. Redrawn from Work, Richardson, Reynolds, and Willis (2008); reprinted with permission of Elsevier.

steps based on Koch's postulates, which are used in biomedicine to generate human disease diagnoses and interventions, can be reframed to fit coral disease and health. This demonstrates the strength of this motivation for reef biologists and the extent to which such translational efforts explicitly underpin and frame their choices of conceptual tools.

This translational motivation, and its underlying goal of putting research at the service of what researchers perceived as a global emergency threatening the very existence of their research objects, is a key historical and epistemic factor. The notion of repertoires as relatively stable arrangements of well-aligned and reproducible elements of research, including institutional and material components, helps us not only identify this factor but also analyze in detail how it accounts for the choices of instruments, concepts, and collaborations to be pursued by researchers and the extent to which these choices affect previously existing commitments and ways of practicing coral biology. For example, a critical advantage brought to coral biology by the infection repertoire is that this approach to research permits more systematic alignment of omic data, observations, and more traditional phenotypic descriptions of damage or lesions (together with symptomatology and possible underlying causes), thus making it possible at least in principle to gather them in databases that are globally accessible. Given the crises faced by coral reefs, the ready availability of data and the opportunity to combine many different types of data in pursuing research on these phenomena are crucial. These efforts allow expanded scale and scope of comparisons of various types of host-pathogen interactions as well as prompt identification of microbiota and their microbiological components.

As we have shown, the repertoires framework helps reveal the complexities often present during episodes of science change, including the continued use of some key concepts and approaches side by side with novel ones that may on the surface appear to be incompatible. For instance, in coral research it rapidly became clear that the notion of assumed stability in fact underestimated the possibility of major changes such as those due to climate change. Subsequently, researchers changed their views to understand such conditions of relative stability and uniformity as potentially negative as in fact they make corals more vulnerable and very sensitive to physical and chemical changes and hence open to mass effects (Woodley et al. 2003, 35).

The repertoire framework also helps us understand recent exchanges between biomedical and marine researchers and the formation of a common set of allegiances and tools that facilitate the circulation of knowledge between these two communities. It could be argued that, thanks to the common ground provided by the infection repertoire, there are now representational claims being transferred not only from biomedical research to coral research but also from coral research back to research on human subjects as some of the conceptual and methodological work on coral disease has recently become used in biomedical work on dysbiosis in relation to gut disease (Bosch and Miller 2016; Ferrer, Méndez-García, Rojo, Barbas, and Moya 2017). Hence, our analysis demonstrates how the adoption of a common repertoire can facilitate interdisciplinary interactions and conceptual transfers across historically distant and distinct research traditions. It also provides an example of how the repertoires framework can be utilized by historians seeking to track the evolution of scientific practices at the community level over longer periods of time, particularly in cases where both change and continuity are simultaneously at issue.

Notes

We wish to acknowledge Javier Suarez for research assistance, Michael Dietrich for useful discussions, Glenn Hawke for assistance with preparation of fig. 10.1, and the Australian Research Council Discovery Project (DP160102989) "Organisms and Us: How Living Things Help Us to Understand Our World" (2016–20) and the ERC Starting Grant "[DATA_SCIENCE]: The Epistemology of Data-Intensive Science" (2014–19) for funding.

1. Although we think it is important to maintain the wording preferred by the scientists whose work we are exploring, we do not believe that these changes in fact represent *paradigm shifts* in a Kuhnian sense as they did not, in fact, reflect a complete overthrow of existing understandings of causes of the changes being observed in coral reefs. We discuss the difference between Kuhnian views and our own repertoires framework in more detail in the next section.
2. It is important to emphasize that this report comes out of a group of researchers based primarily at US institutions and also is clearly a political document aimed at drawing attention to key issues and gaps in research in order to attract funding. We rely on it here because it provides an excellent contemporaneous overview of global coral research and because it is subsequently frequently cited in the research literature in the field as an authoritative source.
3. For instance, the *Oxford English Dictionary* currently defines *repertoire* in two main ways: as the "body of items that are regularly performed" and as the "stock of skills or types of behaviour that a person habitually uses."
4. Here, we provide a brief overview of Kuhn's key ideas and point the reader to the voluminous literature on his work, including various interpretations and criticisms, notably Toulmin (1970), Hoyningen-Huene (1993), and de Langhe (2018).
5. Though the views of Imre Lakatos (1970) on research programs are much less inflexible concerning the degree of change happening within any given program, they are susceptible to similar critiques.
6. This interpretation is one that Kuhn himself would likely endorse (see Kuhn 2000, 287). On Kuhn's internalism, see esp. Wray (2011).
7. The first coral genome to be completely sequenced was a staghorn coral in 2011 (see CoE CRS 2011).
8. These include ReefBase (the Global Information System for Coral Reefs, http://reefbase.org/about.aspx), CoRIS (Coral Reef Information System, https://www.coris.noaa.gov), and, more recently, the International Coral Reef Initiative (ICRI, https://www.icriforum.org).
9. Note that this focus on microbiomes and the role of specific pathogens is not necessarily reductive or mechanistic in nature and, in fact, parallels

the emphasis on complex systems and related disruptions that emerges in this so-called postgenomic period—as well as its reliance on next-generation sequencing and related molecular approaches to the study of microbial environments (Richardson and Stevens 2015; Guttinger and Dupré 2016).

10. Arguably, there have been and continue to be a number of different repertoires at work in this field, including the genome-sequencing repertoire (as documented in historical detail in Hilgartner [2017]) and the model organism repertoire of the 1990s (Ankeny and Leonelli 2011; Leonelli and Ankeny 2015), the microbiome repertoire emerging in the early years of the twenty-first century (Ankeny and Leonelli 2016), the repertoire associated with the "one health" movement (which seeks to improve health and well-being at the interface between humans, nonhuman animals, and their various environments), and a repertoire relating to the investigation of climate change (which arguably developed within climate and environmental science in parallel with discussions within oceanography and marine conservation). However, space limits prevent us from exploring each of these in detail in the context of this essay.
11. Here, again, we contend that this use of *paradigm* is very loose, as is often the case in scientific research, and that the researchers are not intending to make any deep claims about their practices in relation to Kuhn's theories of scientific change.

References

Ankeny, Rachel A., and Sabina Leonelli. 2011. "What's So Special about Model Organisms?" *Studies in the History and Philosophy of Science* 41:313–23.

———. 2016. "Repertoires: A Post-Kuhnian Perspective on Scientific Change and Collaborative Research." *Studies in the History and the Philosophy of Science* 60:18–28.

Antonius, Arnfried. 1981. "The 'Band' Diseases in Coral Reefs." In *The Reef and Man* (Proceedings of the Fourth International Coral Reef Symposium, vol. 2), ed. E. D. Gomez et al., 7–14. Diliman, Quezon City: Marine Sciences Center, University of the Philippines.

Bachelard, Gaston. (1934) 1985. *The New Scientific Spirit.* Boston: Beacon.

Bosch, Thomas C. G., and Margaret J. McFall-Ngai. 2011. "Metaorganisms as the New Frontier." *Zoology* 114:185–90.

Bosch, Thomas C. G., and David J. Miller. 2016. *The Holobiont Imperative: Perspectives from Early Emerging Animals.* Heidelberg: Springer.

Bourne, David G., Melissa Garren, Thierry M. Work, Eugene Rosenberg, Garriet W. Smith, and C. Drew Harvell. 2009. "Microbial Disease and the Coral Holobiont." *Trends in Microbiology* 17:554–62.

Bourne, David G., Kathleen M. Morrow, and Nicole S. Webster. 2016. "Insights into the Coral Microbiome: Underpinning the Health and Resilience of Reef Ecosystems." *Annual Review of Microbiology* 70:317–40.

Bryant, Dirk, and Lauretta Burke. 1998. *Reefs at Risk: A Map-Based Indicator of Potential Threats to the World's Coral Reefs*. Washington, DC: World Resources Institute.

Chang, Hasok. 2012. *Is Water H_2O? Evidence, Realism and Pluralism*. Dordrecht: Springer.

CoE CRS (ARC Centre of Excellence in Coral Reef Studies). 2011. "Gene Secrets of the Reef Revealed: Genome of Staghorn Coral Acropora millepora Sequenced." ScienceDaily, July 7. https://www.sciencedaily.com/releases/2011/07/110705123631.htm.

de Langhe, Rogier. 2018. "An Agent-Based Model of Thomas Kuhn's *The Structure of Scientific Revolutions*." *Historical Social Research* 43:28–47.

Egan, Suhelen, and Melissa Gardiner. 2016. "Microbial Dysbiosis: Rethinking Disease in Marine Ecosystems." *Frontiers in Microbiology* 7:991.

Endean, R. 1977. "Acanthaster planci Infestations on Reefs of the Great Barrier Reef." In *Third International Coral Reef Symposium 1*, ed. D. L. Taylor, 185–91. Miami: University of Miami Press.

Ferrer, Manuel, Celia Méndez-García, David Rojo, Coral Barbas, and Andrés Moya. 2017. "Antibiotic Use and Microbiome Function." *Biochemical Pharmacology* 134:114–26.

Garren, Melissa, and Farooq Azam. 2012. "New Directions in Coral Reef Microbial Ecology." *Environmental Microbiology* 14:833–44.

Garren, Melissa, Laurie Raymundo, James Guest, C. Drew Harvell, and Farooq Azam. 2009. "Resilience of Coral-Associated Bacterial Communities Exposed to Fish Farm Effluent." *PLoS ONE* 4, no. 10:e7319.

Glasl, Bettina, Nicole S. Webster, and David G. Bourne. 2017. "Microbial Indicators as a Diagnostic Tool for Assessing Water Quality and Climate Stress in Coral Reef Ecosystems." *Marine Biology* 164:91.

Glynn, Peter W. 1996. "Coral Reef Bleaching: Facts, Hypotheses and Implications." *Global Change Biology* 2:495–509.

Green, Edmund P., and Andrew W. Bruckner. 2000. "The Significance of Coral Disease Epizootiology for Coral Reef Conservation." *Biological Conservation* 96:347–61.

Guttinger, Stephen, and John Dupré. 2016. "Genomics and Postgenomics." *The Stanford Encyclopedia of Philosophy*, ed. Edward N. Zalta. https://plato.stanford.edu/archives/win2016/entries/genomics.

Hilgartner, Stephen. 2017. *Reordering Life: Knowledge and Control in the Genomics Revolution*. Cambridge, MA: MIT Press.

Hinchliffe, Steve, Nick Bingham, John Allen, and Simon Carter. 2016. *Pathological Lives: Disease, Space and Biopolitics*. Chichester: Wiley-Blackwell.

Hoegh-Guldberg, Ove. 1999. "Climate Change, Coral Bleaching and the Future of the World's Coral Reefs." *Marine and Freshwater Research* 50:839–66.

Hooks, Katarzyna B., and Maureen A. O'Malley. 2017. "Dysbiosis and Its Discontents." *mBio*, vol. 8. 10.1128/mBio.01492-17.

Hoyningen-Huene, Paul. 1993. *Reconstructing Scientific Revolutions: Thomas S. Kuhn's Philosophy of Science*. Chicago: University of Chicago Press.

Hughes, Terence P. 1994. "Catastrophes, Phase Shifts, and Large-Scale Degradation of a Caribbean Coral Reef." *Science* 265:1547–51.

Hughes, Terence P., and J. B. C. Jackson. 1985. "Population Dynamics and Life Histories of Foliaceous Corals." *Ecological Monographs* 55:141–66.

International Coral Reef Initiative (ICRI). 1996. *Report to the United Nations Commission on Sustainable Development*. Washington, DC: ICRI.

Knowlton, Nancy, and Forest Rohwer. 2003. "Multispecies Microbial Mutualisms on Coral Reefs: The Host as a Habitat." *American Naturalist* 162: S51–S62.

Kuhn, Thomas S. 1962. *The Structure of Scientific Revolutions*. Chicago: University of Chicago Press.

———. 2000. *The Road since Structure: Philosophical Essays, 1970–1993, with an Autobiographical Interview*. Edited by James Conant and John Haugeland. Chicago: University of Chicago Press.

Lakatos, Imre. 1970. "Falsification and the Methodology of Scientific Research." In *Criticism and the Growth of Knowledge*, ed. Imre Lakatos and Alan Musgrave, 91–195. Cambridge: Cambridge University Press.

Laudan, Larry. 1977. *Progress and Its Problems*. Berkeley and Los Angeles: University of California Press.

Leonelli, Sabina. 2013. "Integrating Data to Acquire New Knowledge: Three Modes of Integration in Plant Science." *Studies in the History and Philosophy of the Biological and Biomedical Sciences* 4:503–14.

Leonelli, Sabina, and Rachel A. Ankeny. 2015. "Repertoires: How to Transform a Project into a Research Community." *BioScience* 65:701–8.

Mumby, Peter J., and Robert S. Steneck. 2008. "Coral Reef Management and Conservation in Light of Rapidly Evolving Ecological Paradigms." *Trends in Ecology and Evolution* 23:555–63.

Potts, D. C. 1984. "Generation Times and the Quaternary Evolution of Reef-Building Corals." *Paleobiology* 10:48–58

Rajan, Kaushik Sunder, and Sabina Leonelli. 2013. "Introduction: Biomedical Trans-Actions, Post-Genomics and Knowledge/Value." *Public Culture* 25: 463–75.

Reiss, Julian, and Rachel A. Ankeny. 2016. "Philosophy of Medicine." *The Stanford Encyclopedia of Philosophy*, ed. Edward N. Zalta. https://plato.stanford.edu/archives/sum2016/entries/medicine.

Richardson, Laurie L. 1998. "Coral Diseases: What Is Really Known?" *Trends in Ecology and Evolution* 13:438–43.

Richardson, Sarah S., and Hallam Stevens. 2015. *Postgenomics: Perspectives on Biology After the Genome*. Durham, NC: Duke University Press.

Richmond, Robert H. 1993. "Coral Reefs: Present Problems and Future Concerns Resulting from Anthropogenic Disturbance." *American Zoologist* 33:524–36.

Sapp, Jan. 1999. *What Is Natural? Coral Reef Crisis*. Oxford: Oxford University Press.

Tamboli, C. P., C. Neut, P. Desreumaux, and J. F. Colombel. 2004. "Dysbiosis in Inflammatory Bowel Disease." *Gut* 53:1–4.

Toulmin, Stephen. 1970. "Does the Distinction between Normal and Revolutionary Science Hold Water?" In *Criticism and the Growth of Knowledge*, ed. Imre Lakatos and Alan Musgrave, 39–47. Cambridge: Cambridge University Press.

Wilkinson, C. R. 1999. "Global and Local Threats to Coral Reef Functioning and Existence: Review and Predictions." *Marine and Freshwater Research* 50:867–78.

Woodley, C. M., A. W. Bruckner, S. B. Galloway, S. M. McLaughlin, C. A. Downs, J. E. Fauth, E. B. Shotts, and K. L. Lidie. 2003. *Coral Disease and Health: A National Research Plan*. Silver Spring, MD: National Oceanic and Atmospheric Administration.

Work, Thierry M., Laurie L. Richardson, Taylor L. Reynolds, and Bette L. Willis. 2008. "Biomedical and Veterinary Science Can Increase Our Understanding of Coral Disease." *Journal of Experimental Marine Biology and Ecology* 362:63–70.

Wray, K. Brad. 2011. *Kuhn's Evolutionary Social Epistemology*. Cambridge: Cambridge University Press.

ELEVEN

Why Study Sex by the Sea? Marine Organisms and the Problems of Fertilization and Cell Cleavage

MICHAEL R. DIETRICH, NATHAN CROWE, AND RACHEL A. ANKENY

In her exploration of the standardization of organisms for use in physiology, the historian Cheryl Logan has posited that, in the early twentieth century, biologists moved from an assumption of the biological distinctiveness of the organisms they studied to a "presumption of generality" (Logan 2002, 331). In other words, they changed from viewing organisms as diverse and distinct to considering organisms and their traits to be readily and directly generalizable. The impact of this transition—from a presumption of diversity to one of generality—was that many biologists changed the way in which they conducted their research and the types of claims they made about how biological processes worked.

For scientists who saw value in arguing from a diversity of organisms, a general claim would typically be justified inductively via an assembly of instances. If one held the generality presumption, generalized claims could be made on the basis of observations or experiments in a single species and extended without the need for inductive support. This tendency toward the presumption of generality has from the late 1960s forward been recognized in the case of organisms chosen as simple models primarily on the basis of their genetics, such as the nematode *C. elegans* (de Chadarevian 1998; Ankeny 2001b) and others that have come to be recognized as model organisms (Kohler 1994; Ankeny 2001a; Rader 2004; Ankeny and Leonelli 2011). Despite these important examples, the transition to the presumption of generality was not universal in twentieth-century biology. Indeed, as we will see, inductive generalization via diversity was maintained as an appropriate method by some biologists.

In this essay, we explore broader-scale trends in organismal choice during the mid-twentieth century and argue that, among researchers in developmental biology, especially some using marine organisms to study the problems of fertilization and cell cleavage, a presumption of diversity was alive and well. Much of the research on fertilization and cell cleavage during this time was inextricably bound up with the histories of the Marine Biological Laboratory (MBL) at Woods Hole, Massachusetts, and the Misaki Marine Biological Station in Japan. At these stations, biologists conducted groundbreaking work that uncovered the physiochemical mechanisms of fertilization, and they did so using a variety of research organisms rather than making generalizations based on extensive work with one species. It is important to consider why biologists chose what they did for their topics and their research subjects as these reasons often go beyond their interests or the accident of organismal availability at the seaside laboratories. How a set of marine organisms became the subject of study for fertilization and cell cleavage in the mid-twentieth century and what the investigators took to be discoverable using these organisms are the subjects of this essay. Through analysis of a detailed case study of research conducted by Katsuma Dan and Jean Clark Dan, we explore some of the complexities and contingencies involved in these sorts of choices of organisms for research in developmental biology.

The organisms used by developmental biologists have been well documented. Fish and echinoderms, especially sea urchins, have been celebrated by scientists and historians as being integral cogs in the history of embryology (Fantini 1985; Monroy 1986; Dan 1988; Wourms 1997; Ernst 2011; Zeigler, Mirantsev, Angoux, and Kroh 2014). If we are interested in how researchers choose their research organisms in the first place, however, starting with this post hoc claim about the importance of particular organisms leads to deep methodological problems. Such an approach simply reinforces the narratives that have been created over the past century about why they were so important without providing new insights into the use of marine organisms. More significantly, these historical reflections on important marine organisms have reinforced the presumption of generality associated with organism use in the twentieth century as particular species tend to become canalized over time in both scientific practice and historical memory.

To combat this methodological problem, we use a novel approach by beginning with a large and international data set (containing several thousand data points) of developmental biologists in the 1950s and 1960s drawn from the *General Embryological Information Service* (*GEIS*), a periodical that tracked biologists, their fields of research, and the organisms they used for several decades (Crowe et al. 2015). From the *GEIS* data, we construct a representation of the organisms used in different areas of research, something that we call the *organismal landscape*. Using these visualizations, we show how different organisms were used in different research areas, thus providing both a glimpse into general trends in this field and a more data-driven starting point for choice and use of a detailed case study than traditional means for gathering evidence and their accompanying narratives have permitted.

Because *GEIS* is global in scope, we can use its data to examine the worldwide trends in organism use in the postwar period in developmental biology. With reference to the research topics of fertilization and cleavage, we generate a representation of global patterns of organismal use from 1951 to 1963 as well as a wealth of underlying data concerning who was working on which problems, where, and with what organisms.

These trends in the organismal landscape led us to the work of Katsuma Dan and Jean Clark Dan, which is a main focus in this essay. Both made important contributions to the study of fertilization and cell cleavage that were based on careful research using a variety of marine organisms. Their work, and that of many others, contributed to a midcentury trend among researchers of using marine organisms to investigate these topics and draw generalizations from these organisms. Examining in detail why the Dans (and their collaborators) chose and used a diverse range of marine organisms and what they were able to conclude from their research allows us to understand some of the factors underlying this larger historical trend and assess Logan's claims about the turn in this period toward using a limited range of organisms to derive biological generalizations in a new context. We contend that both the broader trends in developmental biology articulated using the *GEIS* data and the detailed case study of the Dans reveal continued use of a relatively broad range of organisms (albeit mostly marine) during this period.

Mapping the Organismal Landscape

To generate a picture of the trends related to organism use and research topic, we analyzed the information reported in a mid-twentieth-century journal that for several decades acted as an important directory for embryologists and their work. *GEIS* was first published in 1949 through the Hubrecht Laboratory in Utrecht, Netherlands. The Hubrecht Laboratory had been founded in 1916 after World War I with the expressed mission of supporting international collaboration between embryologists. When Chris P. Raven and Pieter J. Nieuwkoop took charge of the laboratory after World War II, they sought to embrace this core mission by helping rebuild the scientific community, which had fractured during the war. With this goal in mind, they created a publication, *GEIS*, that would act as both a directory for anyone working on problems related to embryology and a clearinghouse for the understanding of the current state of research in the field.

To collect information for *GEIS*, the Hubrecht Laboratory circulated a yearly survey to as many embryologists as could be located. The survey requested details about the researchers themselves (e.g., their contact information and qualifications) as well as information regarding the current projects that they were actively pursuing. This latter component, and particularly its emphasis on ongoing rather than published work, became central to the periodical's mission. The editors did not want to duplicate the bibliographic information already available to researchers; rather, they hoped to help embryologists across the world identify research trends and find collaborators.

In 1949, the first issue of *GEIS* included contact information and research project details for 550 scientists. Most of those included in the first issue hailed from Europe, but over the next couple of decades Nieuwkoop, who would become the central figure behind the periodical until its demise in 1981, helped expand the geographic reach of the periodical to include the vast majority of embryologists in North America, South America, Asia, Africa, and the Pacific. By the early 1970s, the contact information and research data for nearly three thousand scientists appeared in each issue. In 1980, the last year of its publication run, *GEIS* included the names of over thirty-four hundred biologists at twelve hundred institutions across fifty countries.

Both the contact information and the research data included in *GEIS* offer historians a rich resource for understanding the field of developmental biology after World War II. *GEIS* reported research in two ways. The first was by listing what each scientist described as his or her ongoing research projects along with his or her contact information. The entries often included only one or two projects, but some researchers listed as many as four. For the first fifteen years, the *GEIS* editors also reported research projects via an extensive subject directory that was structured as a nested set of multiple hierarchies. It was broken into major research areas—descriptive and comparative embryology, experimental developmental morphology, developmental physiology, development and genetics, pathology, and regeneration—and each of these major categories was further broken down into subdivisions, which were themselves sometimes further subdivided in order to group similar research projects appropriately. This categorization allowed those using the directory to locate work related to their own and thus find collaborators or perhaps identify gaps in current areas of research that might foster the development of new projects. These major divisions changed slightly over the years, and the subdivisions were periodically reorganized, but the directory generally reflected the broad trends in the field (Crowe et al. 2015). In 1965, the *GEIS* editors decided that the amount of research being reported required a radically different approach to classifying projects, and, thus, they switched from a nested hierarchical system to an alphabetical index of topics.

From the journal's initial publication to its reorganization in 1965, the *GEIS* editors also annotated each reported project with a taxonomic code that reflected the type of organism being used in the research. The *taxonomic subdivision of animal kingdom*, as the *GEIS* editors described it, divided organisms using higher taxonomic ranks, most often according to phyla.[1] These codes were reported alongside the researcher's name in the subject directory, allowing readers to see not only who was working on related problems but also what organisms were being used. After 1964, the reorganization of the subject directory into an index necessitated the removal of the taxonomic-level reporting.

The sixteen-year period during which *GEIS* linked research organisms to subject allows historians to document patterns of organism use over time. To facilitate this analysis, we extracted the information on subjects and organisms used in those areas for each issue of *GEIS* from 1951 to 1963.[2] Our initial analysis sought to identify patterns of organism use across and within research areas, and we used visualization tools to facilitate this process (fig. 11.1). (For our methods for creating the organismal landscape visualizations, see the appendix.)

Historians have long been interested in how researchers choose certain organisms for their research (Kohler 1994; Ankeny 2001a; Rader

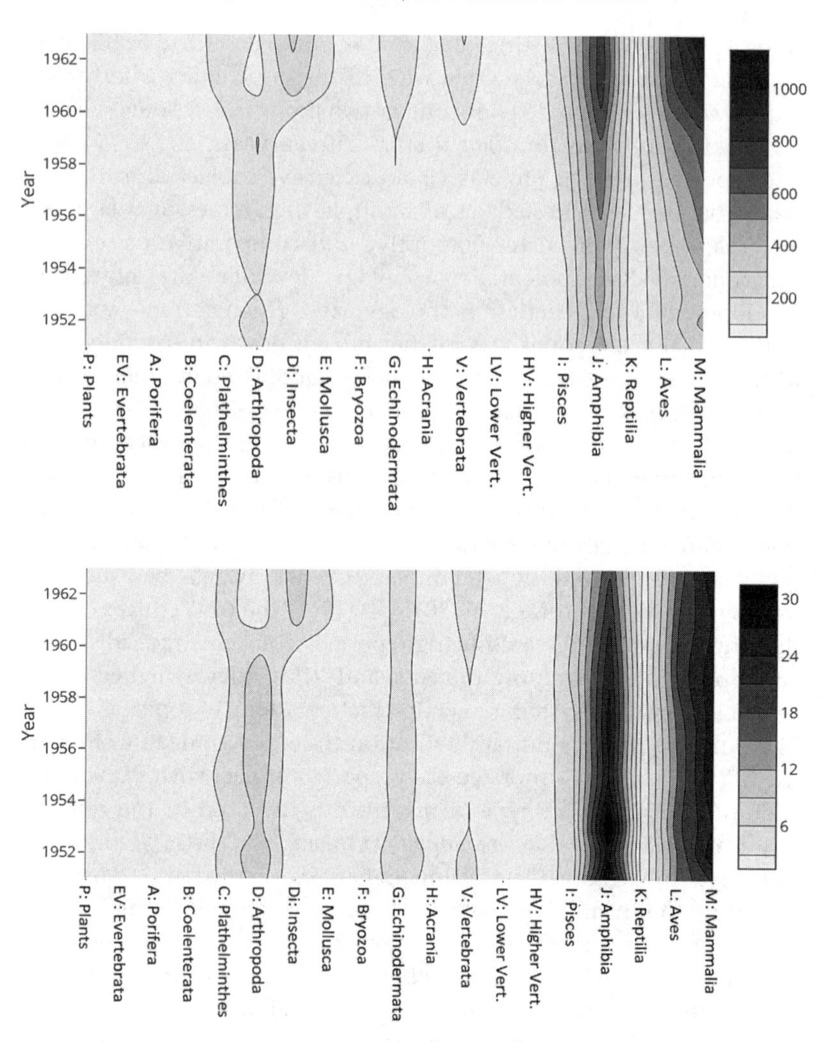

FIGURE 11.1. Organismal landscape diagrams for developmental biology, 1951–63. *Top*: The number of uses of a class of organism in the *General Embryological Information Service* by year. *Bottom*: The percentage per year of uses of a class of organism in the *General Embryological Information Service* by year. Darker areas represent higher usage and lighter areas lower usage.

2004; Leonelli 2007). Data on the work of thousands of scientists derived from *GEIS* allow historians to see broad trends in research choices and then drill down to specific researchers who either illustrate those trends or depart from them. For present purposes, we were interested in the use of marine organisms in developmental biology, so we began looking for trends in marine organism use within organismal landscape diagrams for the major research areas in *GEIS*.

In order to find a developmental biology research topic that preferentially employed marine organisms, we used evidence from the *GEIS* organismal landscape analysis. Developmental physiology, for instance, had a higher concentration of research projects using echinoderms than did other areas. Consequently, we created more fine-grained heat maps depicting annual distributions of organisms across research topics within developmental physiology (fig. 11.2*A* and *B*). For the year 1961, research in activation and fertilization and cleavage stood out as making high use of echinoderms. As research topics, the *GEIS* editors also classified activation and fertilization and cleavage in the larger categories of descriptive and comparative embryology and experimental developmental morphology.

In order to create a profile of organism use in fertilization and cell cleavage research, we combined the data from descriptive and comparative embryology, experimental developmental morphology, and developmental physiology (fig. 11.2*C* and *D*). For both research topics, echinoderms are very strongly represented, but there is also significant research within both topics using mammals. Fish are also heavily used for activation and fertilization research. Because our focus in this essay is on marine systems, we decided to concentrate more deeply on the connection between these research topics and echinoderms. To do so, we went back to the individual volumes of *GEIS* and compiled a list of everyone doing research on these topics from 1951 to 1963. From the list of over one hundred researchers, we focused on those using echinoderms and, from that list, chose two consistently represented researchers, Katsuma Dan (1904–96) and Jean Clark Dan (1910–78). It is important to note that, without doing this analysis, we likely would not have considered researching fertilization and cell cleavage or the work of the Dans. Instead of being guided by the availability of archival sources or the fame or importance of a researcher, we wanted our choice to be guided by data from *GEIS* that indicated relative prevalence of a class of organisms in a research area and researchers with a consistent history of using that class of organisms in those areas. This approach led us to the Dans' research on cell cleavage and fertilization using echinoderms and other marine organisms. Once we began to consider the Dans' research carefully, we were able to document their deliberately inductive approach using a variety of marine organisms, thus providing an excellent case study through which to explore some of the complexities and contingencies involved in choices of organisms for research in develop-

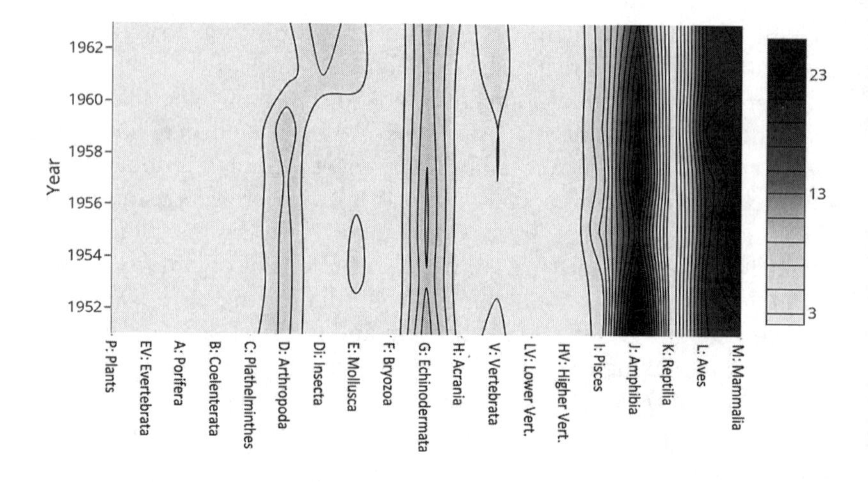

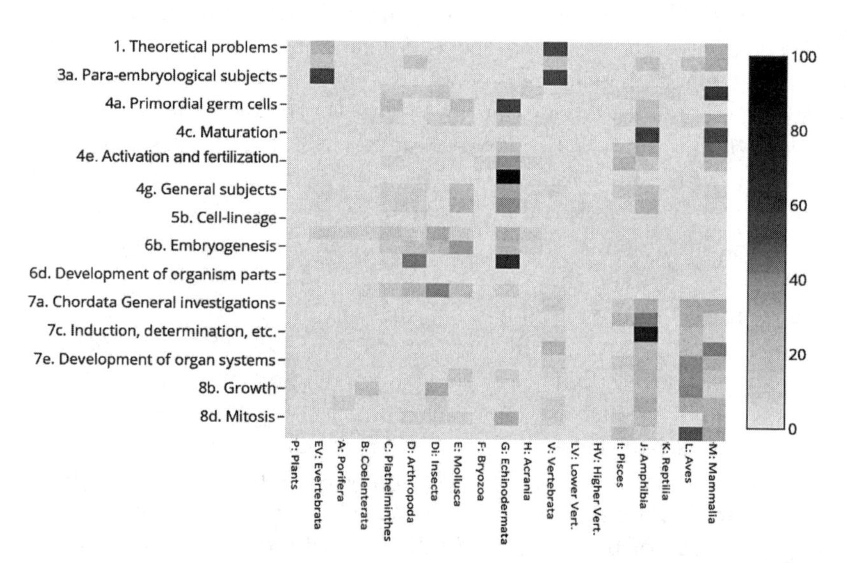

FIGURE 11.2. Organismal use in fertilization and cell cleavage research. *Top*: Organism use in developmental physiology, 1951–63. The percentage by year of uses of a class of organism in the *General Embryological Information Service* for the research area developmental physiology. Darker areas represent higher usage and lighter areas lower usage. *Bottom*: Percentage by topic of organism used within developmental physiology in 1961. *Opposite top*: Organism use in activation and fertilization research from the *General Embryological Information Service*, 1951–63. Numerical values represent the percentage of activation and fertilization projects using a particular class of organisms in a particular year. *Opposite bottom*: Organism use in cleavage research from the *General Embryological Information Service*, 1951–63. Numerical values represent the percentage of cell cleavage projects using a particular class of organisms in a particular year.

	EV. Evertebrata	C. Plathelminthes	D. Arthropoda	Di. Insecta	E. Mollusca	F. Bryozoa, etc	G. Echinodermata	H. Acrania	V. Vertebrata	I. Pisces	J. Amphibia	L. Aves	M. Mammalia
1951	0.0	2.9	8.8	0.0	5.9	2.9	44.1	2.9	0.0	11.8	5.9	0.0	14.7
1953	0.0	7.1	3.6	0.0	7.1	3.6	39.3	3.6	0.0	3.6	7.1	0.0	25.0
1955	0.0	10.5	0.0	0.0	5.3	0.0	36.8	5.3	0.0	15.8	2.6	0.0	23.7
1957	0.0	7.4	0.0	0.0	1.9	0.0	29.6	3.7	0.0	24.1	1.9	0.0	31.5
1959	1.8	7.3	0.0	0.0	0.0	0.0	30.9	3.6	1.8	21.8	7.3	0.0	25.5
1961	1.3	10.0	0.0	0.0	0.0	5.0	26.3	2.5	0.0	23.8	6.3	0.0	25.0
1963	9.0	3.8	0.0	0.0	2.6	0.0	24.4	3.8	1.3	19.2	7.7	0.0	28.2

	EV. Evertebrata	B. Coelenterata	C. Plathelminthes, etc	D. Arthropoda	Di. Insecta	E. Mollusca	G. Echinodermata	H. Acrania	V. Vertebrata	I. Pisces	J. Amphibia	K. Reptilia	L. Aves	M. Mammalia
1951	4.5	9.1	0.0	0.0	0.0	0.0	18.2	0.0	4.5	4.5	36.4	0.0	0.0	22.7
1953	4.0	4.0	4.0	0.0	16.0	0.0	28.0	0.0	4.0	4.0	24.0	0.0	0.0	12.0
1955	0.0	8.7	0.0	0.0	8.7	0.0	47.8	8.7	0.0	0.0	17.4	0.0	0.0	8.7
1957	0.0	0.0	0.0	0.0	3.8	0.0	53.8	3.8	3.8	3.8	11.5	0.0	0.0	19.2
1959	0.0	0.0	0.0	0.0	0.0	0.0	54.2	4.2	4.2	12.5	12.5	0.0	0.0	12.5
1961	2.9	8.6	4.3	2.9	15.7	0.0	27.1	4.3	0.0	7.1	15.7	0.0	1.4	10.0
1963	7.6	5.0	0.8	2.5	11.8	0.0	24.4	4.2	1.7	8.4	16.0	0.0	2.5	15.1

FIGURE 11.2. *(Continued)*

mental biology and particularly debates associated with the best strategies for making generalized claims using single multiple organisms.

Fertilization and Cleavage in Marine Organisms

Considering that the first cellular descriptions of fertilization were produced by Oskar Hertwig and Hermann Fol using sea urchins, the connection between fertilization research and use of echinoderms is not surprising (Farley 1982; Lillie 1916). The association of fertilization research with echinoderms and other marine invertebrates was solidified with the rise of research traditions at the Naples Zoological Station and the MBL, where the experimental tradition in embryology, which Hertwig and Fol's work typified, had been embraced (Maienschein 1991; Pauly 2000).

The Dans came to problems of fertilization and cleavage through American embryological research fostered by the MBL. Katsuma Dan and Jean Clark met in the early 1930s as graduate students at the University of Pennsylvania, where they were both students of Lewis Victor Heilbrunn (Colwin and Colwin 1979; Dan 1978). Like many American biologists, Heilbrunn spent every summer at MBL (Dietrich 2015). At the time, Frank R. Lillie, a leading expert on fertilization, served as president of the board of trustees and was a significant intellectual presence in Woods Hole (Maienschein 1991; Clarke 1998). It would not have been possible to be at the MBL in the 1930s without learning about Lillie's research on fertilization.

Much like Hertwig and Fol, Lillie pursued his investigation of fertilization in marine systems. While he appreciated that fertilization must involve the structural interaction of a sperm cell and an egg cell, his research focused on factors external to the egg and the sperm (Lillie 1919). He was keenly interested in fertilization reactions that occurred between sperm and egg before they came into physical contact. Observations of the clumping of sperm cells in the presence of an egg cell led him to propose that a chemical, fertilizin, was causing an agglutination reaction among the sperm cells. From these observations and other experiments, he and his protégé Ernest Everett Just came to see the egg cell as the initiator of fertilization, claiming that the egg cell released the fertilizin chemical to attract sperm. Lillie and Just's understanding of fertilization, however, was in contrast to the theories of Jacques Loeb, who also spent considerable time at Woods Hole. Loeb, whose work on artificial parthenogenesis had made him famous in the early twentieth century, argued for a more prominent role for the sperm. He believed that the sperm released a chemical, cytolysin, that broke down the outer membrane of the egg and helped initiate development (Maienschein 2014). The MBL acted as the hub for these debates, regularly bringing together Lillie, Just, Loeb, and others every summer despite the fact that they worked at universities around the country (Pauly 2000).

Katsuma Dan and Jean Clark (fig. 11.3) both worked with marine systems in their early research (Ebert 1994; Okada 1996; Sakai 1996). Katsuma was focused on the problem of the fertilization membrane, a structure formed during fertilization that prevents polyspermy, the fertilization of an egg by more than one sperm. Using a range of ma-

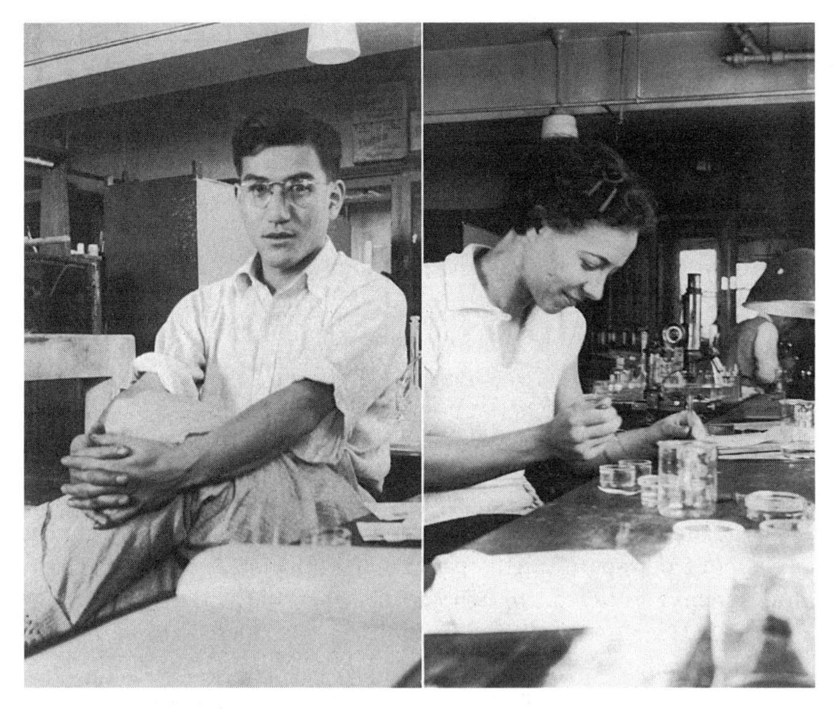

FIGURE 11.3. *Left*: Katsuma Dan, *Embryo Project Encyclopedia* (1931). MBL History Archives, http://embryo.asu.edu/handle/10776/3123. *Right*: Jean Clarke Dan, *Embryo Project Encyclopedia* (1932). MBL History Archives, http://embryo.asu.edu/handle/10776/3122.

rine organisms, including the sea urchin *Arbacia punctulata*, the clam *Cuminga tellinoides*, the starfish *Asterias forbesii*, the sand dollar *Echinarachnius parma*, and the worms *Nereis limbata*, and *Cerebratalus lacteus*, he carefully measured cell membrane charge before and after fertilization in order to evaluate the hypothesis of the University of Minnesota physiologist Jesse F. McClendon that the fertilization membrane was formed by the precipitation of oppositely charged particles (McClendon 1914; Dan 1933, 1934). Katsuma's extensive experimentation with these different organismal systems failed to find any differences in electrical charge that could be involved in the formation of the fertilization membrane.

In her dissertation research Jean also focused on the conditions that led to polyspermy in *Arbacia punctulata, Cuminga tellinoides, Asterias forbesii*, and *Cerebratalus lacteus* but without Katsuma's emphasis on electrokinetics (Clark 1936). Instead, she repeated the original polyspermy experiments that Oskar and Richard Hertwig had published in 1887 using the sea urchin *Strongylocentrotus lividus*. Her experiments investigated the effects of a number of chemical agents, such as nicotine, quinine, strychnine, morphine, and cocaine, on polyspermy. This line of experimentation was then extended to consider the effects of sperm concentration, other chemical solvents, and cation concentration. Heilbrunn's research on the impact of cations, such as calcium ions, clearly informed Jean's research, and her demonstration that sodium, potassium, calcium, and magnesium ions facilitate polyspermy supported his ideas regarding their role in the gelation of the cellular protoplasm (Dan 1936, 382). It is significant that, in direct contrast to the landmark work of the Hertwigs in just one species of sea urchin, the Dans conducted their experiments in species of sea urchins, marine mollusks, marine worms, and starfish. All these species were readily available in Woods Hole, but, most importantly, the decision to engage in working with a broad selection of organisms suggests an inductive approach aimed at creating more generalizable results concerning the problem of polyspermy. This pattern of comparative research in marine systems continued throughout the careers of the Dans.

In 1932, Katsuma's father, Takuma Dan, a Japanese baron and leading industrialist, was assassinated in Japan as part of the League of Blood Incident (Large 2001). As head of Mitsui, Takuma Dan led one of the most powerful industrial conglomerates in Japan. Trained at the Massachusetts Institute of Technology, he was known for his pro-American attitudes and thus became a target of militarists in prewar Japan. In 1934, when he finished his doctorate, Katsuma returned to Japan and began working at the Misaki Marine Station. Katsuma and Jean had fallen in love during their time together, so Katsuma asked his older brother, Ino Dan, the new head of the Dan family, for permission to marry Jean. Ino traveled to the United States to interview Jean and her professors in order to determine whether she would make a suitable wife. He agreed to the marriage, although many in Japan were not supportive. Katsuma and Jean were married in the summer of 1936 in Philadelphia ("Katsuma Dan to Wed Jean McNair Clark" 1936; "Miss Jean Clark Wed to Dr. Katsuma Dan" 1936). They spent the summer in Woods Hole and went to the Naples Zoological Station from January to April 1937 before moving to Japan that May (Colwin and Colwin 1979; Katō 1980; Mizoguchi 2007).

In Japan, the Dans began their new life conducting research at the Misaki Marine Biological Station, which was part of Tokyo Imperial University (Okada 1994; Inaba 2015; Ericson, in this volume). Founded in 1886 on the Miura Peninsula south of Tokyo, the Misaki Marine Biological Station was modeled on the Naples Zoological Station and was the perfect setting for the Dans to develop research programs based on marine organisms (Okada 1994, 137–38). Significantly, neither pursued fertilization research when they arrived at Misaki in the late 1930s. Instead, they worked together to develop various Japanese marine organisms as experimental systems to investigate problems of cell cleavage (Dan and Kubota 1960).

Up to this point, the question of cell cleavage had been addressed largely as a structural issue in the sense that most accounts focused on patterns of cell cleavage: whether they were symmetrical or nonsymmetrical and whether all cells continued dividing equally (see Conklin 1902). For example, in 1912, Just connected the problems of fertilization and cleavage by noting that in *Nereis* the first cell cleavage plane seemed to be determined by the entry point of the sperm (Just 1912). At that time, early theories of mechanisms for cell division included proposals based on changes in surface tension, changes in internal viscosity, and constriction of an equatorial membrane band (Dan 1948).

The move from fertilization to cell cleavage was not a big jump for the Dans. In his doctoral research, Katsuma had investigated the possibility of charge differentials in the cleavage process (Dan 1933). He did not find any, but this work indicated his interest in problems of cell division in terms of cell membrane dynamics. Through the difficult war years—between 1937 and 1948—the Dans and two Japanese collaborators produced a series of eight papers on cell cleavage that championed a mechanism based on the effects of the aster-spindle systems in dividing cells (Dan, Yanagita, and Sugiyama 1937; Dan, Dan, and Yanagita 1938; Dan and Dan 1940, 1942; Dan 1943a, 1943b; Dan and Dan 1947a, 1947b). With one exception, these papers used Japanese marine species: the sea urchins *Mespilia globulus, Strongylocentrotus pulcherrimus*, and *Pseudocentrotus depressus*, the sand dollar *Astriclypeus manni*, and the hydromedusa *Spirocodon saltatrix*. The exception was the American species *Ilyanassa obsoleta*, a sea snail. In these papers, the Dans developed novel techniques to mark and track cell membrane motion during cell cleavage. Using sea urchins, Katsuma realized that very fine particles of kaolin clay would adhere to the cell surface of these organisms. He could then use these clay particles as landmarks to track cell surface motion during cleavage. This method became an essential component in his and Jean's research.

The sea urchins *Mespilia globulus* and *Strongylocentrotus pulcherrimus* and the sand dollar *Astriclypeus manni* were used in the first three papers with the kaolin technique (Dan, Yanagita, and Sugiyama 1937; Dan, Dan, and Yanagita 1938; Dan and Dan 1940; see also Inoué 2016, 14). The eggs of each type of organism were readily available at Misaki. From the first experiment using *Mespilia*, the Dans found that there was gradual expansion of the cell surface in most areas of the cell, except at the furrow between the dividing cells, where there was initial shrinkage followed by massive expansion. These measurements in *Mespilia* were taken after the hyaline plasma layer surrounding the cell had been removed. When it was left intact, the differences between regions of the cell diminished. Reasoning that the thickness of this layer caused this dampening, the Dans repeated their experiments with *Astriclypeus*, which has a very thin plasma layer. In *Astriclypeus*, regional differences were more pronounced, and the dampening effect was ascribed to the effects of the thickness of the plasma layer (Dan 1948).

The Dans' fourth paper on cleavage was published in 1942 but based on data that they had collected years earlier in Woods Hole using the snail *Ilyanassa obsoleta*. In comparison to the sea urchins they had previously used in the United States and Japan, cell cleavage in *Ilyanassa* was slow (Dan 1943a), which made tracking cell membrane movement much easier. In all these experiments, the Dans made a series of camera lucida drawings of the cell membranes and their markers as they changed over time. The distances that the markers moved were then measured across the series of drawings, and the rate of motion was calculated and graphed. In a rapidly dividing system, this process of inscription was more challenging. Although *Ilynassa* cell cleavage was less symmetrical than that observed in sea urchins and the sand dollar, it showed the same pattern of cell membrane shrinkage and expansion. The Dans concluded that, because "it is found in the eggs of different phyla which also differ in cleavage pattern," the initial shrinkage at the cell division furrow "may be considered constant among, at least, egg cells cleaving with this general system of spindle-asters, and in-cutting cleavage furrow" (Dan 1948, 194–95). This generalization regarding the cleavage furrow was not the product of simple induction within species or across closely related species. Instead, it was considered justified because the instances were more phylogenetically distant even though all the species considered were still marine.

In 1942, Katsuma Dan developed a new technique for marking cell membrane dynamics that basically involved puncturing the cell and then observing the deformation of the puncture wound as the cell underwent division. This perforation technique was developed using the sea urchin *Pseudocentrotus depressus* but then extended to fertilization in the sand dollar *Astriclypeus manni* (Dan 1943a, 1943b; Dan and Dan 1947a). The sand dollar was of particular interest because it has eccentric nuclei; thus, instead of division occurring from the center of the cell, it is offset to one side. As Jean Dan put it: "Having achieved the formulation of a mechanism which can, with some success, answer the questions involved in the typical cases, the next step is obviously to apply the concept to progressively more atypical examples of cell cleavage" (Dan 1948, 208). In addition to its eccentric spindles, *Astriclypeus manni* was considered "especially favorable material for such a study because the astral rays are particularly visible and well defined and can be seen crossing the median plane, with longer rays extending to the vegetal pole" (Dan 1948, 208). This natural experiment allowed the Dans to postulate more firmly that connections between the asters and the spindles was relatively rigid, with the result that, during eccentric cell division, the asters rotated and the spindles bent as the furrow from the animal pole pushed into the cell (Dan 1948, 213). Having observed eccentric division in *Astriclypeus manni*, the Dans then confirmed their results with "the extreme case in the spindle eccentricity series," cell division in the hydromedusa *Spirocodon saltatrix* (Dan and Dan 1947b; Dan 1948).

His and Jean's initial experiments on cell division had reinforced Katsuma's belief in a mechanism based on the cell's spindles and asters. In 1943, when he read a report from Wilhelm Schmidt in Giessen, Germany, who had photographed birefringent spindles in sea urchin eggs, Katsuma decided that he had to try it himself (Schmidt 1937). With a group of students, including Shinya Inoué, and a borrowed polarizing microscope, he tried to find similar birefringent structures in his Japanese sea urchins. As Inoué remembers that evening, some people saw the spindles and others did not, so the matter was left undecided. Dan returned to his method of marking with kaolin particles and cell perforation. After the war, however, Inoué brought polarizing microscopy back to the Dans' project with great success. Indeed, his homemade improvements to the available polarizing microscope system allowed him to visualize cell spindles clearly and opened up more direct investigation of their role in cell division (Inoué and Dan 1951; Tani, Shribak, and Oldenbourg 2016; Inoué 2016).

In January 1945, the Misaki Marine Biological Station was taken over by the Japanese navy and converted into a base for miniature suicide submarines meant to prevent any Allied ships from entering Tokyo Bay. Earlier, as the war intensified, Jean had been declared an enemy alien and forced to leave their home in Nagai, which was designated a military zone. She moved with their five children to Kudan in Tokyo and then to the Hakone Mountains. They returned to Nagai only after August 1945. Katsuma was at Misaki on September 28, 1945, when representatives of the US and Japanese navies met to discuss plans for the marine station turned submarine base.

Witnessing the chaotic and poorly translated meeting, Katsuma privately tried to make one US officer appreciate the research mission of Misaki before deciding to write a plea for understanding that he taped to the station's front door. Signed "The Last to Go," this poster compared Misaki to the MBL. When US Navy captain L. S. Parks, who was part of the demilitarization unit, saw it, he took it down and sent it to the MBL, where it remains on display to this day. Somewhere along the line, the poster came to the attention of journalists for *Time* magazine, who published the text under the headline "Appeal to the Goths" (Inoué 2016, 2–5).

After the war, as the Dans wrote up their experiments on eccentric division (Dan and Dan 1947a, 1947b), Katsuma returned to his earliest work on electrokinetic analysis of fertilization. While his initial set of papers used sea urchins from the Woods Hole area, his papers in 1947 used some of the Japanese sea urchins that had been featured in his work on cell cleavage: *Strongylocentrotus pulcherrimus* and *Pseudocentrotus depressus*, with the addition of the sea urchin *Anthocidaris crassispina* (Dan 1947a, 1947b, 1947c). As Hazime Mizoguchi notes, these papers from the Dans constituted five of the twenty papers published in the *Biological Bulletin* in 1947 and marked a period of high productivity for them (Mizoguchi 2007, 30).

During the postwar recovery period, Jean became deeply involved in community work; acting as a liaison between occupying American forces and local farmers, she arranged a cooperative that redistributed American surplus materials and persuaded occupation forces to return much-needed farmland to local control (Katō 1980; Inoué 2016, 41–42). In 1947, she returned to the United States to see her family and spend part of the summer at the MBL. While there, she learned of Bausch and Lomb's new phase contrast light microscope. It is said that she was in the Captain Kidd Bar in Woods Hole when she met Kenneth Cooper, a biologist from Princeton University. She told him about the research she was doing with Katsuma on cell division and the work Inoué had begun with his homemade polarizing microscopes. Cooper was evidently impressed. He invited Shinya to apply to work with him as a graduate student at Princeton and asked Edwin Grant Conklin to have the American Philosophical Society arrange a grant that would buy Jean the first phase contrast microscope that Bausch and Lomb produced. Jean took the microscope back to Japan as a gift for Katsuma, but he gave it to her to use for her own work instead. At Katsuma's suggestion, Jean made the hard choice to leave her work at the Nagai cooperative and return to science full-time (Inoué 2016, 50–52).

While Katsuma continued to work on the role of spindles in cell cleavage in 1948, Jean returned to fertilization and began using the phase contrast microscope to observe the initial contact between sperm and egg in Japanese sea urchins (Dan 1950a). This instrument, and the images that could be produced using it, became an essential component in their research. Jean initially posed an unresolved observation problem for herself: Does the tail of the fertilizing sperm in echinoderms enter the egg or remain outside during fertilization? (Dan 1950b). The phase contrast microscope could, she thought, allow her to settle this question. Using locally available echinoderms, including *Strongylocentrotus pulcherrimus*, *Heliocidaris crassispina*, *Clypeaster japonica*, *Mespilia globulus*, *Pseudocentrotus depressus*, and *Asterina pectinifera*, she found that the entire sperm entered the egg during fertilization in all cases.

Bolstered by her success, Jean turned to another observational enigma, namely, the one associated with the agglutination reaction made famous by Lillie a few decades before. A number of scientists beginning with Fol had observed that sometimes during agglutination thin fibers were visible extending from the acrosome region at the head of the sperm (Dan 1952; Afzelius 2006). In 1927, Grigore Popa had suggested that the acrosome broke open and secreted a sticky substance during agglutination (Popa 1927; Dan 1952). Jean was able to observe the acrosome when sperm from sea urchins and starfish were exposed to egg jelly water. On the basis of her observations, she proposed a crucial role for the acrosome reaction in fertilization. As she put it: "As the spermatozoan actively swims through the loose network of the jelly layer, it responds to the chemical stimulation of the jelly substance by a break down of the membrane covering the front part of the acrosome, so that by the time the sperm reaches the egg surface, a few seconds later, it carries at its tip a mass of freshly exposed lysin with which it effects penetration of the vitelline membrane as the first step in the fertilization process" (Dan 1952, 62). Significantly, when she extended her study of the acrosome reaction to starfish, she added the presence of a filament extending from the sperm acrosome in addition to the release of lysins. These filaments had been recognized as part of the fertilization process but had usually been thought to have arisen from the egg,

Table 11.1 Organism use in Jean Dan's "Studies on the Acrosome" publications

Paper number	Date	Organism	Organism type and number of species
	1950	*Strongylocentrotus pulcherrimus*	Sea urchins (5)
		Heliocidaris crassispina	Starfish (1)
		Clypeaster japonica	
		Mespilia globulus	
		Pseudocentrotus depressus	
		Asterina pectinifera	
I	1954	*Pseudocentrotus depressus*	Sea urchins (2)
		Strongylocentrotus pulcherrimus	
II	1954	*Asterina pectinifera*	Starfish (3)
		Asterias amurensis	
		Astropecten scoparius	
III	1954	*Strongylocentrotus pulcherrimus*	Sea urchins (3)
		Heliocidaris crassispina	
		Pseudocentrotus depressus	
IV	1955	*Mytilus edulis*	Bivalve mollusks (12)
		Lithophaga curta	
		Spondylus cruentus	
		Crassostrea echinata	
		Crassostrea nippona	
		Crassostrea gigas	
		Trapezium sublaevigatum	
		Chama retroversa	
		Petricola japonica	
		Mactra veneriformis	
		Mactra sulcataria	
		Zirfaea subconstricta	
V	1956	*Mytilus edulis*	Bivalve mollusk (1)
VI	1960	*Asterias forbesii*	Starfish (1)
VII	1964	*Pseudocentrotus depressus*	Sea urchins (2)
		Heliocidaris crassispina	
VIII	1967	*Asterina pectinifera*	Starfish (2)
		Asterias amurensis	
IX	1967	*Asterias amurensis*	Starfish (1)
X	1971	*Asterina pectinifera*	Starfish (2)
		Asterias amurensis	

not the sperm. Observing the filaments during agglutination, when no eggs were present, effectively settled the issue of filament origins while highlighting the significance of the acrosome reaction (Dan, Kitahara, and Khori 1954).

In a set of papers describing the acrosome reaction, Dan and her collaborators deliberately extended the range of organisms in order to build a case for the generality of the acrosome reaction (Dan 1952; Dan, Kitahara, and Khori 1954; Wada, Collier, and Dan 1956; Haino and Dan 1961; Niijima and Dan 1965a, 1965b; Dan and Hagiwara 1967; Hagiwara, Dan, and Saito 1967; Dan and Sirakami 1971; see table 11.1).

In a paper with Seiji Wada, for instance, Jean deliberately chose marine bivalves that represented "three of the five orders, and eight families." The organisms that she chose were available at Misaki, had external fertilization that could be experimentally induced, and had easily identifiable acrosomes (Dan and Wada 1955, 40). Her goal was not only to characterize the reaction but also to build an inductive case for its generality in phylogenetically divergent marine organisms. This form of reasoning was characteristic of her and Katsuma's understanding of the appropriate way in which to pursue conclusions that could be generalized beyond the organisms within which they were originally obtained and came to be a cornerstone of their repertoire.

Jean Dan's work on the acrosome reaction was a central feature of the rest of her career and was rightly recognized for its brilliance (Colwin and Colwin 1979; Afzelius 2006). Her return to science and her success with the acrosome reaction garnered her a position as a lecturer (in 1959) and then as a professor (in 1973) at Ochanomizu University, where she not only supervised graduate students but also played an important role in the Tateyama Marine Biological Station, which she directed in 1975 (Colwin and Colwin 1979; Katō 1980).

Conclusion

In the case of the Dans' research in cell biology, the presumption of diversity did not seem to wane in contrast to what has been documented by Logan (2002) primarily in physiology. Indeed, Katsuma and Jean consistently sought inductive support when they wished to make generalizations about cell cleavage and the acrosome reaction. For example, in Matazō Kume and Katsuma Dan's *Invertebrate Embryology* (1968), translated by Jean Dan, Kume and Dan explicitly address what they call the *universality* of fertilization. They summarize the situation as follows: "Lillie's fertilizin began with *Nereis* and extended to sea urchins, but starfish were excluded from the argument because their agglutination is uncertain. The animals in which the acrosome reaction have so far been observed are, at Misaki, six sea urchins, three species of starfish, twelve bi-valve molluscs, two annelids . . . and a sea cucumber. . . . It is thus an important point that the acrosome reaction is found to occur over a clearly wider range than the agglutination reaction" (Kume and Dan 1968, 31).[3]

Thus, this detailed case study reveals that the choice of organisms was not guided by material availability alone. Throughout the Dans' research, issues of epistemic access to phenomena played a crucial role. In addition, the Dans chose systems that they could maintain and manipulate in order to study fertilization, leading to their consistent choices of local marine organisms wherever they were located. That said, they also preferred organisms in which fertilization and cleavage could be easily observed and marked even if those organisms were not easily accessible locally. Their use of a sea snail indigenous to North America for their cleavage experiments (despite working in Japan at the time) serves as an excellent example of how the need to observe particular phenomena outweighed the advantages of local organisms in some instances. Marine systems that are fertilized externally were especially useful for visualizing fertilization, but naturally occurring variation in the phenomena was as important as constancy across species. Hence, another crucial component in the Dans' practice centered on this material choice, which was in contrast to much of existing biological practice during the period, which tended to prioritize use of organisms that were close at hand, particularly in the case of marine organisms.

Of course, even as they sought universality, Kume and Dan limited their inductive reach to marine systems. Jean Dan's review of the acrosome reaction in 1967 extended it to sixty-eight species, of which eight were vertebrates, of which four were marine mammals and four nonmarine mammals (Dan [1967] 1971). All the research on mammals occurred after 1960, suggesting a continuing search for acrosome reactions in other species over time (a trend supported in Piko [1969]). Indeed, returning to *GEIS* reveals that many researchers were investigating fertilization in nonmarine systems. Thus, very similar research on this topic was in fact possible in a wider range of organisms than the Dans chose, and such research was actively pursued by other researchers.

The decision the Dans made to limit the scope of their claims to marine systems may in part represent the constraints of local availability. However, we think it is equally important (and in fact more likely) that they and their collaborators had developed an expertise (particularly in terms of methods and techniques) in marine systems. Moving from sea urchins to rabbits or mice would have been possible but would also have represented a break with an established and effective practice in order to utilize experimental systems where they had no established expertise. These material and social conditions in turn limited the scope of their abilities to infer inductively from the marine systems with which they surrounded themselves. They sought to use a diverse set of marine organisms in order to make claims about the nature of cleavage and fertilization rather than establish generalizations about phenomena based on one or two singular species. In fact, their decisions to compare these processes across a number of different organisms eventually allowed them to solve some of the more vexing questions about the nature of fertilization and cleavage that had plagued scientists since the early twentieth century.

Appendix: Creating an Organismal Landscape

This appendix describes our methods for creating the organismal landscape visualizations used in this essay.

Data for the organismal landscape diagrams were extracted from *GEIS* volumes from 1951 to 1963. Using the research classification and the organismal categories provided in *GEIS,* we created spreadsheets for each year that accounted for the listed scientists, their work, and the nested hierarchical system of research topics that the *GEIS* editors used to sort research. We then counted the numbers of organisms per research topic and filled in the spreadsheets with those values. This gave us organism counts for the most specific research classifications within the nested hierarchy. Numbers of organisms for the higher-level research categories were obtained by summing together the relevant lower-level research categories (see, e.g., fig. 11.1*A* above). In order to control for the effect of growth in the field of embryology/developmental biology, we transformed the numbers of organisms in these spreadsheets into their proportion per year (see, e.g., fig. 11.1*B* above).

It should be noted that text mining and digital analysis tools could not be applied to the creation of the spreadsheets. The formatting of the journals was consistent neither between different volumes nor within individual volumes themselves. Moreover, each entry could be, but was not necessarily, coded in multiple ways, making it difficult to create effective machine learning rules to mine files. For example, a common entry from 1951 includes these codes: "Harvey, E.B—a (4)LV." After the name, the letter after the dash refers to which project of the author's is relevant, the parenthetical notation refers to the investigative approach, and, finally, LV refers to the type of organism, in this case lower vertebrates. Organism codes could be one or two letters. To complicate matters further, the entries are in double columns on every page, making it difficult to do a line-by-line analysis of the text. In other words, the nature of how the *GEIS* volumes were constructed necessitated that all the information be coded by hand into spreadsheets. The average length of each *GEIS* volume was over two hundred pages.

The organismal landscape visualizations were created with Plotly (https://plot.ly) and Microsoft Excel. For visualization in Plotly, a set of data in an Excel spreadsheet that we wanted to visualize was saved as a comma-separated values file (.csv). This file was then imported into Plotly. To create a heatmap visualization, we selected heatmap as a chart type, selected the organismal columns for the *Z*-axis and the year as the *Y*-axis values. To create a contour map visualization, we selected contour as a chart type, selected the organismal columns for the *Z*-axis and the year as the *Y*-axis values. Labels, colors, and other features of the visualization were created or selected with other tools in the Plotly program.

The heatmap diagrams in figure 11.2*C* and *D* above were generated in Excel by applying conditional formatting to a table of values in an Excel spreadsheet.

Notes

Research for this project has partially been supported via the Australian Research Council Discovery Project (DP160102989) "Organisms and Us: How Living Things Help Us to Understand Our World" (2016–20).

1. These codes and taxonomic ranks are as follows: EV. Evertebrata, C. Plathelminthes, D. Arthropoda, Di. Insecta, E. Mollusca, F. Bryozoa, etc., G. Echinodermata, H. Acrania, V. Vertebrata, I. Pisces, J. Amphibia, L. Aves, M. Mammalia.
2. For most of this period, *GEIS* published its subject index only every other year.
3. Note that, although they are exceedingly deferential to Lillie and generously acknowledge the influence and impact of his research on fertilizin and agglutination, Kume and Dan suggest that perhaps he picked the wrong phenomena as the key to fertilization.

References

Afzelius, Björn. 2006. "Preformed Acrosome Filaments: A Chronicle." *Brazilian Journal of Morphological Science* 23:279–85.

Ankeny, Rachel A. 2001a. "Model Organisms as Cases: Understanding the Lingua Franca at the Heart of the Human Genome Project." *Philosophy of Science* 68:S251–S261.

———. 2001b. "The Natural History of *Caenorhabditis elegans* Research." *Nature Reviews Genetics* 2: 474–79.

Ankeny, Rachel A., and Sabina Leonelli. 2011. "What's So Special about Model Organisms?" *Studies in the History and Philosophy of Science* 42:313–23.

Clark, Jean. 1936. "An Experimental Study of Polyspermy." *Biological Bulletin* 70:361–84.

Clarke, Adele E. 1998. *Disciplining Reproduction: Modernity, American Life Sciences, and "the Problems of Sex."* Berkeley and Los Angeles: University of California Press.

Colwin, Laura, and Arthur Colwin. 1979. "Jean Clark Dan." *Nature* 278:492.

Conklin, Edwin Grant. 1902. *Karyokinesis and Cytokinesis in the Maturation, Fertilization and Cleavage of Crepidula and Other Gasteropoda*. Philadelphia: Academy of Natural Sciences Press.

Crowe, Nathan, Michael R. Dietrich, Beverly Alomepe, Amelia Antrim, Bay Lauris ByrneSim, and Yi He. 2015. "The Diversification of Developmental Biology." *Studies in the History and Philosophy of Biological and Biomedical Sciences* 53:1–15.

Dan, Jean C. 1948. "On the Mechanism of Astral Cleavage." *Physiological Zoology* 21:191–218.

———. 1950a. "Fertilization in the Medusan, *Spirocodon saltatrix*." *Biological Bulletin* 99:412–15.

———. 1950b. "Sperm Entrance in Echinoderms, Observed with the Phase Contrast Microscope." *Biological Bulletin* 99:399–411.

———. 1952. "Studies on the Acrosome, I: Reaction to Egg-Water and Other Stimuli." *Biological Bulletin* 103:54–66.

———. (1967) 1971. "Acrosome Reaction and Lysins." In *Fertilization: Comparative Morphology, Biochemistry, and Immunology* (vol. 1), ed. Charles B. Metz and Alberto Monroy, 237–93. New York: Academic.

Dan, Jean C., and Y. Hagiwara. 1967. "Studies on the Acrosome, IX: Course of Acrosome Reaction in the Starfish." *Journal of Ultrastructure Research* 18:562–79.

Dan, Jean C., A. Kitahara, and T. Kohri. 1954. "Studies on the Acrosome, II: Acrosome Reaction in Starfish Spermatozoa." *Biological Bulletin* 107:203–18.

Dan, Jean C., and Seiji K. Wada. 1955. "Studies on the Acrosome, IV: Acrosome Reaction in Bivalve Spermatozoa." *Biological Bulletin* 109, no. 1:40–55.

Dan, Jean C., and A. Sirakami. 1971. "Studies on the Acrosome, X: Differentiation of the Starfish Acrosome." *Development, Growth and Differentiation* 13, no. 1:37–52.

Dan, Katsuma. 1933. "Electrokinetic Studies of Marine Ova, I: *Arbacia punctulata*." *Journal of Cellular and Comparative Physiology*. 3, no. 4:477–92.

———. 1934. "Electrokinetic Studies of Marine Ova, II: *Cuminga tellinoides, Asterias forbesii, Echinarachnius parma, Nereis limbata*, and *Cerebratalus lacteus*." *Biological Bulletin* 66:247–56.

———. 1936. "Electrokinetic Studies of Marine Ova, III: The Effect of Dilution of Sea Water, and of Sodium and Calcium upon the Surface Potentials of *Arbacia* Eggs." *Physiological Zoology* 9:43–57.

———. 1943a. "Behavior of the Cell Surface during Cleavage, V: Perforation Experiment." *Journal of the Faculty of Science of Tokyo Imperial University, Section IV* 6:297–322.

———. 1943b. "Behavior of the Cell Surface during Cleavage, VI: On the Mechanism of Cell Division." *Journal of the Faculty of Science of Tokyo Imperial University, Section IV* 6:323–68.

———. 1947a. "Electrokinetic Studies of Marine Ova, V: Effect of pH-Changes on the Surface Potentials of Sea-Urchin Eggs." *Biological Bulletin* 93, no. 3: 259–66.

———. 1947b. "Electrokinetic Studies of Marine Ova, VI: The Effect of Salts on the Zeta Potential of the Eggs of *Strongylocentrotus pulcherrimus*." *Biological Bulletin* 93, no. 3:267–73.

———. 1947c. "Electrokinetic Studies of Marine Ova, VII: Relation between the Zeta Potential and Adhesiveness of the Cell Membrane of Sea-Urchin Eggs." *Biological Bulletin* 93, no. 3:274–86.

———. 1978. "Katsuma Dan on His First Meeting with Victor Heilbrunn." *The Embryo Project Encyclopedia*. http://embryo.asu.edu/handle/10776/1432.

———. 1988. *Uni to kataru* (Dialog with sea urchins). Tokyo: Gakkai Shuppan Sentā.

Dan, Katsuma, and Jean Clark Dan. 1940. "Behavior of the Cell Surface during Cleavage, III: On the Formation of New Surface in the Eggs of *Strongylocentrotus pulcherrimus*." *Biological Bulletin* 78:486–501.

———. 1942. "Behavior of the Cell Surface during Cleavage, IV: Polar Lobe Formation and the Cleavage of Eggs of *Ilyanassa obsolete Say*." *Cytologia* 12: 246–61.

———. 1947a. "Behavior of the Cell Surface during Cleavage, VII: On the Division Mechanism of Cells with Excentric Nuclei." *Biological Bulletin* 93: 139–62.

———. 1947b. "Behavior of the Cell Surface during Cleavage, VIII: On the Cleavage of Medusan Eggs." *Biological Bulletin* 93:163–88.

Dan, Katsuma, Jean Clark Dan, and T. Yanagita. 1938. "Behaviour of the Cell Surface during Cleavage, II." *Cytologia* 8:521–31.

Dan, Katsuma, and Hiroshi Kubota. 1960. "Data on the Spawning of *Comanthus japonica* between 1937 and 1955." *Embryologia* 5, no. 1:21–37.

Dan, Katsuma, T. Yanagita, and M. Sugiyama. 1937. "Behavior of the Cell Surface during Cleavage, I." *Protoplasma* 28:66–81.

de Chadarevian, S. 1998. "Of Worms and Programmes: *Caenorhabditis elegans* and the Study of Development." *Studies in the History and Philosophy of Biology and Biomedical Science* 29:81–105.

Dietrich, Michael R. 2015. "Explaining the 'Pulse of Protoplasm': The Search for Molecular Mechanisms of Protoplasmic Streaming." *Journal of Integrative Plant Biology* 57:14–22.

Ebert, James D. 1994. "Katsuma Dan and the Quiet Revolution." *Biological Bulletin* 187:125–31.

Ernst, Susan. 2011. "Offerings from an Urchin." *Developmental Biology* 358: 285–94.
Fantini, Bernadino. 1985. "The Sea Urchin and the Fruit Fly: Cell Biology and Heredity, 1900–1910." *Biological Bulletin* 168:99–106.
Farley, John. 1982. *Gametes and Spores*. Baltimore: Johns Hopkins University Press.
Hagiwara, Y., J. C. Dan, and A. Saito. 1967. "Studies on the Acrosome." *Journal of Ultrastructure Research* 18, nos. 5–6:551–61.
Haino, Kazu, and Jean C. Dan. 1961. "Some Quantitative Aspects of the Acrosomal Reaction to Jelly Substance in the Sea Urchin." *Embryologia* 5:376–83.
Inaba, Kazuo. 2015. "Japanese Marine Biological Stations: Preface to the Special Issue." *Regional Studies in Marine Science* 2:1–4.
Inoué, Shinya. 2016. *Pathways of a Cell Biologist: Through yet Another Eye*. New York: Springer.
Inoué, Shinya, and Katsuma Dan. 1951. "Birefringence of the Dividing Cell." *Journal of Morphology* 89, no. 3:423–55.
Just, E. E. 1912. "The Relation of the First Cleavage Plane to the Entrance Point of the Sperm." *Biological Bulletin* 22:239–52.
Katō Kyōko. 1980. *Nagisa no uta: Aru joryū seibutsugakusha no shōgai*. Tokyo: Kōdansha.
"Katsuma Dan to Wed Jean McNair Clark." 1936. *Washington Post*, July 3, X18.
Kohler, Robert E. 1994. *Lords of the Fly: Drosophila Genetics and the Experimental Life*. Chicago: University of Chicago Press.
Kume, Matazō, and Katsuma Dan. 1968. *Invertebrate Embryology: Musekitsui dobutsu hasseigaku*. Belgrade: National Library of Medicine, Public Health Service, US Department of Health, Education and Welfare, and National Science Foundation.
Large, S. S. 2001. "Nationalist Extremism in Early Showa Japan: Inoue Nissho and the 'Blood-Pledge Corps' Incident, 1932." *Modern Asian Studies* 35:553–64.
Leonelli, Sabina. 2007. "*Arabidopsis*, the Botanical *Drosophila*: From Mouse Cress to Model Organism." *Endeavour* 31:34–38.
Lillie, Frank R. 1916. "The History of the Fertilization Problem." *Science* 43:39–53.
———. 1919. *Problems of Fertilization*. Chicago: University of Chicago Press.
Logan, Cheryl A. 2002. "Before There Were Standards: The Role of Test Animals in the Production of Empirical Generality in Physiology." *Journal of the History of Biology* 35:329–63.
Maienschein, Jane. 1991. *Transforming Traditions in American Biology, 1880–1915*. Baltimore: Johns Hopkins University Press.
———. 2014. *Embryos under the Microscope: The Diverging Meanings of Life*. Cambridge, MA: Harvard University Press.
McClendon, J. F. 1914. "On the Nature and Formation of the Fertilization Membrane of the Echinoderm Egg." *Internationale Zeitschrift für physikalisch-chemische Biologie* 1:163–68.

"Miss Jean Clark Wed to Dr. Katsuma Dan." 1936. *New York Times*, July 5, N5.
Mizoguchi, Hazime. 2007. "Japanese Biologists at the Marine Biological Laboratory, Woods Hole." *Historia scientarum* 17:20–37.
Monroy, Alberto. 1986. "A Centennial Debt of Developmental Biology to the Sea Urchin." *Biological Bulletin* 171:509–19.
Niijima, L., and J. Dan. 1965a. "The Acrosome Reaction in *Mytilus edulis*, I: Fine Structure of the Intact Acrosome." *Journal of Cell Biology* 25:243–48.
Niijima, L., and J. Dan. 1965b. "The Acrosome Reaction in *Mytilus edulis*, II: Stages in the Reaction, Observed in Supernumerary and Calcium-Treated Spermatozoa." *Journal of Cell Biology* 25:249–59.
Okada, Masukichi. 1996. "Katsuma Dan." *Development Growth and Differentiation* 38, no. 5:449.
Okada, T. S. 1994. "Experimental Embryology in Japan, 1930–1960: A Historical Background of Developmental Biology in Japan." *International Journal of Developmental Biology* 38:135–54.
Pauly, Philip. 2000. *Biologists and the Promise of American Life: From Meriwether Lewis to Alfred Kinsey*. Princeton, NJ: Princeton University Press.
Piko, Lajos. 1969. "Gamete Structure and Sperm Entry in Mammals." In *Fertilization: Comparative Morphology, Biochemistry, and Immunology* (vol. 2), ed. Charles B. Metz and Alberto Monroy, 325–403. New York: Academic.
Popa, G. T. 1927. "The Distribution of Substances in the Spermatozoon (Arbacia and Nereis)." *Biological Bulletin* 52:238–57.
Rader, Karen A. 2004. *Making Mice: Standardizing Animals for American Biomedical Research, 1900–1955*. Princeton, NJ: Princeton University Press.
Sakai, H. 1996. "Katsuma Dan." *Development, Growth and Differentiation* 38:450.
Schmidt, W. J. 1937. *Die Doppelbrechung von Karyoplasma, Zytoplasma und Metaplasma*. Berlin: Gebrüder Borntraeger.
Tani, T., M. Shribak, and R. Oldenbourg. 2016. "Living Cells and Dynamic Molecules Observed with the Polarized Light Microscope: The Legacy of Shinya Inoué." *Biological Bulletin* 231:85–95.
Wada, S. K., J. R. Collier, and J. C. Dan. 1956. "Studies on the Acrosome." *Experimental Cell Research* 10:168–80.
Wourms, John. 1997. "The Rise of Fish Embryology in the Nineteenth Century." *American Zoologist* 37:269–310.
Ziegler, Alexander, Georgy V. Mirantsev, Michael Angoux, and Andreas Kroh. 2014. "Historical Aspects of Meetings, Publication Series, and Digital Resources Dedicated to Echinoderms." *Zoosystematics and Evolution* 90: 45–56.

TWELVE

Hagfish and Vascular Biology: Why the Marine Model Matters

MARIANNE A. GRANT AND WILLIAM C. AIRD

Hagfish are unusual animals, and you are not likely to forget them if you have seen them. SEBASTIAN SHIMELD AND PHILLIP DONOGHUE

Not even a mother could love a hagfish. CAREY GOLDBERG

We have long been fascinated by the observation that blood vessels and their cell lining (the endothelium) differ in structure and function between different parts of the human body. These vascular bed–specific phenotypes are important for meeting the unique needs of diverse organs. For example, endothelial cells from different organs use distinct formulas of procoagulants and anticoagulants to maintain blood in a fluid state within the circulation. Importantly, from a medical perspective, dysfunction or failure of site-specific properties explains why vascular disease, whether stroke, heart attack, vasculitis, or pulmonary hypertension, always affects discrete and predictable segments of the vascular tree (Rosenberg and Aird 1999, 1555–64). As our research in mice and humans uncovered increasing layers of complexity in the vasculature, we began to wonder whether we were examining an advanced property of the mammalian system or whether phenotypic heterogeneity represents a more fundamental design feature of the endothelium.

One way to address this question is through studies in comparative biology. Previous investigations have shown that, while invertebrates, cephalochordates (lancelets), and tunicates (sea squirts) have blood vessels, none has a true endothelial cell lining. In contrast, all vertebrates examined to date have a closed system of blood vessels lined by endothelial cells (Monahan-Earley, Dvorak, and Aird 2013, 46–66). These basic observations suggest that the endothelium evolved over a remarkably short time span following the divergence of the vertebrate lineage. When exactly did this occur? And why? What selective advantage did this new cell layer provide? And was phenotypic heterogeneity a built-in feature from the start or a later addition that met evolving needs of the organism?

Why does any of this matter? After all, physicians and their patients seem to get by just fine without considering such esoteric questions. For example, a cardiologist treating Mr. Smith for coronary artery disease is keenly interested in the patient's risk factors for atherosclerosis: Does he smoke? Is his blood cholesterol high? Does he have diabetes or hypertension? In identifying risk factors for coronary artery disease, the doctor is able to prescribe active intervention for his patient with the explicit goal of reducing that individual's cardiovascular morbidity and mortality.

A physician with an interest in the evolutionary origins of health and disease would ask a somewhat different question about Mr. Smith's condition, namely, Why do *Homo sapiens*, of all species in the world, develop atherosclerotic heart disease in the first place? He might consider how the closed-loop geometry of the cardiovascular system of necessity creates microdomains of disturbed blood flow (at branch points and curvatures), which in turn lead to endothelial dysfunction and local buildup of atherosclerotic plaque. This is an example of an evolutionary trade-off whereby the selective advantage of a closed circulatory system—as distinct from the open circulation present in many invertebrates—seems to have outweighed the disadvantages associated with regions of disturbed flow. The Darwinian physician would also point to the fact that natural selection operates at the level of reproductive success (traits have been selected to transmit genes to future generations) and that chronic diseases of the elderly, including atherosclerosis, escape rigorous quality control. And, finally, he (not to mention paleo-diet enthusiasts) would emphasize that our cardiovascular system was designed for a much earlier era and that a mismatch between our genome and our environment lies at the root of the atherosclerosis endemic.

A consideration of evolutionary questions matters because it provides an explanatory framework for understanding human health and disease, whether it be the endothelium as an organ, phenotypic heterogeneity as a structural and functional entity, or atherosclerosis as a disease construct. Physicians and medical researchers are in the business of explaining disease to their patients and their families, to the public, and to each other, and those who focus their lens exclusively on proximate mechanisms of disease ignore important evolutionary legacies that render us vulnerable to disease, including design constraints, path dependence, trade-offs, and selective pressures. In 1973, Theodosius Dobzhansky famously observed that "nothing in biology makes sense except in the light of evolution" (Dobzhansky 1973, 125–29). In 1994, Randolph Nesse and George Williams published their *Why We Get Sick: The New Science of Darwinian Medicine* (Nesse and Williams 1994). This was the first attempt to formalize Dobzhansky's principle in the realm of medicine and helped launch a new field that has become known as *evolutionary medicine*.[1] Simply put, evolutionary medicine broadens our horizons, changes the way we think about our patients and their symptoms, and opens new avenues of research. And, finally, evolutionary and comparative perspectives provide—at least for some of us—a spiritual dimension to our work by enhancing our sense of connection to the rich tapestry of life.

To gain insights into the evolutionary origins of the endothelium and endothelial heterogeneity, we set our sights on the invertebrate-vertebrate divide. There was little information in the published literature about the endothelium in the earliest vertebrates, the cyclostomes (jawless fishes), a group that includes hagfish and lamprey. We saw this knowledge gap as an opportunity because a comparison of features between hagfish or lamprey and gnathostomes (jawed vertebrates) might provide powerful insights into core properties of this vital cell layer.[2]

To pursue our interests, we spent eight summers at Mount Desert Island Biological Laboratory (MDIBL) in Maine, procuring, harvesting, dissecting, and analyzing the cardiovascular system of the Atlantic hagfish, *Myxine glutinosa*. We opened our specimens up, observed their two hearts beating asynchronously, and followed their arteries and veins to all reaches of the body. We cut sections from organs and incubated them with various dyes and lectins. We prepared ultrathin sections and processed them for electron microscopy to examine the fine structure of the endothelium. We employed intravital microscopy to observe blood cells moving in the arteries and veins of living specimens both under normal conditions and in response to inflammatory mediators. We removed blood from hagfish and measured the time it took to clot. And we applied modern molecular biology techniques to clone genes from the endothelium.

The goal of this essay is to take the reader on our journey from the bench to the beach and back to the bench. We begin with a consideration of place, the MDIBL, a small marine biology facility nestled in Salisbury Cove, Maine. We then introduce our model, the hagfish, and briefly discuss its role in research, both past and present. Finally, we describe our own studies with this creature. In the end, we hope to weave a story about how an unlikely marine model organism—unsavory to most, fascinating and elegant to us—can teach us about our own biology.

The Research Setting

Marine research stations have played an important role in the development of the biological sciences for well over a century (Dexter 1988, 3–6; Maienschein 1990, 399–403; Epstein and Epstein 2005, 1–7). The MDIBL had its beginnings in 1898 when John Sterling Kingsley, chair and professor of Biology at Tufts College, established a teaching and research laboratory at South Harpswell, Maine, that housed Tufts College's Summer School of Biology. Kingsley was no stranger to marine biology. He had served as an instructor (and eventually director) of the Annisquam Laboratory on the north shore of Cape Ann near Gloucester, Massachusetts, as well as the Marine Biological Laboratory at Woods Hole, Massachusetts. However, he was interested in studying more northern waters, and Casco Bay provided an abundance of boreal fauna and flora.[3]

In the meantime, Mount Desert Island (MDI), a large island off the coast of Maine, was drawing an ever-increasing number of rusticators and had become a favorite summer retreat for a handful of affluent high-society families, including the Hamiltons, the Vanderbilts, the Pulitzers, and the Rockefellers.[4] In the early twentieth century, George Dorr, whose Boston-based family had long summered on MDI, together with Charles W. Elliot, president of Harvard College and a summer resident of Northeast Harbor on the island, organized a small group of year-round island residents who had a common interest in protecting the island from overdevelopment and preserving it in its natural state (Evans 2015, 44). The group sought donations of private land and money and by 1912 had secured over five thousand acres of property. The land was declared a national monument in 1916 and would be renamed Acadia National Park in 1929.

Speaking at a celebration of the original national monument in 1916, Dorr said: "It struck me what a splendid and useful thing it would be if we could provide down here, in a spot so full of biologic interest and unsolved biologic problems, so rich in various beauty and locked around by the cool northern sea, a summer home, however simple, for men of science working in the government bureau, in the museums and universities. They would come down to work, as Henry Chapman and Charles Sedwick Minot used to do, on a fresh field of life, bird or plant or animal, and then go back invigorated, ready to do more valuable work the whole winter through in consequence of this climate boon and stimulating change" (Evans 2015, 45). Dorr realized his dream when he and his friends purchased land in Salisbury Cove, on the northern coast of the island, a few miles from the town of Bar Harbor, and convinced members of the Harpswell Laboratory to move to MDI, "where the conditions were better and more permanently assured than they possibly could be at Casco Bay" (Evans 2015, 47). Thus was born MDIBL, which opened its doors to summer investigators in 1921.[5] A year later, in a review of the new station, Roy Waldo Miner wrote: "From a biologist's standpoint, its situation is exceptional. The sheltered coast line on this side of the island is indented by a succession of coves, floored with sandy mud, and backed by a rocky rampart of cliffs, which jut out at intervals as picturesque headlands covered with spruce growth. The tide rises and falls a distance of twelve feet so that a considerable stretch of mud flat is laid bare at low tide, where marine worms, clams, crabs, and gastropods abound. The waters . . . after good dredging, and the wharf piles of the extensive United States coaling station on the shore of the opposite mainland are crowded with marine algae, ascidians, sea anemones, and sea stars. The deep waters of Frenchman's Bay are alive with various marine dishes and Crustacea, inclining lobsters" (Evans 2015, 57).

Every summer, researchers from various medical school and basic biology departments throughout the United States and around the world converged on MDI to study their favorite marine organisms. In the early years, most of these investigators were PhD scientists committed to basic studies in marine biology. However, the lab soon gained a reputation for its biomedical research. In a fund-raising letter written in 1928, the MDIBL was described as attracting "a score of investigators from our leading universities [who] have been quietly at work upon subjects, many of which bear directly upon human welfare. There are six investigators now working upon cancer and allied problems" (Evans 2015, 74). In the president's report for 1988, Frank Epstein, a seasonal investigator whom we will meet later in this essay, wrote: "The Mount Desert Island Biological Laboratory conducts basic research in marine biology, with many implications for human health and for the environment. . . . As the laboratory enters its second century, its goal is to continue as a center of excellence in marine and biomedical research. . . . We anticipate that, in its second century as in its first, advances in the diagnosis, prevention and treatment of disease will result from insights gained at this laboratory into the most fundamental processes of life" (Evans 2015, 497).[6]

In the pre–molecular biology era, researchers at MDIBL made significant headway into the problem of epithelial transport and renal physiology. Some of these individuals were towering figures in their fields, including Homer Smith, Eli (E. K.) Marshall, and James Shannon. Depending on the questions they were asking (and the availability of specimens), these investigators turned to any number of organisms, including harbor seals, beavers, frogs, salamanders, hagfish, dogfish shark, eel, sculpin, toadfish, lungfish, guppies, little skate, goosefish, flounder, and rock gunnel. Although the research landscape was dominated by transport studies, many other areas were explored over the decades, including cardiovascular physiology (e.g., splenic blood flow in dogfish, nervous control of cardiovascular function in dogfish and sculpin, cardiac muscle contraction in the sea potato and dogfish, and control of blood flow in gills using an isolated perfused head preparation from sculpin), toxicology (including the effects of carcinogens on embryogenesis), hematology (e.g., red blood cell volume regulation in dogfish, flounder, and skates; metabolism and ion transport in the erythrocyte of the dogfish, skate, and harbor seal; the Bohr effect on the dogfish; and red cell carbonic anhydrase function in red blood cells), liver function (e.g., bile formation and secretion in perfused livers from dogfish and skates), tissue regeneration (e.g., regeneration of forelimbs and tail in amphibians and morphogenesis of regenerating fins in killifish), and physiology of the eye (e.g., ion transport and oxygen consumption of corneas from frogs and a variety of fish species) (Evans 2015).

As technology progressed in the twentieth century, MDIBL attempted to keep pace with modern biomedical research. With the advent of molecular biology in the 1990s, researchers turned to the production and screening of cDNA libraries, expression of mRNA in oocytes, and the cloning of genes, particularly those that encoded transporters, channels, and cotransporters. Many of these newer approaches involve time-intensive, iterative protocols, and investigators now found themselves collecting specimens during their summer months at MDIBL and completing their experiments at their home institutions. With the introduction of cell culture, modern microscopic and imaging techniques, and DNA sequencing, investigators moved beyond physiological studies of whole organisms and perfused organs to address more focused questions concerning cell dynamics, calcium fluxes, protein and mRNA expression, and ligand-receptor interactions. In 1994, Frank Epstein, then the chairman of the Board of Trustees, noted: "[The MDIBL] was in a unique position to exploit the great strength of the comparative approach in relating genetic information to function. On the brink of our second 100 years, our aim is to be the premier place in the world where creatures of the sea are used to relate molecular biology to cellular activity, where the tools of classical biology and those of the newest aspects of molecular genetics can be used in concert, where outstanding scientists from many countries and their students can ponder and clarify and illuminate the meaning and mechanism of life, here on the seashore, down by the sea, where it all began" (Evans 2015, 649).

We joined the summer faculty in the years 2005–8 and 2011–14 as part of the New Investigator Program, which was aimed at "bringing interested scientists to the laboratory for purposes of stimulating new ideas and areas of investigations and participating in the educational life of the community" (Evans 2015, 273). We had heard about MDIBL from Frank Epstein, who at the time was a clinical colleague at the Beth Israel Deaconess Medical Center, a Harvard Medical School–affiliated hospital in Boston. Like so many of his contemporaries at MDIBL, Epstein was concerned about the prospects of attracting the next generation of clinician-scientists to the shores of Salisbury Cove. Since two-income families had become the rule, it was no longer feasible for spouses to spend several weeks in Maine during the summer. Moreover, mounting pressures to pay for one's own salary, either through external grant support or clinical revenue, created a disincentive for clinician-scientists to indulge in a one- or two-month sabbatical at MDIBL. Not one to be deterred, Epstein, who was aware of our interest in evolutionary medicine, was relentless in his recruitment efforts. In our many discussions, he would often echo the words of his colleague Richard Hays from the Albert Einstein College of Medicine when describing the MIDBL experience: "All of this stands in contrast to the hectic, compartmentalized life we lead during most of the year, and this chance to share thoughts and work in fellowship is, I think, the reason many of us got into the game in the first place" (Evans 2015, 353). There was something in Epstein's sales pitch that deeply resonated with us: his passion for the comparative method, his love of the sea, and his unwavering commitment to the MDIBL mission were contagious and inspiring. So, after securing the necessary financial resources and the support of our families, we ventured to MDIBL in the summer of 2005. By this time, the station had grown to include twelve summer research laboratories, two administrative buildings, a cooperative dinning-hall, housing for students and faculty, and a conference hall. Just the year before, in the summer of 2004, the lab had hosted seventy-seven principle investigators, eighty-six undergraduate students, and fifty-four medical students (Evans 2015, 894).

We were assigned a small, spartan two-hundred-square-foot lab, the Epstein Lab, originally built in the 1920s by William Procter, an avid naturalist and patron of the MDIBL. The lab, a single-story wood shed with a large window that provided a beautiful view of Frenchman Bay, was outfitted with one small refrigerator, two benches for carrying out experiments, and a single desk. We were housed in rustic no-frills cabins that were owned by the MDIBL, we shared our meals with other scientists, students, and administrative staff in the dining hall, and we socialized on the beach at night with our colleagues. We attended and participated in weekly Monday Morning Seminars (a tradition at MDIBL since 1976) and Friday Brown Bag Seminars, both of which were conducted outdoors with the aid of a chalkboard.[7] And we supervised a large number of high school students, college undergraduates, and medical students in our lab over the years. As physicians and scientists leading busy lives in Boston, we always looked forward to our annual sojourn in MDIBL—for the change in pace, for the connection with nature and the sea, and for the opportunity to work with our model organism, *Myxine glutinosa*, otherwise known by fisherman as the *slime eel*.

Introduction to the Slime Eel

MDIBL houses all its fish in an outdoor facility consisting of large circulating seawater tanks with water maintained at 17 degrees Celsius. The largest of these tanks contains dogfish shark. These animals seem to be in constant motion, their dorsal fin frequently knifing through the water's surface. Other open-topped tanks contain an assortment of fish species, including killifish, rays, and sculpin. Hidden among this panoply of open seawater aquariums is a large, circular green-walled tank—perhaps 6 feet in diameter and 4 feet deep—covered by a square piece of plywood, weighted down and held in place by a chipped cinderblock.[8] A hose threaded through a small circular opening in the plywood delivers fresh cold seawater to the tank as old water is siphoned off from below. The first hint that this tank contains the lab's supply of hagfish is the sign above it that reads: "RINSE OFF all slime from the fishnets after use." There is a good reason for this: hagfish are notorious for the copious amounts of slime they produce (a 150-gram Pacific hagfish can produce up to 900 milliliters of slime) (Fudge, Schorno, and Ferraro 2015, 947–67). Just last year, *National Geographic* ran an online piece entitled "'Slime Eels' Explode on Highway After Bizarre Traffic Accident" that recounts how a truck carrying thousands of fish wrecked on Oregon's highway, covering the road and cars in slime (Bittel 2017). Reassuring the reader that "this isn't a scene from the set of *Ghostbusters*," the writer captures the otherworldliness of the sight: "In a truly sci-fi scenario, thousands of mucus-spewing hagfish—destined for dinner plates in Asia—coated a road in Oregon." Another version of the same accident mused: "It is not every day that the mucus of a living fossil destroys a Prius. . . . No one was injured physically, at least. The psychological damage can't be quantified" (Paoletta 2017). These stories and pictures tell most readers all they will ever know about hagfish: they are slimy, fear-provoking monstrosities that have found a niche market in certain Asian countries (especially South Korea and Japan).

To learn more, we need to remove the makeshift top from the rounded tank—first the cinderblock, then the heavy piece of plywood. Both are placed on the ground next to the aquarium. Peering into the open top, there is nothing to see but deep currents of dark water that is freezing cold to the touch. If there are fish inside, they are certainly not anxious to be seen. At this point, we take advantage of a nearby fishnet with a long handle to "stir the pot." Moving the net back and forth along the bottom of the tank, we rouse the inhabitants, which begin to swim toward the surface of the water.

Within no time, dozens of these creatures—each about 1.5–2 feet in length—are frantically circling the top of the tank. For the uninitiated, as we were in our first year at MDIBL, these creatures look like some cross between a headless snake and an eel (fig. 12.1). They are long, slender, and pinkish. They can move quickly, in spurts, and they do so with waves of lateral undulation. Their head region is not very distinct

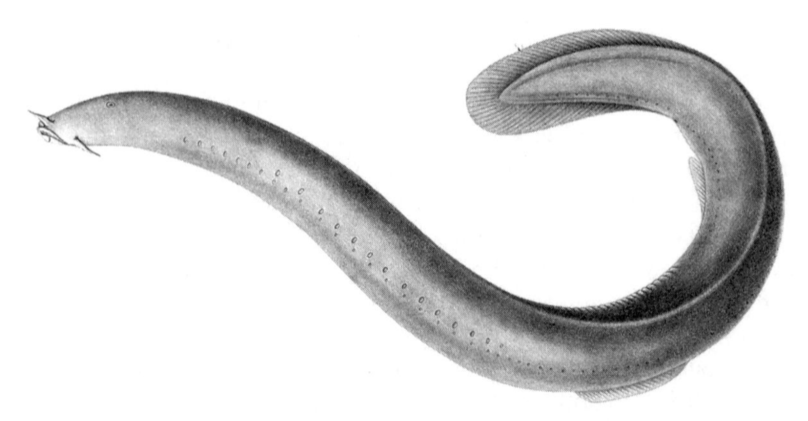

FIGURE 12.1. Hagfish—also called *slime eels*—are eel-shaped, slime-producing marine fish. They are the only known living animals that have a cartilaginous skull and do not have a vertebral column, although they do have rudimentary vertebra-like cartilages. Along with lampreys, hagfish are jawless vertebrates (Agnatha) and are ancestral to the jawed vertebrates (Gnathostomata). Hagfish have no paired fins; instead, a single fin fold runs along the ventral (lower) surface of the body. They have sensory barbels around the mouth, a single nostril, and no external eyes. Instead of vertically articulating jaws, they have a pair of horizontally moving structures with tooth-like projections. Hagfish feed chiefly on dead and dying fish of varying species by boring into the body and consuming viscera and musculature. Drawing by J. H. Richard from Gillis, MacRae, Phelps, and Ruel (1855–56).

since they lack jaws and external eyes. On closer inspection, the snout is recognizable by its short, paired tentacles. Capturing them with the bare hand as they skim the water's surface is hardly a consideration for novices but over time becomes the method of choice if only to keep the fishnets clean of slime. Whether scooped up by hand or by net, the hagfish immediately begin to release slime, a process that continues unabated after they are placed in a small multipurpose plastic bucket containing seawater for transport to the lab. By the time we make the short trek from aquarium to lab, the fish are typically caught up in their own slime, unable to move. Some exhibit knotting behavior, in which they tie themselves into a knot and work the knot from head to its tail, scraping off the slime as they go.[9] They are able to perform this maneuver because they have no vertebrae.

Once in the lab, we get to work. We add a fish anesthetic called MS-222 (its chemical name is tricaine methanesulfonate) to the bucket of saltwater. After the anesthetic takes effect, we lay the immobilized specimens on the lab bench—head to the left, tail to the right—and begin our inspection. There are no paired fins; instead, a single fin fold runs along the ventral (lower) surface of the body (fig. 12.1 above).

The scaleless skin is smooth, shiny, loose, and very slimy. There are two rows of segmentally arranged slime glands distributed along the ventrolateral surface. These glands have the appearance of whitehead pimples and, when squeezed, squirt out their milky contents. The fish's mouth, which contains two pairs of tooth-like rasps on the top of a tongue-like projection, is unlike anything we have seen. It is literally the stuff of nightmares. At this point of the examination, we would have to concur with the Smithsonian Channel description of the hagfish as "a slime-emitting ocean-dweller that's remained unchanged for three hundred million years—and it shows. It has a skull (but no spine), velvet smooth skin, and a terrifying pit of a mouth that's lined with rows of razor-sharp teeth."[10]

Next, we make a median ventral incision with a sharp pair of dissecting scissors. When cutting through the body wall, the loose skin easily pulls away from the underlying muscle because the innermost surface of the hagfish skin and the axial musculature are separated by a large, blood-filled subcutaneous sinus. Once through to the abdominal cavity, we are struck by the seeming simplicity of the internal structures. The gut is not separated into a stomach and intestine; rather, it constitutes one long, straight, uniform tube that is suspended by the mesentery. When the intestine is displaced to one side, four large linear structures are seen attached to the body wall in the midline, extending down the length of the body: two posterior cardinal veins and two pronephric ducts (the dorsal aorta is hidden from view). Anterior in the body cavity lies the liver, which consists of two discernible lobes. A green-colored gall bladder is seen to project between the two lobes. Lifting up the liver reveals a visibly pulsating brachial heart. Next to this heart and connected to the liver is a smaller second heart that is beating at its own rhythm. Dissecting caudally, we encounter a series of internal gill pouches, six on each side. At this point, we encounter extensive connective tissue that seems to get firmer and thicker as we approach the head region, completely obscuring our view.

As physician-scientists, steeped in the ins and outs of human (and to a lesser extent mouse) anatomy and physiology, we felt in over our heads as we continued our dissection of this unusual creature. We also experienced a sense of wonder and excitement, as though we were treading new, uncharted territory. But, of course, we were not. As we will discuss in the next section, biologists in the nineteenth century and the early twentieth routinely opened up hagfish and studied their gross anatomy. Their written descriptions of organs and tissues, including the vasculature, are breathtaking in their detail, their powers of observation so clearly superior to our own. Much of this early work has withstood the test of time, and we are indebted to these pioneering investigators for their foundational contributions. As we finished our novice work that first day, we were reminded of the inspiring words from a 1966 *Scientific American* paper we had read in preparation for our first visit: "The reader may find it difficult to conceive of a fish that has four hearts [two of these are real hearts], only one nostril and no jaws or stomach; that can live for months without feeding; that performs feats of dexterity by literally tying itself in knots. The organism nonetheless exists; it is called the hagfish. . . . Its anatomy and its habits are revealing. As the lowest form of true fish and vertebrate, the 'hag' offers a matchless opportunity to investigate a stage in the evolution of vertebrates at a most primitive level" (Jensen 1966, 82–90).

Myxine: Worm or Fish?

Pehr Kalm, a Swedish naturalist and apostle of Carl Linnaeus's, was the first to describe hagfish in his 1753 travel book *En resa til Norra America* (A journey to northern America) (Kalm 1753). Kalm had set out on a journey to the northern part of America in October 1747, at the behest of the Royal Swedish Academy of Sciences, to study the area's wildlife, collect samples, and write descriptions for Linnaeus. Natural historians like Kalm and Linnaeus were motivated by a desire to describe and classify existing and new species precisely and, in so doing, catalog the whole of God's creation.

Shortly after setting out, Kalm's ship was severely damaged by a series of violent storms. It crawled into Grimstad harbor on the southern coast of Norway, where it was laid up for several weeks undergoing repairs. Never one to miss an opportunity, Kalm made the most of his unplanned stay in Grimstad, documenting his observations in a daily journal. An entry on Friday, January 8, 1748, tells the story of a certain "Pehr in the Garden" bringing home three specimens of a particular species of fish to show him. The locals called the fish *piral* or *pilor* and viewed it as a menace because of "the damage it causes": "When one lays out lines or nets and catches various kinds of fish, such as cod, haddock etc., these sucking-fish come and fasten themselves on to the cod or whatever fish is there, then suck away the flesh from the fish so that within a few hours there is nothing left but skin and bones; they generally bite a hole in the fish and crawl into it, whereupon they suck it up in the manner described" (Kalm 1771, 80). Kalm noted the strange appearance of these creatures, including the absence of eyes, and he was quite certain that they had never been described. He placed the specimens in a large basin with fresh seawater: "[After an hour] the water became quite foul with a white and viscous slime or rather jelly, which looked like a thin and pellucid glue. . . . [I]t finally became so sticky that when one pulled it up the fish itself came up surrounded by it" (Kalm 1771, 80).

Kalm believed he had discovered a new species of blind *Petromyzon* or lamprey. As was his custom, he shared his findings with Linnaeus. In 1758, Linnaeus named the species *Myxine glutinosa* and, in a departure from Kalm, classified it among the nonarthropod invertebrates. Indeed, in the tenth edition of his *Systema naturae, M. glutinosa* appears as a member of the first order of the sixth class, vermes intestinalis, alongside the leech, Toredo shipworm, and several intestinal worms (Janvier 2007). Perhaps Linnaeus was influenced by the reports that fisherman routinely found hagfish inside the bodies of their daily catch.[11] In all events, his short description of the new species still resonates today: "Intrat et devorat pisces; aquam in gluten mutat" (enters into and devours fishes; turns water into glue) (Linnaeus, 1758, 650).

The first anatomical descriptions of Myxine were published in the Scandinavian and German literature, for which no English translations exist. However, as Ragnar Fange, a biologist from Kristineberg Marine Biological Station, in Fiskebäckskil, Sweden, recounts, a Norwegian bishop and naturalist named Johan Ernst Gunnerus was the first to report the results of hagfish dissection in 1763 (Fange 1998, xiii–xix). Gunnerus referred to his specimens as *Sleep-Mark* or *slime worm*. Anders Jahan Retzius, a natural historian and chemist from Lund, Sweden, and member of the Royal Swedish Academy of Sciences, wrote a commentary in 1790 questioning Linnaeus's classification of Myxine among the order of worms and proposed that it was more highly related to lampreys or snakes. By the late 1790s, it was widely accepted that Myxine was not an invertebrate but rather a kind of fish.

Anatomical Research in the Nineteenth Century

In the nineteenth century, Retzius's son Anders Adolph Retzius and grandson Magnus Gustaf Retzius provided cursory descriptions of hagfish anatomy.[12] However, the early English literature credits Johannes Peter Müller as the undisputed father of the field. As a physician and professor of anatomy and physiology at Berlin, Müller was impassioned by comparative anatomy and would spend each August and September traveling to the Baltic Sea, the North Sea, or the Mediterranean Sea to study marine organisms and collect specimens for his anatomical museum (Otis 2004). Between 1836 and 1845, he published a series of illustrated articles on the comparative anatomy of Myxine in German.

Frank J. Cole, professor of zoology, University College, Reading, authored *A Monograph on the General Morphology of the Myxinoid Fishes, Based on a Study of Myxine*, published in six parts between 1905 and 1925.[13] In the opening paragraphs of part 1, he refers to Müller as "the illustrious founder of knowledge of myxinoid anatomy" and relates his own findings to those of Müller throughout the monograph. Ayres and Jackson, writing in 1901, began their article on the morphology of myxinoid skeleton and musculature: "It is now 64 years since Johannes Müller published the first part of his 'Anatomie der Myxinoiden'. Since that time almost nothing has been added to our knowledge of the anatomy of these interesting and important forms of ancestral vertebrates. Not a few papers have dealt with the comparative anatomy of the myxinoid fishes, but the authors of these papers, almost without exception have drawn their anatomical knowledge from Müller's monograph instead of going to nature for facts, the assumption being that Müller saw everything worth knowing and saw it right" (Ayers and Jackson 1901, 185–186). Warning against slavish adherence to the "authority of the eminent German anatomist," they go on to describe differences between their observations and Müller's. In this passage, we see acknowledgment of Müller as the authority on hagfish anatomy, a desire on the part of the authors to see for themselves, and a notable shift from the traditional naturalist's goal of cataloguing God's creation to an interest in evolutionary relationships between ancestral and later forms.

Physiological Research in the Twentieth Century

In the twentieth century, research on hagfish continued at a number of marine research stations. As new research tools and experimental methods emerged in the early to mid-twentieth century, investigators turned to physiological studies, at the level of both whole organism and perfused organs. They were no longer just passively observing their specimens; they were now manipulating them in controlled experiments. For example, they injected fish with phenol red and measured its elimination in the bile and urine. They evaluated the effect of injected endotoxin and snake venom on survival, and they determined whether antigens, including infectious agents, could induce antibody production (Papermaster, Condie, and Good 1962, 355–57; Faenge and Krog 1963, 713). Some went so far as to perform skin autografts and allografts in hagfish to test for immunological memory and graft rejection (Hildemann and Thoenes 1969, 506–21). There are also descriptions of metabolic rates at different ambient temperatures, body weight and blood electrolytes under different salt concentrations, and cardiac performance under low-oxygen conditions (McFarland and Munz 1965, 383–98; Munz and Morris 1965, 1–6; Hansen and Sidell 1983, R356–R362). Researchers catheterized ureters to measure urine output and glomerular filtration rate (Riegel 1978, 261–77). They injected hagfish with mammalian hormones and in one case surgically removed the hypophysis to evaluate endocrine function over the next several months (Kerkof, Boschwitz, and Gorbman 1973, 231–40; Matty, Tsuneki, Dickhoff, and Gorbman 1976, 500–516; Inui and Gorbman 1977, 423–27). Blood pressure measurements were made in different arteries, and mircopuncture techniques were used to evaluate transmembrane potentials of isolated spontaneously beating hearts (Jensen 1965, 443–58; Riegel 1986, 359–71). In addition to these physiological data, new methods for chemical analyses led to the characterization of hagfish hemoglobins, bile salts, and hormones (Quast, Paleus, Ostlund, and Bloom 1967, 2849–54; Bannai, Sugita, and Yoneyama 1972, 505–10; Li, Tomita, and Riggs 1972, 344–54). And, finally, the introduction of electron microscopy in the 1960s afforded high-resolution analysis of tissues and cells and provided a powerful new tool for correlating structure with function.

What motivated these twentieth-century investigators? Why did they choose to spend their time and resources on studying such an esoteric, poorly accessible, and ancient organism? A survey of the papers in question reveals a common explanation: the authors considered the position of the hagfish on the phylogenetic tree as an opportunity to explore evolutionary origins of vertebrate systems. Consider, for example, a paper published in *Nature* in 1962 on the immune system of vertebrates that begins: "The hagfish, an agnathan cyclostome, is the most primitive of the true vertebrates. Since it is the first species in the phylogenetic scale with formed, circulating red blood cells and haemocytopoietic tissue in the submucosa of the gut tract, definition of its immune response seemed particularly relevant to our investigation of the phylogeny of immune responsiveness" (Papermaster, Condie, and Good 1962, 355–57). A report from 1983 on the metabolic basis of hagfish tolerance to hypoxia states: "Because of the unusual working conditions of hagfish cardiac muscle and the pivotal position of these animals in vertebrate phylogeny, we reasoned that a careful metabolic study of the tissue might reveal mechanisms of anoxic tolerance and also yield conceptual insight into the evolution of obligate aerobic function in the myocardium of higher vertebrates" (Hansen and Sidell 1983, R356–R362). These and many other examples underscore the perceived importance of hagfish as a valuable evolutionary model.

Hagfish Research Today

With advances in molecular cloning and the availability of increasingly sophisticated anatomical and histological methods, the use of hagfish as an evolutionary model has intensified over the past twenty years. Shimeld and Donoghue, from the Department of Zoology, University of Oxford, summarized the situation in 2012: "[The] attraction [of the cyclostomes] stems from where they branch in animal phylogeny. The only surviving lineages from a once diverse and disparate evolutionary grade of jawless fishes, they provide an experimental window into the developmental biology and genomic constitution of the ancestral vertebrate. It is important to note, however, that neither lampreys nor hagfish can be taken as literal proxies for the ancestral vertebrate: both lineages have acquired characteristics specific to cyclostomes, and both have transformed or lost ancestral vertebrate characters. However, in comparison to vertebrate outgroups, such as the urochordates, and in-groups, such as sharks and bony fishes, lampreys and hagfish can provide insights into the molecular and genomic changes that underlie the assembly and subsequent evolution of the vertebrate and gnathostome body plans" (Shimeld and Donoghue 2012, 2091–99).[14] Indeed, hagfish have been used to study the ancestral origins of several systems, ranging from the vertebrate brain, to craniofacial development, vertebrate gills, adaptive immunity, phototransduction, and, in our own studies, the endothelium and hemostasis (Oisi, Ota, Kuraku, Fujimoto, and Kuratani 2013, 175–80; Im et al. 2016, 203–10; Lamb et al. 2016, 2064–87; Gillis and Tidswell 2017, 729–32; Sugahara, Murakami, Pascual-Anaya, and Kuratani 2017, 163–74).

A second area of current research interest revolves around a long-standing debate over the phylogenetic relationship between hagfish, lampreys, and gnathostomes (jawed vertebrates). Although there is general consensus that each "cyclostome" order represents a single clade (in other words, all hagfish belong to one clade, all lamprey to another), the exact relationship between the three lineages has been more controversial. Traditional comparative morphology and physiology support hagfish as a sister group to the vertebrates (especially given their lack of vertebrae).[15] This is referred to as the *cyclostome/gnathostome theory* (for a review, see Ota and Kuratani [2008, 999–1011]). However, molecular data (including analyses of mitochondrial, nuclear ribosomal RNA genes and nuclear protein–coding genes as well as microRNAs) suggest that hagfish and lamprey form a monophyletic sister group that diverged from the basal position on the vertebrate tree, with gnathostomes representing an outgroup (for a review, see [Ota and Kuratani 2007, 329–37]). According to this latter phylogenetic scheme (termed the *craniate/vertebrate theory*), some of the unique features of hagfish previously deemed to be ancestral in origin are in fact derived.[16]

A third and related focus of research is the study of hagfish embryogenesis. Most hagfish live in difficult-to-access deep sea habitats, buried under the ocean floor. Where and how hagfish fertilize and then lay their eggs in the wild remains a mystery, and, until recently, the rarity of fertilized eggs significantly hampered progress in our understanding of hagfish development. A major breakthrough in hagfish embryology occurred in 2005 when Japanese investigators from the Evolutionary Morphology Group at RIKEN successfully obtained embryos from *E. burgeri* in an artificial spawning tank.[17] The premise of their work, like that of so many before them, was that "a comparison of embryonic development between hagfishes, lamprey and gnathostomes should provide hints as to their phylogenetic relationships." Using a combination of in situ hybridization and histology of freshly fixed hagfish embryos, they showed that hagfish neural crest cells follow a typical vertebrate pattern of delamination and migration (Ota, Kuraku, and Kuratani 2007, 672–75). In subsequent studies, they demonstrated that hagfish embryos display additional traits, including vertebrae, that were previously believed to be ancestrally absent (Ota, Fujimoto, Oisi, and Kuratani 2011, 373; Oisi, Ota, Fujimoto, and Kuratani 2013, 944–61; Oisi, Ota, Kuraku, Fujimoto, and Kuratani 2013, 175–80). The presence of these traits during development suggests that they were lost during evolution and led the authors to conclude that "it is now clear that cyclostomes constitute a monophyletic group, with gnathostomes as an outgroup" (Kuratani, Oisi, and Ota 2016, 229–38).[18]

Advances in molecular biology and an interest in biomimetics have resulted in growing interest in the area of hagfish slime, the goal being to leverage its properties to develop a renewable protein-based biofiber (see, e.g., Fudge, Schorno, and Ferraro 2015, 947–67). Finally, hagfish are important contributors to the benthic ecosystem, and, in recent years, they have been increasingly targeted for their flesh and skin (tanned hagfish skin is a popular source of leather products, particularly wallets and belts). Gaining insights into spawning habits, population dynamics, relationships of length, age, and maturity, age of recruitment, extent of resource, and reproductive potential is a necessary foundation for maintaining a sustainable fishery (Powell, Kavanaugh, and Sower 2005, 158–65). These types of studies involve a different style of population-level fieldwork that leverages technological advances, including tethered mobile Internet-operated vehicles and baited camera deployments.

Hagfish and the Endothelium

Our studies of hagfish over the past ten years have helped inform us about the evolutionary origins of the endothelium. We found that hagfish do indeed possess an endothelial lining throughout their vasculature and that the endothelium displays remarkable site-specific heterogeneity in ultrastructure, function, and antigen expression (for a review, see Monahan-Earley, Dvorak, and Aird [2013, 46–66]). For example, we found that brain capillaries are lined by a continuous endothelium (Dvorak and Aird 2016, 161–206). The endothelial cells were remarkable for their abundance of intracellular vesicles and their overlapping and interdigitating lateral borders, features that are consistent with a functional blood brain barrier. In contrast, the endothelium that lines the capillary-like sinusoids of the hagfish liver contained many gaps that allow for direct access of substances between blood and hepatocytes (Yano et al. 2007, 613–15). In the liver of other vertebrates, this type of endothelium is termed *discontinuous* and is believed to play an important role in filtering substances from the blood as it passes from the mesenteric circulation (which absorbs food substances in the gastrointestinal tract) to the systemic circulation.

Electron microscopic examination of endothelial cells from other blood vessels, including the aorta and those from the skin, heart, skeletal muscle, and kidney, revealed unique ultrastructural properties in each vascular bed (Yano et al. 2007, 613–15). To demonstrate molecular heterogeneity, we stained various tissue sections with fluorescently labeled lectins. Lectins are naturally occurring carbohydrate-binding proteins that recognize cell surface determinants. Previous studies in mice and humans have shown that they differentially bind to one or another type of endothelial cell. Consistent with these data, we showed marked variation in the efficacy of vascular labeling of different lectins in hagfish organs, suggesting the existence of vascular bed–specific endothelial glycoproteins (Yano et al. 2007, 613–15). Finally, in intravital microscopy studies, we were able to demonstrate that neutrophils adhere to the endothelium of capillaries and postcapillary venules but not arterioles, following the local injection of histamine (Yano et al. 2007, 613–15). These findings are similar to those in mice and provide evidence for functional differentiation of the endothelium in hagfish. Separately, we asked whether the endothelium plays a role in mediating vasomotor tone in hagfish and, if so, whether this function is differentially regulated across the vascular tree. We found that hagfish arteries from the mesentery and skeletal muscle demonstrated site-specific mechanisms of endothelium-dependent vasomotor relaxation (Feng et al. 2007, R894–R900). Collectively, our results suggest that the endothelium appeared in an ancestral vertebrate following the divergence from the urochordates and cephalochordates between 540 and 510 million years ago and that endothelial heterogeneity appeared during the same narrow window of time as a core feature of this cell layer.

Our observations raise an interesting question. Given that endothelial cell heterogeneity is so highly conserved in vertebrates, what fitness advantage does it provide? We can only speculate, of course, but we presume that phenotypic heterogeneity reflects the remarkably pleiotropic role of the endothelium in meeting the varying needs of diverse tissue types. Moreover, some site-specific properties reflect a local self-preserving adaptation of the endothelial cells to their particular microenvironment. A third consideration is that phenotypic plasticity provides the endothelium with the flexibility to adjust temporally to a wide range of physiological and pathophysiological influences, including pathogens and toxins.

In our respective labs in Boston, we have spent many years employing basic tools of molecular and cell biology and mouse genetics to dissect the proximate mechanisms of endothelial cell heterogeneity. On the basis of our studies, we recently employed dynamical systems theory to reframe endothelial cell heterogeneity in terms of multistability (Regan and Aird 2012, 110–30). Multistability, or the ability of single cells to maintain distinct phenotypes under identical extracellular conditions, is generated through positive regulatory feedback and/or epigenetic modifications that generate barriers between distinct states. These states can be represented as valleys on a landscape, with the internal state of an endothelial cell represented as a marble on its surface. Endothelial cells constantly navigate this landscape in response to extracellular signals as well as intracellular noise. When exposed to appropriate signals, cells can cross barriers and enter other valleys. We demonstrated that cell-to-cell variation in the expression of an endothelial-specific gene—von Willebrand factor (VWF)—is controlled by a low-barrier, noise-sensitive bistable switch that involves random transitions in the DNA methylation status of the *VWF* promoter (Yuan et al. 2016, 10160).

In iterative fashion, we went back to the beach to clone and characterize the von Willebrand factor in hagfish. Our goal was to determine whether the hagfish gene was similarly heterogeneous in its expression and, thus, potentially subject to noise-induced *on-off* state transitions. We identified a single *vwf* transcript that encodes a simpler protein compared with higher vertebrates but is nonetheless functional (Grant et al. 2017, 2548–58). However, in contrast to mammals, Vwf was not patchy in its expression but detected throughout the endothelium of all vascular beds examined. These findings suggest that vascular bed–specific expression of VWF evolved following the divergence of hagfish and that its expression in hagfish is locked into a single *on* state, impervious to the effects of biological noise or organ-specific extracellular cues. In other words, while phenotypic heterogeneity is a highly conserved feature of the endothelium, the nature of the endothelial mosaic and the mechanisms by which it is generated are likely to differ widely across vertebrate species.

Conclusion

Twelve years ago, we set out to complement our traditional research program on endothelial cell heterogeneity with beachside studies of hagfish. Our choice of model organism was based on its basal position on the vertebrate phylogenetic tree. Our choice of destination, namely, MDIBL, was influenced by our mentor, Frank Epstein. As luck would have it, we had the financial means to support our research during those years at MDIBL, and we had the blessing of our chairman of medicine. However, our decision to set aside what amounted to about twenty months of our time to study marine life on the shore of Salisbury Cove was not without its detractors. When preparing to leave one summer for Maine, one of us was confronted by a preeminent cardiologist at the Brigham and Women's Hospital: "Why are you wasting your time studying a lowly fish? Why don't you do something that will help patients?" We faced similar (though far more diplomatic) resistance from the National Institutes of Health, which consistently questioned the clinical relevance of hagfish research in our grant applications. Such criticisms gave us pause. It is admittedly a stretch to imagine any imminent clinical application emerging from our work. That being said, we believe that there is a place for less traditional approaches—including evolutionary biology—in the attempt to understand human health and disease. What's more, as physicians and biomedical researchers, we are by nature intellectually curious about the living world, and, as such, we are not averse to seeking knowledge for knowledge's sake. Finally, for those of us who are moved by nature and the sea, beachside research is invigorating and inspirational, a perfect counterbalance to the daily pressures and stresses of city life and academe. The MDIBL offered an ideal environment where physicians and marine biologists could interact on a daily basis, where the sights and sounds of the ocean were conducive to spurts of creativity, and where a period of rejuvenation prepared us for the rigor of the upcoming academic year.

Notes

1. The term *evolutionary medicine* ultimately prevailed over *Darwinian medicine* to avoid any confusion with social Darwinism. There are no fewer than five books on this subject—including Gluckman, Beedle, Buklijas, Low, and Hanson (2016), Trevathan, Smith, and McKenna (2008), Stearns and Koella (2008), and Perlman (2013)—and two journals, *Evolution, Medicine and Public Health* (*EMPH*), founded by Stephen Stearns in 2013, and the *Journal of Evolutionary Medicine,* initiated in 2012 by Paul W. Ewald. In introducing the inaugural issue of *EMPH*, Stearns wrote: "Evolutionary medicine can help to reduce suffering and save lives. It is also full of new ideas, unexpected connections and scientific surprises" (Stearns 2013, 2). Herein lies the crux in the field: so far, evolutionary principles have served as a powerful explanatory tool in medicine, but the value of evolutionary medicine as an active research displine whose results translate into improved patient care remains to be seen.
2. More specifically, a character that is shared between cyclostomes and gnathostomes is likely to represent a trait established in the most recent common ancestor of all vertebrates. If this feature is not shared by the nonvertebrate chordates, such as the tunicates and amphioxus, it likely represents a key to understanding the evolutionary origin of vertebrate body plans.

3. By 1915, seventy-nine different investigators from forty-three institutions had spent time at the Harpswell Laboratory and collectively published over sixty papers in areas ranging from natural history to embryology and comparative anatomy (Evans 2015, 27).
4. Evans describes MDI as "a summer playground for the rich and famous" (Evans 2015, 44). Frank Epstein and his son, Jon Epstein, point out: "These institutions [marine research stations such as MDIBL] began as a natural outgrowth of the intense interest in natural history that characterized high society in the eighteenth and nineteenth centuries" (Epstein and Epstein 2005, 1).
5. The station was initially named the Weir Mitchell Station of the Harpswell Laboratory and renamed MDIBL in 1923.
6. In 2000, Edward Benz, the Osler Professor and chairman of medicine at the Johns Hopkins University School of Medicine, echoed these words in a talk on the human genome project and what it meant to MDIBL: "MBIDL is the only marine biological laboratory whose mission statement clearly states that advancing human health is a priority. During its first 100 years, MDIBL distinguished itself by the scientific contributions its investigators have made both to basic marine biology and to the advance of biomedical research, using marine organisms as models for human physiology and pathophysiology" (Evans 2015, 823).
7. Epstein described these seminars as being "crowded with students ranging from high school seniors to Full Professors, in which the sun shines in the eye of the speaker" (Evans 2015, 667).
8. The cover is designed to provide hagfish—who are photophobic bottom dwellers—with darkness and prevent them from escaping over the side of the tank.
9. The knotting behavior has been described as follows: "When the hagfish handles large food items (e.g., whole fish carcasses with intact, puncture-resistant skins), a knot forms in the posterior body, and then slides towards the head so that it can be pressed against the food surface. The body knot then makes a stable platform that effectively antagonizes the toothplate movements when it is in contact with the food item. In addition to feeding behaviors, the knotting movements of hagfish can be elicited when the animals are discarding excess slime from their body surfaces and when attempting to evade predators (Jensen 1966, 82–90). In all cases, hagfish knotting is executed with an impressive combination of speed, agility, and precision; however, these behaviors are rarely observed in extant fishes, including elongate, gape-limited species. The formation and manipulation of body knots requires coordinated bending and twisting movements, which, we theorize, are facilitated in hagfish by the presence of a flexible, elongate body devoid of vertebrae, a relatively complex arrangement of axial musculature, and loose skin" (Clark, Crawford, King, Demas, and Uyeno 2016, 243–56).

10. "Crazy Monster—Sea Creatures," Smithsonian Channel, https://www.smithsonianmag.com/videos/category/science/the-hagfish-is-the-slimy-sea-creature-of-you.
11. In the early twentieth century, Bashford Dean, whom we will discuss later in the essay, weighed in on the confusion with classifying hagfish: "Because the myxinoids have been so little understood, they have been more misrepresented and maligned than any other vertebrates. The earliest investigators classified them with the worms. Even after this error had been corrected, and the slime eels accorded a place with the vertebrates, complete enjoyment of their new position was denied them on account of their round, leech-like mouths, indicated by the word 'Bdellostoma.' The shape of the mouth coupled with the fact that the eels were often brought to the surface of the water with their heads, and even their entire bodies, buried in the flesh of fish caught on hooks or in traps, misled many investigators to believe that the myxinoids are parasites, and therefore are degenerate. I am convinced, however, that the myxinoids are not parasites" (Conel 1930, 69).
12. Anders A. Retzuis, a Swedish professor of anatomy and a supervisor at the Karolinska Institute in Stockholm, carried out detailed anatomical studies on the hagfish; notably, he investigated the circulatory system and first described the subcutaneous blood sinus and other vascular structures. Gustaf Retzuis, a Swedish physician and anatomist also at Karolinska, intensively studied the morphology of *Myxine glutinosa*, made careful anatomical observations of the nervous and sensory systems and structures, and described the function of the caudal heart.
13. See https://books.google.com/books?id=B25BAQAAMAAJ.
14. An outgroup is a species or group of species from an evolutionary lineage that diverged before the lineage that includes the species of interest (the ingroup).
15. The lack of true vertebrae has led to claims that hagfish are not vertebrates but rather join the vertebrates to form the taxon *craniates*.
16. According to Ota: "The cyclostome/gnathostome theory was widely accepted among paleontologists and morphologists until the late 1970s, when the craniate/vertebrate theory, based on morphological and physiological traits shared by lampreys and gnathostomes, was proposed. . . . Unlike the cyclostome/gnathostome theory, the craniate/vertebrates theory suggests that gnathostomes and lampreys represent sister groups, with hagfishes taxonomically basal to the rest of the vertebrates (gnathostomes plus lampreys). . . . As the latter theory fits the idea of graded evolution, and a large number of phenotypic characters, including morphological, histological and physiological traits common to gnathostomes and lampreys apparently exclude hagfishes from the vertebrate clade, it had been accepted among morphologists and paleontologists until recently. The craniate/vertebrate theory still survives, especially

among morphologists. Actually, even in the recently reconstructed phylogenetic trees based on morphological data, including newly discovered fossil vertebrates, the hagfishes are still located in the basal position, without exception. The monophyly of lampreys and hagfishes (cyclostome/gnathostome theory) came to light again at the end of the previous century, when molecular phylogenetic analysis was applied to the reconstruction of the phylogenetic relationships of living organisms. Because of the increasing availability of genomic data from vertebrates, the molecular phylogeny was reconstructed with much higher accuracy than before, and the cyclostome/gnathostome theory has now come to be accepted widely among molecular phylogeneticists" (Ota and Kuratani 2008, 999–1001).

17. *E. burgeri* was chosen because it is one of the few coastal shallow-water-dwelling species of hagfish. The study investigators tried to recapitulate the environment of the hagfish by keeping them in the dark in a seawater-filled aquarium equipped with a high-power filtering system and with temperature regulation at 16°C, a protein skimmer, and fine-grained sand and oyster shells on the bottom (Ota and Kuratani 2008, 999–1011).
18. This conclusion is supported by others, including Caputo et al., who wrote: "The monophyly of cyclostomes now seems to be supported by molecular, morphological and developmental data" (Caputo Barucchi, Giovannotti, Nisi Cerioni, and Splendiani 2013, 80–89).

References

Ayers, H., and C. M. Jackson. 1901. "Morphology of the Myxinoidei, I: Skeleton and musculature." *Journal of Morphology* 17, no. 2:185–226.

Bannai, S., Y. Sugita, and Y. Yoneyama. 1972. "Studies on Hemoglobin from the Hagfish Eptatretus burgeri." *Journal of Biological Chemistry* 247, no. 2: 505–10.

Bittel, J. 2017. "'Slime Eels' Explode on Highway After a Bizarre Traffic Accident." *National Geographic*, July 14, https://news.nationalgeographic.com/2017/07/hagfish-slime-oregon-highway.

Caputo Barucchi, V., M. Giovannotti, P. Nisi Cerioni, and A. Splendiani. 2013. "Genome Duplication in Early Vertebrates: Insights from Agnathan Cytogenetics." *Cytogenetic and Genome Research* 141, nos. 2–3:80–89.

Clark, A. J., C. H. Crawford, B. D. King, A. M. Demas, and T. A. Uyeno. 2016. "Material Properties of Hagfish Skin, with Insights into Knotting Behaviors." *Biological Bulletin* 230, no. 3:243–56.

Conel, J. L. 1930. "The Genital System of the Myxinoidea: A Study Based on Notes and Drawings of These Organs in Bdellostoma Made by Bashford Dean." In *The Bashford Dean Memorial Volume: Archaic Fishes*, ed. Eugene W. Gudger, 63–100. New York: Order of the Trustees of the American Museum of Natural History.

Dexter, R. W. 1988. "History of American Marine Biology and Marine Biology Institutions Introduction: Origins of American Marine Biology." *American Zoologist* 28, no. 1:3–6.

Dobzhansky, T. 1973. "Nothing in Biology Makes Sense Except in the Light of Evolution." *American Biology Teacher* 35, no. 3:125–29.

Dvorak, A. M., and W. C. Aird. 2016. "Endothelium in Hagfish." In *Hagfish Biology*, ed. S. L. Edwards and G. G. Gross, 161–206. Boca Raton, FL: Taylor & Francis.

Epstein, F. H., and J. A. Epstein. 2005. "A Perspective on the Value of Aquatic Models in Biomedical Research." *Experimental Biology and Medicine* 230, no. 1:1–7.

Evans, D. H. 2015. *Marine Physiology Down East: The Story of the Mt. Desert Island Biological Laboratory*. Perspectives in Physiology. New York: Springer.

Faenge, R., and J. Krog. 1963. "Inability of the Kidney of the Hagfish to Secrete Phenol Red." *Nature* 199:713.

Fange, R. 1998. "Introduction: Early Hagfish Research." In *The Biology of Hagfishes*, ed. J. M. Jorgensen, J. P. Lombolt, R. E. Weber, and H. Malte, xiii–xix. London: Chapman & Hall.

Feng, J., K. Yano, R. Monahan-Earley, E. S. Morgan, A. M. Dvorak, F. W. Sellke, and W. C. Aird. 2007. "Vascular Bed–Specific Endothelium-Dependent Vasomomotor Relaxation in the Hagfish, Myxine glutinosa." *American Journal of Physiology—Regulatory, Integrative and Comparative Physiology* 293, no. 2:R894–R900.

Fudge, D. S., S. Schorno, and S. Ferraro. 2015. "Physiology, Biomechanics, and Biomimetics of Hagfish Slime." *Annual Review of Biochemistry* 84:947–67.

Gillis, J. A., and O. R. Tidswell. 2017. "The Origin of Vertebrate Gills." *Current Biology* 27, no. 5:729–32.

Gillis, J. M., A. MacRae, S. L. Phelps, and E. Ruel. 1855–56. *The U.S. Naval Astronomical Expedition to the Southern Hemisphere during the Years 1849–1852*. Washington, DC: A. O. P. Nicholson.

Gluckman, P. D., A. Beedle, T. Buklijas, F. Low, and M. A. Hanson. 2016. *Principles of Evolutionary Medicine*. 2nd ed. Oxford: Oxford University Press.

Goldberg, C. 2001. "Students Pursue One of the Ocean's Slimy Mysteries." *New York Times*, January 2.

Grant, M. A., D. L. Beeler, K. C. Spokes, J. Chen, H. Dharaneeswaran, T. E. Sciuto, A. M. Dvorak, G. Interlandi, J. A. Lopez, and W. C. Aird. 2017. "Identification of Extant Vertebrate Myxine glutinosa VWF: Evolutionary Conservation of Primary Hemostasis." *Blood* 130, no. 23:2548–58.

Hansen, C. A., and B. D. Sidell. 1983. "Atlantic Hagfish Cardiac Muscle: Metabolic Basis of Tolerance to Anoxia." *American Journal of Physiology* 244, no. 3:R356–R362.

Hildemann, W. H., and G. H. Thoenes. 1969. "Immunological Responses of Pacific Hagfish, I: Skin Transplantation Immunity." *Transplantation* 7, no. 6:506–21.

Im, S. P., J. S. Lee, S. W. Kim, J. E. Yu, Y. R. Kim, J. Kim, J. H. Lee, and T. S. Jung. 2016. "Investigation of Variable Lymphocyte Receptors in the Alternative Adaptive Immune Response of Hagfish." *Developmental and Comparative Immunology* 55:203–10.

Inui, Y., and A. Gorbman. 1977. "Sensitivity of Pacific Hagfish, Eptatretus stouti, to Mammalian Insulin." *General and Comparative Endocrinology* 33, no. 3:423–27.

Janvier, P. 2007. "Evolutionary Biology: Born-Again Hagfishes." *Nature* 446, no. 7136:622–23.

Jensen, D. 1965. "The Aneural Heart of the Hagfish." *Annals of the New York Academy of Sciences* 127, no. 1:443–58.

———. 1966. "The Hagfish." *Scientific American* 214, no. 2:82–90.

Linnaeus, C. 1758. *Systema naturae per regna tria naturae, secundum classes, ordines, genera, species, cum characteribus, differentiis, synonymis, locis*. Pt. I. 10th ed. Stockholm: Salvius.

Kalm, P. 1753. *En resa til Norra America*. Stockholm: Tryckt på L. Salvii kostnad.

———. 1771. *Travels into North America; Containing Its Natural History, and a Circumstantial Account of Its Plantations and Agriculture in General, with the Civil, Ecclesiastical and Commercial State of the Country, the Manners of the Inhabitants, and Several Curious and Important Remarks on Various Subjects*. London: T. Lowndes.

Kerkof, P. R., D. Boschwitz, and A. Gorbman. 1973. "The Response of Hagfish Thyroid Tissue to Thyroid Inhibitors and to Mammalian Thyroid-Stimulating Hormone." *General and Comparative Endocrinology* 21, no. 2:231–40.

Kuratani, S., Y. Oisi, and K. G. Ota. 2016. "Evolution of the Vertebrate Cranium: Viewed from Hagfish Developmental Studies." *Zoological Science* 33, no. 3:229–38.

Lamb, T. D., H. Patel, A. Chuah, R. C. Natoli, W. I. Davies, N. S. Hart, S. P. Collin, and D. M. Hunt. 2016. "Evolution of Vertebrate Phototransduction: Cascade Activation." *Molecular Biology and Evolution* 33, no. 8:2064–87.

Li, S. L., S. Tomita, and A. Riggs. 1972. "The Hemoglobins of the Pacific Hagfish, Eptatretus stoutii, I: Isolation, Characterization, and Oxygen Equilibria." *Biochimica et biophysica acta* 278, no. 2:344–54.

Maienschein, J. 1990. "Neurobiology a Century Ago at the Marine Biological Laboratory, Woods Hole." *Trends in Neurosciences* 13, no. 10:399–403.

Matty, A. J., K. Tsuneki, W. W. Dickhoff, and A. Gorbman. 1976. "Thyroid and Gonadal Function in Hypophysectomized Hagfish, Eptatretus stouti." *General and Comparative Endocrinology* 30, no. 4:500–516.

McFarland, W. N., and F. W. Munz. 1965. "Regulation of Body Weight and Serum Composition by Hagfish in Various Media." *Comparative Biochemistry and Physiology* 14:383–98.

Monahan-Earley, R., A. M. Dvorak, and W. C. Aird. 2013. "Evolutionary Origins of the Blood Vascular System and Endothelium." *Journal of Thrombosis and Haemostasis* 11 (suppl. 1): 46–66.

Munz, F. W., and R. W. Morris. 1965. "Metabolic Rate of the Hagfish, Eptatretus stoutii (Lockington) 1878." *Comparative Biochemistry and Physiology* 16, no. 1:1–6.

Nesse, R. M., and G. C. Williams. 1994. *Why We Get Sick: The New Science of Darwinian Medicine.* New York: Times Books.

Oisi, Y., K. G. Ota, S. Fujimoto, and S. Kuratani. 2013. "Development of the Chondrocranium in Hagfishes, with Special Reference to the Early Evolution of Vertebrates." *Zoological Science* 30, no. 11:944–61.

Oisi, Y., K. G. Ota, S. Kuraku, S. Fujimoto, and S. Kuratani. 2013. "Craniofacial Development of Hagfishes and the Evolution of Vertebrates." *Nature* 493, no. 7431:175–80.

Ota, K. G., S. Fujimoto, Y. Oisi, and S. Kuratani. 2011. "Identification of Vertebra-Like Elements and Their Possible Differentiation from Sclerotomes in the Hagfish." *Nature Communications* 2:373.

Ota, K. G., S. Kuraku, and S. Kuratani. 2007. "Hagfish Embryology with Reference to the Evolution of the Neural Crest." *Nature* 446, no. 7136: 672–75.

Ota, K. G., and S. Kuratani. 2007. "Cyclostome Embryology and Early Evolutionary History of Vertebrates." *Integrative and Comparative Biology* 47, no. 3:329–37.

———. 2008. "Developmental biology of Hagfishes, with a Report on Newly Obtained Embryos of the Japanese Inshore Hagfish, Eptatretus burgeri." *Zoological Science* 25, no. 10:999–1011.

Otis, L. 2004. "Johannes Müller (1801–1858)." *The Virtual Laboratory.* http://vlp.mpiwg-berlin.mpg.de/essays/data/enc22.

Paoletta, R. 2017. "Hagfish Slime Is Wonderful." *Gizmodo,* July 14, 2017. https://gizmodo.com/hagfish-slime-is-wonderful-1796918961.

Papermaster, B. W., R. M. Condie, and R. A. Good. 1962. "Immune Response in the California Hagfish." *Nature* 196:355–57.

Perlman, R. L. 2013. *Evolution and Medicine.* Oxford: Oxford University Press.

Powell, M. L., S. I. Kavanaugh, and S. A. Sower. 2005. "Current Knowledge of Hagfish Reproduction: Implications for Fisheries Management." *Integrative and Comparative Biology* 45, no. 1:158–65.

Quast, R., S. Paleus, E. Ostlund, and G. Bloom. 1967. "Some Spectrophotometric Characteristics of Hemoglobin in the Atlantic Hagfish (Myxine glutinosa L.)." *Acta chemica scandinavica* 21, no. 10:2849–54.

Regan, E. R., and W. C. Aird. 2012. "Dynamical Systems Approach to Endothelial Heterogeneity." *Circulation Research* 111, no. 1:110–30.

Riegel, J. A. 1978. "Factors Affecting Glomerular Function in the Pacific Hagfish Eptatretus stouti (Lockington)." *Journal of Experimental Biology* 73: 261–77.

———. 1986. "Hydrostatic Pressures in Glomeruli and Renal Vasculature of the Hagfish, Eptatretus stouti." *Journal of Experimental Biology* 123:359–71.

Rosenberg, R. D., and W. C. Aird. 1999. "Vascular-Bed-Specific Hemostasis and Hypercoagulable States." *New England Journal of Medicine* 340, no. 20: 1555–64.

Shimeld, S. M., and P. C. Donoghue. 2012. "Evolutionary Crossroads in Developmental Biology: Cyclostomes (Lamprey and Hagfish)." *Development* 139, no. 12:2091–99.

Stearns, S. C. 2013. Editorial. *Evolution, Medicine, and Public Health* 1:1–2.

Stearns, S. C., and J. C. Koella. 2008. *Evolution in Health and Disease*. 2nd ed. Oxford: Oxford University Press.

Sugahara, F., Y. Murakami, J. Pascual-Anaya, and S. Kuratani. 2017. "Reconstructing the Ancestral Vertebrate Brain." *Development, Growth and Differentiation* 59, no. 4:163–74.

Trevathan, W., E. O. Smith, and J. J. McKenna. 2008. *Evolutionary Medicine and Health: New Perspectives*. New York: Oxford University Press.

Yano, K., D. Gale, S. Massberg, P. K. Cheruvu, R. Monahan-Earley, E. S. Morgan, D. Haig, U. H. von Andrian, A. M. Dvorak, and W. C. Aird. 2007. "Phenotypic Heterogeneity Is an Evolutionarily Conserved Feature of the Endothelium." *Blood* 109, no. 2:613–15.

Yuan, L., G. C. Chan, D. Beeler, L. Janes, K. C. Spokes, H. Dharaneeswaran, A. Mojiri, W. J. Adams, T. Sciuto, G. Garcia-Cardena, G. Molema, P. M. Kang, N. Jahroudi, P. A. Marsden, A. Dvorak, E. R. Regan, and W. C. Aird. 2016. "A Role of Stochastic Phenotype Switching in Generating Mosaic Endothelial Cell Heterogeneity." *Nature Communications* 7:10160.

Epilogue: The Future of Biological Research Will Be Found in the Oceans

ALEJANDRO SÁNCHEZ ALVARADO

Throughout history, the enormity and mystery of the ocean have always occupied an important place in human consciousness. Life likely arose in its waters. The vast numbers and diversity of marine species we know and the many more we have yet to discover are the result of billions of years of experiments in evolution. Such richness of mostly untapped biological information stands the best chance we may have as biologists and as a species to understand life. Yet, twentieth- and twenty-first-century biology has focused on just a few randomly selected and possibly nonrepresentative species to unravel the still mysterious mechanisms that make life possible. This epilogue will explore how we may have arrived at the present status quo and argue why and how we must change it in order to spur significant leaps in our understanding of biology.

Twentieth-century biology and more recent biomedical research can be said to be the history of the emergence of what are broadly and interchangeably referred to as *model systems* or *model organisms*. Fruit flies, mice, *Xenopus*, nematodes, and zebrafish were all essentially established last century as standards to interrogate complex biological processes in multicellular organisms. Notable exceptions are the chicken, whose embryos have been studied since Aristotelian times, and the frog, which has been a favorite since Luigi Galvani hung frog legs outdoors to observe how they would react to lightning (Stern 2005). The present-day hegemony of a few organisms would have been difficult to predict. The paradigm is simple enough: study a few animals in great detail and granularity, and the principles of life that make humans possible can be identified, analyzed, and understood. At first blush, this is a reasonable expectation as it acknowledges the common evolutionary origin that is most likely shared by all animals. What is lost in this widely accepted paradigm, however, is the random, almost stochastic nature by which these animals were chosen.

Our current and dominant model systems were not selected because they occupied such remarkably important positions in the history of life that understanding their biology was of critical importance to unravel the Gordian knot of evolution. Rather, our present and minimal set of model organisms rose to prominence simply because they exaggerated particular attributes that aided scientists of the day to resolve problems they were keen to understand. They were available and had desired qualities. Chief among these qualities, for those studying development in particular, were lots of eggs and embryos, transparency, short life cycles, and ease of adaptation to the laboratory (Bolker 1995).

Although the approach of studying a handful of organisms in great detail has served us well for a century, it ultimately ignores many organisms whose biology may ultimately prove to illuminate much more brightly what we still fail to understand. Problems like aging, stem cell biology, regeneration, and scale and proportion are but a few of the central problems of which our present understanding is rudimentary at best. Consequently, it is appropriate to question the status quo.

Consider that a key attribute of the animals we have chosen to study in great molecular detail is that they can all be readily domesticated and thus brought into the lab. Is it not possible, therefore, that the genetic principles we have extracted from their study may ultimately be but the genetics of domestication? We may, in fact, be unaware of many genetic processes that are essential to our understanding of the inheritance of traits and the evolution of species.

Take, for example, the persistence of genome heterozygosity in wild populations of the planarian *Schmidtea mediterranea* from Sardinia (Guo, Zhang, Rubinstein, Ross, and Sánchez Alvarado 2016). The genome of these animals retains heterozygosity even after ten generations of inbreeding, violating expected Hardy-Weinberg predictions based on genetic studies of primarily domesticated genetic systems (Nagylaki 1992). Or take the case of the sea lamprey *Petromyzon mari-*

nus, a representative of an ancient vertebrate lineage that diverged from our own nearly 500 million years ago. *P. marinus* subjects the genome of its somatic cells to programmed rearrangement, resulting in animals that are effectively chimeric with germ cells possessing a full complement of genes and all other cell types possessing a smaller, reproducible fraction of the germline genome (Smith et al. 2018). Or consider an even closer relative, the naked mole rat *Heterocephalus glaber*, the longest-lived rodent known, which exhibits delayed and/or attenuated age-associated physiological declines, defying Gompertzian mortality laws predicting that death rate increases exponentially with age (Ruby, Smith, and Buffenstein 2018).

Such non-Mendelian processes and many other phenomena likely occur in nature with unsuspected frequency in less docile but possibly evolutionarily very important animals and their understanding may ultimately help us solve many vexing puzzles of our own biological existence. Given the remarkable technological advances of the past decade, it seems completely appropriate to turn our present paradigm on its head and, instead of bringing nature to science by domesticating animals in our laboratories, look beyond and bring our indoor science outdoors into the laboratory of nature.

All Animals Are Model Systems

The end of the nineteenth century found biologists struggling to understand the first principles underpinning the inheritance of traits (Richards 2003). Darwin's theory of evolution argued for successive changes driven by selection in the history of animal and plant species, changes that had to be somehow transmitted from one generation to the next. Looking for answers in fertilized oocytes, Edward van Beneden discovered and provided the first comprehensive description of meiosis (van Beneden 1883), while August Weismann put forward the germ plasm theory in 1892 with its "ids" and distinction between somatic and germ lines in an effort to explain how traits are inherited (Weismann 1893). Simultaneously, Edmund Beecher Wilson began to elaborate the concept of lineage and cell fate (Wilson 1892), and, just a few years later, Tschermak (Tschermak 1900), Correns (Correns 1900a, 1900b), and de Vries (de Vries 1900) rediscovered Mendel's work. In fact, the literature of the period was populated by many and mostly unsuccessful efforts and discussions attempting to integrate the seemingly disparate concepts of inheritance and evolution (Richards 2003).

Yet the absence of integration did not prevent the unprecedented number of scientific discoveries and the attendant progress in understanding biological complexity made at the end of the nineteenth century and the beginning of the twentieth. It is important to underscore two central attributes likely responsible for this outcome: technological advances in microscopy and histology and the absence of the modern concept of model systems or even model organisms.

I would argue that the great advances made during this historical period in our understanding of cell biology, embryogenesis, and genetics sprang from the general expectation by the field at large that the answers to the questions of inheritance were to be found in nature and that biologist must rigorously engage in querying the natural world to discover such answers. This approach to biology was first postulated by the French experimentalist Claude Bernard, who wrote: "Dans l'investigation scientifique, les moindres procédés sont de la plus haute importance. Le choix heureux d'un animal, d'un instrument construit d'une certaine façon, l'emploi d'un réactif au lieu d'un autre, suffisent souvent pour résoudre les questions générales les plus élevées. Chaque fois qu'un moyen nouveau et sûr d'analyse expérimentale surgit, on voit toujours la science faire des progrès dans les questions auxquelles ce moyeu peut être applique"[1] (Bernard 1865, 27). This concept of choosing a specific animal to address a specific biological problem was later restated—and in quite similar terms—by the Danish physiologist August Krogh. In fact the concept is known today as Krogh's principle. It states: "For a large number of problems there will be some animal of choice or a few such animals on which it can be most conveniently studied" (Krogh 1929, 202). Thus, scientists of the nineteenth century and the early twentieth explored—with a freedom almost unimaginable today—a broad range of organisms. For example, van Beneden determined how the process of meiosis works using ascaris eggs (van Beneden 1883), Weismann cemented his germ plasm continuity hypothesis working with hydrozoa (Weismann 1883), Wilson developed the concept of lineage and fate using the polychaete *Nereis* (Wilson 1892), and Conklin developed the notion of cytoplasmic inheritance using ascidians (Conklin 1905). In fact, it was not unusual to see articles illuminating our understanding of fundamental biological principles using hemichordates, clams, beetles, cockroaches, and many different species of plants and fungi (Wilson 1900). The remarkably broad collection of organisms used at a time of great growth of knowledge and understanding is in blunt contrast to the few animal species populating the vast majority of biomedical research laboratories in the world today.

It can be argued, therefore, that the frequency and number of probing and transformative questions in biology have a higher likelihood of being asked when the mind is exposed to and challenged by life's astonishing diversity. It is only through such exposure that we can learn firsthand what we did not know was already possible in biology. Exploration of the natural world allowed scientists in the late nineteenth century and the early twentieth to form better questions based on an understanding of the broad gamut of phenomenology displayed by animal and plant diversity. The desire to answer these questions led them to settle on specific animals chosen almost entirely for utilitarian reasons to advance our understanding of embryology, cell biology, and genetics.

Thomas Hunt Morgan, known best for his Nobel Prize–winning work on genetics, provides a good example. From 1888 until 1916, his bibliography encompasses the study of a remarkably ample and diverse collection of research organisms that he studied to ask a wide range of different questions about how organisms develop. Amphibians, ascidians, pycnogonids (sea spiders), hemichordates, sea urchins, fish (teleosts), amphioxi, earthworms, planarians, hermit crabs, cnidarians, aphids, shrimps, and at least a dozen more animals received his attention (Sturtevant 1959). In other words, the fruitfly *Drosophila melanogaster*, the one organism that will be forever attached to his name, was merely another species in a large collection of animals being studied at the time. But Morgan chose *Drosophila* over others not because it occupied some interesting position in the evolution of animals but simply because it was a prodigious egg layer, compact, and easy to rear in captivity and exhibited variations in the lab. Throughout the twentieth century, other biologists also took the same approach, and, with time, organisms other than the selected few became largely invisible or ignored by biologists.

Because their predecessors chose to focus on a few organisms and eventually elevated them to the pedestal of model systems, twenty-first-century biologists have, unwittingly or not, inherited a perspective of the natural world in which it is implied that only a few animals are worthy of investigation. This is not logical. All animals are the result of a set of contingencies and evolutionary principles that combined and adapted in ways unique to them alone and are, therefore, model systems in their own right. Given our common ancestry, such principles should apply not only to our own biology but also and more importantly to our understanding of life as a natural phenomenon. Hence, a time existed, not long ago, when all animals were potential subjects of biological investigation, and the time to retake that thread of history, fueled as before by current and astonishing technological advances, may once again be upon us.

A Legacy Reexamined

From the last half of the twentieth century to the present, the life sciences have strived for and rewarded the building of abstractly beautiful, logically impeccable, and comprehensively simplified systems. Consider that most of the approximately $30 billion annual budget of the National Institutes of Health has been and continues to be invested in the investigation of the biology of a handful of species. These organisms were selected out of convenience and, on statistical bases alone, are unlikely to encompass the gamut of what is possible and what we must understand in biology. Yet the many genome sequences deciphered to date, both plant and animal, underscore the remarkable fact that all known living species share common evolutionary ancestry. Given that evolution follows a syncopated rhythm of complex and contingent histories shaped by the vagaries and vicissitudes of time, place, and environment, it follows that simple laws with predictable outcomes are highly unlikely to describe the fortuitous nature of life fully. Hence, the likelihood of the present expectation that the study of as few as seven species in the biomedical sciences—a number accounting for the main set of model organisms—will uncover all the principles necessary to understand the underpinnings of life and, by extension, our own biology is essentially zero (Sánchez Alvarado 2018).

Our penchant for simplification and specialization is confining our ability to interrogate life to unacceptably narrow confines. This is reflected in the terminology we have chosen in the field of developmental biology to describe embryonic processes and the temporal transformations of tissues. Students of this central discipline in modern biomedical research pride themselves in their ability to invoke a rather large list of terms such as *holoblastic* and *meroblastic cleavages, animal* and *vegetal poles, gastrulation, epiboly, archenteron* and *blastocoels, germ layers, protostomes* and *deuterostomes,* among others, to describe changes in cellular composition, shape, and events underpinning development.

These terms, undoubtedly, have brought harmony in our ability to communicate experimental findings.

Yet consider for a minute the origin of these terms that so readily populate modern developmental biology: most, if not all, were coined in the early to mid-nineteenth century, a time when scientists had no knowledge of evolution, cell biology, or genetics. How could it be, then, that, knowing what we know today, these terms have made it to the present time and continue to be utilized unquestioningly? Terms that were coined as descriptive approximations of then mysterious biological processes have become absolutes with the passage of time. This, of course, comes at a price: to force the discovery of new biology to fit neatly into these limiting and antiquated terms, ultimately driving the belief that processes not conforming to established dogma are outliers and therefore strange and less important, as reflected by the use of terms such as *nonmodel* and *model* organisms. Given that the vast majority of known species remain by comparison starkly understudied, it may very well be that the current model organisms used to advance biomedical research may be more the exception than the rule. It is imperative, therefore, to determine whether our specialization has been impeding progress or, worse yet, leading us astray.

Why Put Words into Animals' Mouths? Let Animals Tell Us Their Own Stories

It is generally agreed that the oceans are the largest depository of biodiversity on our planet, a belief underscored by the regular discovery of numerous new marine species every time probes are sent into its waters (Costello et al. 2013). Yet the National Oceanographic and Atmospheric Agency estimates that nearly 95 percent of our oceans remain unexplored (NOAA 2017). While this might strike us as hyperbole, the following fact illustrates that we in biomedical research in particular and biology in general are not engaged in either a massive or a systematic exploration of our oceans. When the first extrasolar planet was detected orbiting around Gamma-Cephei in 1988 (Campbell, Walker, and Yang 1988), the distance alone (approximately 45 light years away) was sufficient reason to explain why it took us nearly 500 years from the first time Galileo aimed a telescope at the moon to make this momentous discovery. But distance can hardly explain why we would also discover in that same year what is arguably the most abundant life-form in our planet: *Prochlorococcus marinus* (Chisholm et al. 1988). So abundant, in fact, is this species that it can be seen from the international space station. Paleontological evidence indicates that our species has been collecting food from the oceans for at least 165,000 years (McBrearty and Brooks 2000; Henshilwood and Marean 2003), which means that we have literally been bathing in and probably swallowing *P. marinus* for just as long, yet its discovery occurred only a few decades ago. If one extrapolates from this experience, less abundant species may as well be completely invisible. The truth is that we do not even know how many different species exist in the oceans.

Thus, the mysteries of the oceans remain vast and deep. Remarkable species abound in their waters, some already discovered, and many yet to be found. Take, for example, the protozoan *Syringammina fragilissima*, discovered off the coast of Scotland and described by Brady in 1883 (Brady 1883). Although we normally associate protozoans with microscopes because most are very small, *S. fragilissima* is presently the largest single-celled organism known to science at up to 20 centimeters across. Its relative the ciliated protozoan *Tetrahymena thermophila* is, by comparison, only approximately 50 micrometers in diameter. This is approximately a twenty-five-thousand-fold difference in scale alone. If one considers that ribozymes (Cech 1987), telomeres (Greider and Blackburn 1989), histone modifications (Allis, Glover, and Gorovsky 1979), and molecular motors like dynein (Gibbons and Rowe 1965) were all first discovered in *Tetrahymena*, yielding several Nobel Prizes along the way, imagine what new biology awaits to be discovered in the vast and mostly unexplored genotypic and phenotypic space occupied by single-celled organisms and represented at one extreme by the marine protozoan *S. fragilissima*.

Just as there are likely countless surprises waiting for us in the marine unicellular world, there may likely be many others, unsuspected and perhaps even more astonishing, lying in wait to be discovered in the multicellular life inhabiting the oceans. In fact, it is not difficult to find in already-discovered marine species biological attributes that are in stark contrast to those displayed by traditional laboratory animals. Take, for example, a recently discovered sponge off the coast of Hawaii that is the size of a minivan (3.5 meters in length, 2.0 meters in width, and 1.5 meters high) and likely thousands of years old, thus challenging our present understanding of longevity and scale and proportion regulation in simpler organisms (Wagner and Kelley 2017). Or consider the hydrozoan species *Turritopsis dohrnii, Turritopsis nutricola, Laodicea undulate*, and *Podocoryne carnea*, which, upon thermal, chemical, or physical stress or merely aging (senescence), have been reported to revert their life cycle by back transformation of the adult state into juvenile, earlier developmental stages (Piraino, Boero, Aeschbach, and Schmid 1996; Carla, Pagliara, Piraino, Boero, and Dini 2003; Piraino, Schmich, Bouillon, and Boero 2004; Schmich et al. 2007). Then there are animals that partition their respiration and derive their energy by establishing symbioses with algae, such as in the cnidarian-dinoflagellate symbioses of *Symbiodinium* (Hawkins, Hagemeyer, Hoadley, Marsh, and Warner 2016). A closer relative to vertebrates, the marine planktonic chordates known as salps, displays an obligatory alternation between sexual (blastozooid or aggregate) and asexual (oozooid or solitary) life-history stages or generations (Godeaux, Bone, Braconnot, and Bone 1998; Henschke, Everett, Richardson, and Suthers 2016).

Altogether, the remarkable longevity of sponges, the exaggerated potential for dedifferentiation of hydrozoans, the metabolic integration of photo- and chemosynthesis in *Symbiodinium*, and the alternation of meiotic and mitotic reproduction in ascidians provide unique experimental paradigms through which to understand key and as-of-yet-unresolved problems: (1) how regulatory networks of gene expression and their attendant cell behaviors may control aging and senescence, (2) the directionality of ontogeny (i.e., normal vs. reverse development), (3) the physiological integration of metabolism, and (4) developmental potency.

If our collective efforts to date have helped us illuminate many aspects of biology through the study of a handful of organisms, what could we accomplish if hundreds of carefully selected species were to be subjected to rigorous molecular, genetic, and cellular interrogation? Advances in genomics, gene editing, microscopy, computational biology, and aquaculture provide us with a unique opportunity to start a systematic exploration of our oceans in order to identify species harboring new biology. Given that most of biology has not been studied at such discrete levels, such an effort will surely yield not only industries, remedies, and drugs we cannot imagine today but also quite possibly entirely new biological disciplines. Why not attempt, then, an effort to generate the new experimental paradigms of biology by studying the life of our oceans in unprecedented levels of detail?

And what better place to launch such an effort than our seaside laboratories, such as the Marine Biological Laboratory (MBL) in Woods Hole? Right now, the MBL's marine resources department provides investigators with organisms as varied as octopuses, lampreys, ribbon worms, bryozoans, sea urchins, and toadfish. Through its location next to the ocean, it also gives easy access to myriad largely unstudied microbes and other marine invertebrates. An enterprise that exploits the oceans through our marine laboratories will surely revolutionize our understanding of life and define new systems and approaches that may ultimately become the centerpiece of biomedical research in the twenty-first century. Let us learn from history to shape the future.

Note

1. "In scientific research, the smallest processes are of the utmost importance. The happy choice of an animal, an instrument built in a certain way, the use of a reagent instead of another, is often enough to solve the highest general questions. Whenever a new and safe means of experimental analysis arises, science is always seen to make progress in the questions to which these new approaches may be applied."

References

Allis, C. D., C. V. Glover, and M. A. Gorovsky. 1979. "Micronuclei of Tetrahymena Contain Two Types of Histone H3." *Proceedings of the National Academy of Sciences of the United States of America* 76, no. 10:4857–61.

Bernard, C. 1865. *Introduction à l'étude de la médecine expérimentale*. Edited by J. B. Baillière et Fils. Paris: Libraires de l'Académie impériale de médecine.

Bolker, J. A. 1995. "Model Systems in Developmental Biology." *BioEssays* 17, no. 5:451–55.

Brady, H. B. 1883. "Note on Syringammina, a New Type of Arenaceous Rhizopoda." *Proceedings of the Royal Society of London* 35:155–61. doi 10.1098/rspl.1883.0031.

Campbell, B., G. A. H. Walker, and S. Yang. 1988. "A Search for Substellar Companions to Solar-Type Stars." *Astrophysical Journal, Part 1* 331:902–21.

Carla, E. C., P. Pagliara, S. Piraino, F. Boero, and L. Dini. 2003. "Morphological and Ultrastructural Analysis of Turritopsis Nutricula during Life Cycle Reversal." *Tissue and Cell* 35, no. 3:213–22.

Cech, T. R. 1987. "The Chemistry of Self-Splicing RNA and RNA Enzymes." *Science* 236, no. 4808:1532–39. doi 10.1126/science.2438771.

Chisholm, S. W., R. J. Olson, E. R. Zettler, R. Goericke, J. B. Waterbury, and N. A. Welschmeyer. 1988. "A Novel Free-Living Prochlorophyte Abundant in the Oceanic Euphotic Zone." *Nature* 334:340. doi 10.1038/334340a0.

Conklin, E. G. 1905. "Mosaic Development in Ascidian Eggs." *Journal of Experimental Zoology* 2:145–223.

Correns, C. E. 1900a. "G. Mendel's Law on the Behaviour of Progeny of Variable Hybrids." *Botanischen Gesellschaft* 8:156–68.

———. 1900b. "Über Levkojenbastarde: Zur Kenntniss der Grenzen der Mendelschen Regeln." *Botanisches Centralblatt* 84:97–113.

Costello, M. J., P. Bouchet, G. W. Boxshall, K. Fauchald, D. P. Gordon, et al. 2013. "Global Coordination and Standardisation in Marine Biodiversity through the World Register of Marine Species (WoRMS) and Related Databases." *PLOS ONE*, January 9. doi 10.1371/journal.pone.0051629.

de Vries, H. 1900. "Sur la loi de disjunction des hybrides." *Comptes rendus de l'Académie des sciences* 130:845–47.

Gibbons, I. R., and A. J. Rowe. 1965. "Dynein: A Protein with Adenosine Triphosphatase Activity from Cilia." *Science* 149, no. 3682:424–26. doi 10.1126/science.149.3682.424.

Godeaux, J., Q. Bone, J.-C. Braconnot, and Q. Bone. 1998. "Anatomy of Thaliacea." In *The Biology of Pelagic Tunicates*, ed. Q. Bone, 1–24. Oxford: Oxford University Press.

Greider, C. W., and E. H. Blackburn. 1989. "A Telomeric Sequence in the RNA of Tetrahymena Telomerase Required for Telomere Repeat Synthesis." *Nature* 337:331. doi 10.1038/337331a0.

Guo, L., S. Zhang, B. Rubinstein, E. Ross, and A. Sánchez Alvarado. 2016. "Widespread Maintenance of Genome Heterozygosity in Schmidtea mediterranea." *Nature Ecology and Evolution* 1, no. 1:19. doi 10.1038/s41559-016-0019.

Hawkins, T. D., J. C. Hagemeyer, K. D. Hoadley, A. G. Marsh, and M. E. Warner. 2016. "Partitioning of Respiration in an Animal-Algal Symbiosis: Implications for Different Aerobic Capacity between Symbiodinium spp." *Frontiers in Physiology* 7:128. doi 10.3389/fphys.2016.00128.

Henschke, N., J. D. Everett, A. J. Richardson, and I. M. Suthers. 2016. "Rethinking the Role of Salps in the Ocean." *Trends in Ecology and Evolution* 31, no. 9:720–33. doi 10.1016/j.tree.2016.06.007.

Henshilwood, C. S., and C. W. Marean. 2003. "The Origin of Modern Human Behavior: Critique of the Models and Their Test Implications." *Current Anthropology* 44:627–51.

Krogh, A. 1929. "The Progress of Physiology." *Science* 70:200–204. doi 10.1126/science.70.1809.200.

McBrearty, S., and A. Brooks. 2000. "The Revolution That Wasn't: A New Interpretation of the Origin of Modern Humans." *Journal of Human Evolution* 39:453–563.

Nagylaki, T. 1992. *Introduction to Theoretical Population Genetics*. Vol. 21, *Biomathematics*. Berlin: Springer.

National Oceanic and Atmospheric Administration (NOAA). 2017. "How Much of the Ocean Have We Explored?" https://oceanservice.noaa.gov/facts/exploration.html.

Piraino, S., F. Boero, B. Aeschbach, and V. Schmid. 1996. "Reversing the Life Cycle: Medusae Transforming into Polyps and Cell Transdifferentiation in *Turritopsis nutricula* (Cnidaria, Hydrozoa)." *Biological Bulletin* 190:303–12.

Piraino, S., D. De Vito, J. Schmich, J. Bouillon, and F. Boero. 2004. "Reverse Development in Cnidaria." *Canadian Journal of Zoology* 82:1748–54.

Richards, R. J. 2003. "Biology." In *From Natural Philosophy to the Sciences*, ed. D. Cahan, 14–48. Chicago: University of Chicago Press.

Ruby, J. G., M. Smith, and R. Buffenstein. 2018. "Naked Mole-Rat Mortality Rates Defy Gompertzian Laws by Not Increasing with Age." *eLife* 7:e31157. doi 10.7554/eLife.31157.

Sánchez Alvarado, A. 2018. "To Solve Old Problems, Study New Research Organisms." *Developmental Biology* 433, no. 2:111–14. doi 10.1016/j.ydbio.2017.09.018.

Schmich, J., Y. Kraus, D. De Vito, D. Graziussi, F. Boero, and S. Piraino. 2007. "Induction of Reverse Development in Two Marine Hydrozoans." *International Journal of Developmental Biology* 51, no. 1:45–56. doi 10.1387/ijdb.062152js.

Smith, J. J., N. Timoshevskaya, C. Ye, C. Holt, M. C. Keinath, H. J. Parker, M. E. Cook, J. E. Hess, S. R. Narum, F. Lamanna, H. Kaessmann, V. A. Timoshevskiy, C. K. M. Waterbury, C. Saraceno, L. M. Wiedemann, S. M. C. Robb, C. Baker, E. E. Eichler, D. Hockman, T. Sauka-Spengler, M. Yandell, R. Krumlauf, G. Elgar, and C. T. Amemiya. 2018. "The Sea Lamprey Germline Genome Provides Insights into Programmed Genome Rearrangement and Vertebrate Evolution." *Nature Genetics*. doi 10.1038/s41588-017-0036-1.

Stern, C. D. 2005. "The Chick: A Great Model System Becomes Even Greater." *Developmental Cell* 8, no. 1:9–17. doi 10.1016/j.devcel.2004.11.018.

Sturtevant, A. H. 1959. "Thomas Hunt Morgan. A Biographical Memoir." In *Biographical Memoirs*, 281–325. Washington, DC: National Academy of Sciences, USA.

Tschermak, E. 1900. "Ueber künstliche kreuzung bei Pisum sativum." *Berichte der Deutschen Botanischen Gesellschaft* 18:232–49.

van Beneden, E. 1883. "Recherches sur la maturation de l'oeuf et la fecondatlon: *Ascaris megalocephala*." *Archives de biologie* 4:265–640.

Wagner, D., and C. D. Kelley. 2017. "The Largest Sponge in the World?" *Marine Biodiversity* 47:367.

Weismann, A. 1883. *Die entstehung der sexualzellen bei den hydromedusen: Zugleich ein Betrag zur Kenntniss des Baues und der Lebenserscheinungen dieser Gruppe*. Jena: Gustav Fischer.

———. 1893. *The Germ-Plasm: A Theory of Heredity*. New York: Scribner.

Wilson, E. B. 1892. "The Cell-Lineage of *Nereis*." *Journal of Morphology* 6: 361–480.

———. 1900. *The Cell in Development and Inheritance*. London: Macmillan.

Acknowledgments

This project was made possible by the welcoming environment of the Marine Biological Laboratory (MBL) in Woods Hole, Massachusetts, which hosted the 2016 annual History of Biology Seminar "Why Marine Studies?" The seminar series is funded by Arizona State University (ASU) and organized by Jessica Ranney and Andrea Cottrell of the ASU Center for Science and Society. The related MBL History Project is supported by grants from the National Science Foundation and the Edwin S. Webster Foundation, for which we are most grateful.

Jane Maienschein's research is funded by grants from the National Science Foundation and the James S. McDonnell Foundation. Rachel A. Ankeny's research associated with this project has been funded through the Australian Research Council Discovery Project "Organisms and Us: How Living Things Help Us to Understand Our World" (DP160102989).

Contributors

William C. Aird, MD
Professor of Medicine
Beth Israel Deaconess Medical Center
Harvard Medical School
Research North 246
99 Brookline Ave.
Boston, MA 02215
617-667-1031
waird@bidmc.harvard.edu

Rachel A. Ankeny
Professor
Departments of History and Philosophy
University of Adelaide
Adelaide 5005 SA
Australia
rachel.ankeny@adelaide.edu.au

Nathan Crowe
Assistant Professor
University of North Carolina Wilmington
601 South College Rd.
Wilmington, NC 28403
910-962-3309
crowen@uncw.edu

Michael R. Dietrich
Professor
University of Pittsburgh
1101 Cathedral of Learning
4200 Forbes Ave.
Pittsburgh, PA 15260
412-624-5896
mdietrich@pitt.edu

Kjell David Ericson
Assistant Professor (Program Specific)
Center for the Promotion of Interdisciplinary Education and Research
Kyoto University
Yoshida-honmachi
Sakyo-ku
Kyoto 606-8501
Japan
+81-75-012-2837
ericson.kjelldavid.6a@kyoto-u.ac.jp

Marianne A. Grant
Assistant Professor of Medicine
Beth Israel Deaconess Medical Center
Harvard Medical School
Research North 270E
99 Brookline Ave.
Boston, MA 02215
978-356-6500 x2444
mgrant@ebsco.com

Christiane Groeben, MA
Senior Archivist and Independent Scholar (Ret.)
Stazione Zoologica Anton Dohrn
Via Enea Zanfagna 52
80125 Naples
Italy
+39 081 19566329
christiane.groeben@fastwebnet.it

Sabina Leonelli
Professor in Philosophy and History of Science
Codirector, Exeter Centre for the Study of the Life Sciences (Egenis)
Department of Sociology, Philosophy, and Anthropology
University of Exeter
Byrne House
St. Germans Rd.
EX4 4PJ Exeter
United Kingdom
+44 1392725157
s.leonelli@exeter.ac.uk

Christine Yi Lai Luk
Assistant Professor
Department of the History of Science
Tsinghua University
Beijing
China
+86 18410700703
chrisluk@tsinghua.edu.cn

Kate MacCord, PhD
Program Administrator and McDonnell Foundation Fellow
Marine Biological Laboratory
7 MBL Street
Woods Hole, MA 02543
508-289-7513
kmaccord@mbl.edu

Jane Maienschein
University Professor, Regents' Professor, and President's Professor
Arizona State University
School of Life Sciences
427 East Tyler Mall
Tempe, AZ 85287-4501
480-965-6105
maienschein@asu.edu

Karl S. Matlin
Professor Emeritus
Committee on Conceptual and Historical Studies of Science
Whitman Investigator, Marine Biological Laboratory
University of Chicago
16 Sippewissett Rd.
Falmouth, MA 02540
kmatlin@uchicago.edu

Kathryn Maxson Jones, MA
McDonnell Foundation Scholar, Marine Biological Laboratory
PhD Candidate, Program in History of Science
Department of History
Princeton University
Dickinson Hall, Room 129
Princeton, NJ 08544
kmaxson@princeton.edu

Samantha Muka
Assistant Professor
Stevens Institute of Technology
College of Arts and Sciences
1 Castle Point Terrace
Hoboken, NJ 07030
201-216-3401
smuka@stevens.edu

Nipam H. Patel, PhD
Director
Marine Biological Laboratory
7 MBL St.
Woods Hole, MA 02543
508-289-7300
director@mbl.edu

Alejandro Sánchez Alvarado, PhD
Investigator
Stowers Institute for Medical Research
1000 East 50th St.
Kansas City, MO 64110
816-926-4530
asa@stowers.org

Katharina Steiner, PhD
Marie Skłodowska-Curie Postdoctoral Fellow
History of Science, Medicine, and Technology Program
University of Wisconsin–Madison
5118 Mosse Humanities Building
455 N. Park St.
Madison, WI 53706
608-890-2612
ksteiner4@wisc.edu

Index

The letter *t* following a page number denotes a table; page numbers in italics refer to figures.